Importance of Microbiology Teaching and Microbial Resource Management for Sustainable Futures

Importance of Microbiology Teaching and Microbial Resource Management for Sustainable Futures

Edited by

İpek Kurtböke
Centre for Bioinnovation and the School of Science, Technology and Engineering, University of the Sunshine Coast, QLD, Australia
World Federation of Culture Collections

ELSEVIER

ACADEMIC PRESS
An imprint of Elsevier

Academic Press is an imprint of Elsevier
125 London Wall, London EC2Y 5AS, United Kingdom
525 B Street, Suite 1650, San Diego, CA 92101, United States
50 Hampshire Street, 5th Floor, Cambridge, MA 02139, United States
The Boulevard, Langford Lane, Kidlington, Oxford OX5 1GB, United Kingdom

ISBN: 978-0-12-818272-7

For information on all Academic Press publications visit our website at https://www.elsevier.com/books-and-journals

Publisher: Stacy Masucci
Acquisitions Editor: Linda Versteeg-Buschman
Editorial Project Manager: Timothy Bennett
Production Project Manager: Omer Mukthar
Cover Designer: Greg Harris

Cover image credits - Dr. İpek Kurtböke, personal collection (images were captured at Central Analytical Research Facility (CARF) at the Queensland University of Technology (QUT), a Microscopy Australia linked lab, 2012).

Typeset by TNQ Technologies

Contents

4. Multimedia in microbiology education: a critical sustainable resource

Danilla Grando

5. Improving public understanding of microorganisms by integrating microbiology concepts into science teaching throughout the education system

Megan L. Lloyd and Jemma A. Berry

11. Yeast in a megadiverse country for sustainable futures

Enrique Javier Carvajal Barriga, Patricia Portero-Barahona and Alexandra Narváez-Trujillo

12. Yeast as a model organism for teaching biotechnology and human cell biology leading to sustainable futures

Ian G. Macreadie

Contributors

Pravech Ajawatanawong, Division of Bioinformatics and Data Management for Research, Faculty of Medicine Siriraj Hospital, Mahidol University, Bangkok, Thailand

Jemma A. Berry, Edith Cowan University, Joondalup, WA, Australia

Chooseel Bunsuwansakul, Division of Bioinformatics and Data Management for Research, Faculty of Medicine Siriraj Hospital, Mahidol University, Bangkok, Thailand

Thelma Caine, CABI, Bakeham Lane Egham, Surrey, United Kingdom

Enrique Javier Carvajal Barriga, Pontificia Universidad Católica del Ecuador, Centro Neotropical para la Investigación de la Biomasa, Colección de Levaduras Quito-Católica (CLQCA), Quito, Ecuador

Graça S. Carvalho, CIEC (Research Centre on Child Studies), University of Minho, Braga, Portugal

Suwanee Chunhametha, Thailand Bioresource Research Center (TBRC), National Center for Genetic Engineering and Biotechnology (BIOTEC), National Science and Technology Development Agency (NSTDA), Pathum Thani, Thailand

Lily Eurwilaichitr, Internaional Collaboration Office (INCO), National Sceince and Technology Development Agency, (NSTDA), Pathum Thani, Thailand

Danilla Grando, School of Science, RMIT University, Melbourne, VIC, Australia

Piyanun Harnpicharnchai, Thailand Bioresource Research Center (TBRC), National Center for Genetic Engineering and Biotechnology (BIOTEC), National Science and Technology Development Agency (NSTDA), Pathum Thani, Thailand

Supawadee Ingsriswang, Thailand Bioresource Research Center (TBRC), National Center for Genetic Engineering and Biotechnology (BIOTEC), National Science and Technology Development Agency (NSTDA), Pathum Thani, Thailand

Supattra Kitikhun, Thailand Bioresource Research Center (TBRC), National Center for Genetic Engineering and Biotechnology (BIOTEC), National Science and Technology Development Agency (NSTDA), Pathum Thani, Thailand

D.İ. Kurtböke, Centre for Bioinnovation and the School of Science, Technology and Engineering, University of the Sunshine Coast, Maroochydore BC, QLD, Australia; World Federation of Culture Collections

Somsak Likhitrattanapisal, Thailand Bioresource Research Center (TBRC), National Center for Genetic Engineering and Biotechnology (BIOTEC), National Science and Technology Development Agency (NSTDA), Pathum Thani, Thailand

Nelson Lima, CEB (Centre of Biological Engineering), University of Minho, Braga, Portugal

Megan L. Lloyd, The University of Western Australia, Crawley, WA, Australia

Ian G. Macreadie, Honorary Professor of Biotechnology, School of Science, RMIT University, Melbourne, VIC, Australia

Anabela Martins, CEB – Biological Engineering Centre, Universidade do Minho, Braga, Portugal

Joaquim F. Martins da Silva, Colégio Luso-Internacional de Braga, Braga, Portugal

Alexandra Narváez-Trujillo, Pontificia Universidad Católica del Ecuador, Centro Neotropical para la Investigación de la Biomasa, Colección de Levaduras Quito-Católica (CLQCA), Quito, Ecuador

Phongthana Pasookhush, Division of Bioinformatics and Data Management for Research, Faculty of Medicine Siriraj Hospital, Mahidol University, Bangkok, Thailand

Jim Philp, Organisation for Economic Cooperation and Development (OECD), Paris, France

Patricia Portero-Barahona, Pontificia Universidad Católica del Ecuador, Centro Neotropical para la Investigación de la Biomasa, Colección de Levaduras Quito-Católica (CLQCA), Quito, Ecuador

Lorenz Christian Reimer, Leibniz Institute DSMZ-German Collection of Microorganisms and Cell Cultures, Braunschweig, Germany

Matthew J. Ryan, CABI, Bakeham Lane Egham, Surrey, United Kingdom

Paulo Sampaio, Systems and Production Department, University of Minho, Braga, Portugal

David Smith, CABI, Bakeham Lane Egham, Surrey, United Kingdom

Andrey Yurkov, Leibniz Institute DSMZ-German Collection of Microorganisms and Cell Cultures, Braunschweig, Germany

Preface

Current global realities clearly indicate that there is an urgent need to implement environmentally friendly solutions to improve health, agriculture, nutrition..., whilst minimising disease and environment destruction. Multidisciplinary approach in research and combined actions involving eco-innovations is getting increasingly necessary to be able to achieve the **Sustainable Development Goals** (SDGs) of the United Nations, which are intended to be achieved by 2030. The SDGs include (1) no poverty, (2) zero hunger, (3) good health and well-being, (4) quality education, (5) gender equality, (6) clean water and sanitation, (7) affordable and clean energy, (8) decent work and economic growth, (9) industry, innovation and infrastructure, (10) reducing inequality, (11) sustainable cities and communities, (12) responsible consumption and production, (13) climate action, (14) life below water, (15) life on land, (16) peace, justice and strong institutions and (17) partnerships for goals.

Global citizens will play a significant role in increasing public awareness, as well as the target-directed education of responsible global citizens to ensure sustainable futures. Microbiology has been one of the key disciplines that served the mankind through harnessing beneficial activities of microorganisms ranging from soil fertility and productivity to detoxification of man-made and natural pollutants. The extraordinary metabolic activities of microorganisms stem from their fascinating metabolic diversity and genetic adaptability. However, the importance and relevance of this discipline has not been fully understood and often overlooked and microbial resources, as one of the key role players in the salvation of the planet, have not been fully tapped. Although the use of microorganisms in pharmaceutical industry for drug development has been one of the successful examples of the utilisation of microbial power since the 1940s, Earth's vast microbial genetic pool still remains undiscovered. Environmental microbiology is still in its infancy; however, advancing molecular techniques will no doubt fill the information gaps and will play a key role in revealing functional diversity of microorganisms and the crucial role they play in sustaining life functions. Such understanding in turn will facilitate function-based environmental applications of microorganisms in particular for agricultural and waste management purposes. Information on microbial diversity and creation of effective tools to access and recovery of such diversity is crucial for biotechnology. Moreover, microbial culture collections and preservation of key features of the microbial genetic pool are imperative for many different sectors ranging from health to environmental management. Culture collections play a capacity-building role, to help countries to better understand and utilise their microbial diversity. They are also bodies that the public and the policy makers can call upon for objective help in developing regulations and guidelines for the safe and ethical use of biological resources.

In a sequel to *Microbial Resources: From Functional Existence in Nature to Applications* (ISBN: 9780128047651, Academic Press, Elsevier), *Importance of Microbial Teaching and Microbial Resource Management for Sustainable Futures* covers importance of microbiology education and teaching in relation to creation of sustainable futures. It also provides examples of key microbial resources ranging from amoebae to yeast as well as with further examples from the CABI's unique culture collection in the United Kingdom and their importance for sustainability. The book also includes the topics related to the importance of microbial preservation, data management and quality control in culture collections.

I thank Emeritus Prof. John Bartlett (Australia), Prof. Nelson Lima (Portugal), Prof. Graça Carvalho (Portugal) and Dr. David McKay (Australia) for providing peer-reviews and feedbacks on the chapters.

As stated by the UN, the SDGs 'are the blueprint to achieve a better and more sustainable future for all, they address the global challenges we face, including those related to poverty, inequality, climate and environmental degradation, prosperity and peace and justice'. 'The goals interconnect and in order to ensure a just transition that leaves no one behind'; accordingly, 'it is important that the goals and targets are achieved by 2030'. Eight years is a short span for scientific and biotechnological outputs and eco-innovations; thus, there is no time to waste. Increase in multidisciplinary understanding and bringing the forces together will play a key role and importance of microbiology and the vital roles of microorganisms will be increasingly understood and appreciated. The timely publication of the book I believe will strengthen information generation on the roles of microorganisms for sustainability and it will be one of the key resources highlighting the importance and potential contributions of microbiology that can aid into the timely achievement of the SDGs.

İpek Kurtböke

Chapter 1

Pasteurisation for sustainable futures

D.İ. Kurtböke[1,2]

[1]*Centre for Bioinnovation and the School of Science, Technology and Engineering, University of the Sunshine Coast, Maroochydore BC, QLD, Australia;* [2]*World Federation of Culture Collections*

Chapter outline

Sustainability and sustainable development goals

Sustainability refers to the 'ability to be maintained at a certain rate or level' with 'avoidance of the depletion of natural resources in order to maintain an ecological balance' as well as relating to how 'the biosphere and human civilisation can co-exist in a sustainable way' (https://www.lexico.com/definition/sustainability). Sustainable development refers to the 'development that meets the needs of the present without compromising the ability of future generations to meet their own needs' (WCED, 1987; Scoones, 2007). Examples include 'sustainable cities, economies, resource management, businesses and livelihoods'

Importance of Microbiology Teaching and Microbial Resource Management for Sustainable Futures
https://doi.org/10.1016/B978-0-12-818272-7.00012-2

(Clark, 2007; Scoones, 2007). As defined by Scoones (2007), effective sustainable practices involve 'building epistemic communities of shared understanding of and common commitment to' sustainable practices. Such practices also involve 'creation of networks, formation of alliances, constructing or involving relevant institutions and organisations to formulate projects with funds made available'. Sustainability is also defined as a 'boundary term' to describe where 'science meets politics, and politics meets science' (Gieryn, 1999; Scoones, 2007). Scoones (2007) also notes that 'it is at this complex intersection between science and politics where boundary work takes place, and where words, with often ambivalent and contested meanings, have an important political role in processes of policy making and development'.

In the above context, one of the globally relevant approaches at the policy makers level has been the establishment of **sustainable development goals** (SDGs) by the United Nations (UN) in 2015, which are intended to be achieved by 2030. The SDGs include (1) no poverty, (2) zero hunger, (3) good health and well-being, (4) quality education, (5) gender equality, (6) clean water and sanitation, (7) affordable and clean energy, (8) decent work and economic growth, (9) industry, innovation and infrastructure, (10) reducing inequality, (11) sustainable cities and communities, (12) responsible consumption and production, (13) climate action, (14) life below water, (15) life on land, (16) peace, justice and strong institutions and (17) partnerships for goals (https://www.un.org/sustainabledevelopment/).

However, Fagunwa and Olanbiwoninu (2020) have recently stated that, although the above listed SDGs are well structured to address the global challenges mankind faces, they are unrealistic unless strategies to reach the targets within the next 10 years are defined in detail. Cash et al. (2002, 2003) have also pointed out that 'the challenge of meeting human development needs whilst protecting the earth's life support systems, confronts scientists, technologists, policy makers and communities from local to global levels' Science and Technology (S&T) thus must play a more central role in sustainable development, 'yet little **systematic scholarship** exists on how to create institutions that effectively harness S&T for sustainability. The efforts to mobilise S&T for sustainability are more likely to be effective when they manage boundaries between knowledge and action in ways that simultaneously enhance the salience, credibility and legitimacy of the information they produce' (Gross et al., 2014; Cash and Belloy, 2020).

Henry and Vollan (2014) also highlight that 'social relationships (and hence networks) are important to many issues of sustainability', for example, 'transfers of knowledge, cooperation on the management of shared resources, and formulation of policies that are meant to influence behaviour'. They also add that 'the literature on networks and sustainability is vast and amorphous and illustrates that, although in early stages, theoretical development on networks can offer practical insights into solving problems of sustainability'.

Science advances through tentative answers to a series of more and more subtle questions which reach deeper and deeper into the essence of natural phenomena.

Louis Pasteur.

van Kerkhoff and Lebel (2006) stressed that 'actions towards sustainable development require a mix of scientific, economic, social and political knowledge and judgments. The role of research-based knowledge in this complex setting is ambiguous and diverse, and it is undergoing rapid change both in theory and in practice' They also note that 'research, politics, researchers and publics are intertwined in a constant struggle of justifications, explanations and decisions in an uncertain and complex world'. "Relationships between research-based knowledge and action can be better understood as arenas of shared responsibility, embedded within larger systems of power and knowledge that evolve and change over time". This conceptualisation they claim 'offers a more appropriate starting point for understanding the role of research in sustainable development than the conventional models of trickle-down, transfer and translation'.

It is characteristic of experimental science that it opens ever widening horizons of our vision

Louis Pasteur.

Biomimicry and sustainability

The more I study nature, the more I stand amazed at the work of the Creator. Science brings men nearer to the Creator

Louis Pasteur.

The terms '**Biomimicry**' or '**Biomimetics**' coined by Otto Schmidt in the 1950s (Harkness, 2002) refer to 'biologically inspired design or adaptation or derivation from nature' (Bhushan, 2009). It is 'technical emulation of biological forms, processes, patterns and systems' with a focus on 'green product innovation' (Benyus, 2013; Grove et al., 2016; Kennedy and Marting, 2016); thus, biomimicry can provide a platform of 'sustainable-inspired eco-innovations' such as for energy savings and waste reduction (Kirkewoog, 2017). Kennedy et al. (2015) highlighted that bio-inspired designs are 'resilient, adaptable, multifunctional, regenerative and generally zero-waste' and 'when deeply informed by biology, design, the thinking shifts away from an anthropocentric model and considers product life cycles and earth system limitations'. They also stated that sustainability needs may form a new frontier for innovation and biomimicry can be the merger between sustainability and innovation. Designers can thus work alongside the biologists to support human civilisation within the ecological limits of the planetary support system. Collado-Ruano (2015) highlighted 'biomimicry as a necessary eco-ethical dimension for a future human sustainability' and recommended its integration in the **'Global Citizenship Education' program of the UNESCO** (see Section Global citizenship education).

Role of microbiology in sustainability

Messieurs, c'est les microbes qui auront le dernier mot (Gentlemen, it is the microbes who will have the last word)

Louis Pasteur.

Microbiology deals with the effects and practical uses of microorganisms that influence our day-to-day living. It has direct links to many daily life aspects such as food security, health and well-being, clean energy, environmental health, waste degradation and climate change indicators (Fagunwa and Olanbiwoninu, 2020; Sorlini, 2020). If microbial indicators and microbiological data are carefully integrated into the global action plan, microbiology can accelerate the process required to achieve SDGs by policy makers (Fagunwa and Olanbiwoninu, 2020; Sorlini, 2020).

As stated by De Vrieze et al. (2020) for thousands of years, microbial processes were used to produce useful products that served mankind, e.g., alcoholic beverages, bread and cheese. Following the discovery of penicillin and advancements in understanding of microbial metabolism after the World War II, explosive discoveries were made in the fields of pharmaceuticals, food and enzymology for biotechnological and industrial applications. They were then followed by the application of microorganisms in the environment such as for waste management and renewable energy generation (De Vrieze et al., 2020). Further insights were recently gained through advances in molecular microbiology that revealed the true potential of microorganisms (genome studies) as well as their diversity and functions in the environment (metagenomics and functional diversity). These advances opened up new frontiers for the application of microorganisms in diverse fields ranging from bioremediation to bioenergy and to agriculture. Examples include the genetically modified *Caldicellulosiruptor bescii*-mediated conversion of switchgrass to ethanol (Chung et al., 2014).

Without the presence of pure cultures and only with DNA information, metagenomics now provides information on microbial communities in the environment. Functional metagenomics combines information obtained from high-resolution genomic analysis of such uncultured microbial communities with their functions in a specific environment allowing meaningful analysis of the existence of such communities. Further functional analysis of specific microbial communities can subsequently be achieved via integrative approaches using metatranscriptomics and metaproteomics by obtaining comprehensive information on the genes, nucleic acids, proteins and metabolites of the microbial communities (Ghosh et al., 2019). Specific examples include the use of genomics and metagenomics to identify specific characteristics of deep-sea microbial communities such as their adaptation to high-pressure and low-temperature conditions as well as their interactions with complex mixtures of organic matter whilst responding to changes in the ocean's biogeochemical state (Nagata et al., 2010).

Screening of environmental metagenomic libraries can easily identify novel genes that encode enzymes required to synthesise new antibiotics. Metagenomics can also target products with unique properties, such as the heat stability of specific enzymes. Extreme environments can be screened for targeted characteristics where extreme environmental parameters shape microbial metabolism (Madigan et al., 2021). Microbial detergent enzymes from such environments can reduce synthetic chemical use and water by 'optimally functioning near the boiling point of water in the presence of dilute acids' (Madigan et al., 2021). Colonial green algae can be the future sources of biofuels, such as *Botryococcus braunii* which excreted long chain hydrocarbons with the consistency of crude oil. As a result, a fraction of the world's oil supply might feasibly be produced via photosynthetic algae in near future (Madigan et al., 2021). Besides their use in the industrial context, metagenomic approaches can also reveal microbial processes such as by providing insights into soil C, N, P, S and other elemental cycles (Myrold et al., 2013). Again, the impact of global warming and defrosting on microbial communities can be measured using metagenomics at specific sites that can provide information on the disappearing or new functions of shifting microbial communities (Mackelprang et al., 2011).

Herndorn (2018) has recently noted that the methane released by the presently thawing permafrost might become the major contributor to global warming, outweighing carbon dioxide. However, 'transport of methane from deep soils to the atmosphere might also be controlled by freeze-thaw dynamics and the activities of **CH_4-oxidising microorganisms** and CH_4-transporting plants'. Thus, 'improved understanding of carbon cycling across different redox regimes' in different defrost zones may lead to preventative measures. One effective example for the control of methane emissions in this context has been information generation on the relevant processes regulating CH_4 emission from rice fields by combined analysis of methanogenic and methanotrophic microbial activities by Krüger et al. (2001). Their findings defined suitable mitigation options to control CH_4 emission via late flooding or intermittent drainage.

All above listed examples agree with Stein (2020) who recently noted that 'although climate change presents the largest challenge ever to human society, it also offers opportunities to restructure our practices and lifestyles towards sustainability. Understanding how microbial metabolism evolved and continues to function in a geochemical context is thus crucial for our continued adaptation and survival'.

Microorganisms can also play a significant role as bioindicators for climate change; thus, if noted in a timely manner, can result in preventative measures implementations (Cavicchioli et al., 2019). Microbiology can also assist decision-making processes by providing key information defined by Scott (1991) (Box 1.1) which can be of significance to policy makers for timely delivery of the SGDs.

BOX 1.1 Microbiology assisting information generation

Factual information

- Bibliographic information
- Names of organisms
- Codes of organisms
- Culture collections
- Nucleic acid data

Interpretive information

- Taxonomic information
- Distribution maps
- Multimedia databases

Decision-making information

- Expert systems are computer programs that emulate decision-making processes based on factual knowledge and sets of logical rules for applying that knowledge to solve problems. They are used for agricultural applications, including pest diagnosis, pest management, environmental control, irrigation design, financial analysis, advisory services ...

Predictive information

- Mathematical models that have been built to predict, e.g., incidence of plant disease, in which parameters can be varied and the predicted outcome observed, or to make more elaborate predictions such as the rate at which pesticide tolerance may evolve in a population of pathogens.

Microbially-mediated solutions to current global problems are limitless, ranging from bio-cement, biopolymer production to energy generation and bioconversion of lignocellulosic waste. Current real-life examples include (1) ecoLogicStudio Business Ecosystem https://www.ecologicstudio.com/about AirBubble, children's playground integrated with air-purifying microalgae (https://www.ecologicstudio.com/projects/airbubble-playground-and-exhibition), (2) use of bio-luminesce to change the way light is produced (https://en.glowee.com/), (3) microbially induced calcium carbonate precipitation for bio-cement production (Choi et al., 2017) and even (4) using mushrooms to build biodegradable coffins, which 'gets absorbed into the earth within 4–6 weeks and total process of decomposition is completed within 2–3 years' (https://loop-biotech.homerun.co/).

Demain (1992) highlighted that 'microbial secondary metabolism was going to be a new theoretical frontier for academia and a new opportunity for industry'. He was proven to be correct over the last 30 years and the diversity of microbial metabolic compounds has fascinated scientists with a diverse range of applications. Microbial biotechnology has already and will continue to significantly contribute towards economic growth and employment creation (Timmis et al., 2017).

Pasteur and the prepared mind

> ***In the field of observation chance favours only the prepared mind, intuition is given only to him who has undergone long preparation to receive it***
>
> Louis Pasteur.

'The mind which is trained to observe the details of natural phenomena, and to reason concerning the bearing of known laws on such phenomena is the **prepared mind**'. It is a class of mind that grasps the 'significance of a new observation, or of a variation from a known sequence of events, and thus establishes a new law or invents a practical procedure' (Pearce, 1912). Pasteur and many of his co-workers due to their sound background training and scholarly understanding possessed the minds 'prepared to utilise scientific imagination, that enabled them to grasp the opportunity offered by **chance observation**' (Pearce, 1912). Incremental build up and cumulative learnings and experiences were required to understand the value of that particular '**chance observation**'. Only the prepared mind can place the 'proper block exactly where it belongs' which then can establish 'a new landmark for future progress' (Pearce, 1912).

Pearce (1912) highlighted that Pasteur's background in physical chemistry including crystallography was an example of the 'prepared mind' when he correctly interpreted the significance of the chance observation involving *Penicillium glaucum* in tartaric acid salt solution which ultimately changed the nature of the fluid due to the fermentative activity of the fungus. Throughout his career, Pasteur was able to switch between disciplines conducting investigations ranging from chemistry, to biochemistry, to biology, to medicine. Such interdisciplinary understandings also brought him into contact with problems that concerned larger and larger groups and resulted in finding solutions to many devastating ones (Hendrick, 1991). Examples include Pasteur, despite never having handled a silkworm before, agreeing to investigate the silkworm disease that was ravaging France's silk industry and identifying two associated diseases: silkworm nosema and flacherie. His investigation resulted in identification of hereditary and contagious disease (nosema, caused by a microsporidian) as well as implementation of hygiene rules, good ventilation and quarantine of the suspect batches (flacherie). As a result, for the first time, problems of hereditary and contagion were scientifically proven and prophylaxis rules were established (The middle years 1862–1877 | Institut Pasteur, https://www.pasteur.fr/en/institut-pasteur/history/middle-years-1862-1877).

Hendrick (1991) noted that Pasteur lived in a time of political instability, changing monarchs, falling empire and emerging republics and revolutions of 'brutal ferocity'. France was then faced with significant medical and agricultural problems and Pasteur deliberately chose scientific projects which, if successful, would alleviate basic problems of the era. Examples include elimination of the puerperal fever following childbirth to the development of vaccines for anthrax and rabies to saving the silkworm industry.

Ullmann (https://www.britannica.com/biography/Louis-Pasteur) noted that during his career, 'Pasteur touched on many problems, he never accepted defeat and he always tried to convince sceptics, though his impatience and intolerance were notorious when he believed that truth was on his side. Throughout his life he was an immensely effective observer and readily integrated relevant observations into conceptual schemes'.

All above listed qualities are reasons for his '**prepared mind**' stemming from years of painstakingly and meticulously conducted investigations with critical and objective thinking and careful observations and analysis. Examples include the **chance observation** on the lost pathogenicity of the chicken cholera culture that still retained pathogenic characteristics, which resulted in the immunisation and application of the concept to many other diseases (https://www.britannica.com/biography/Louis-Pasteur/Spontaneous-generation, 2021).

Calkins (1923) cited Duclaux (1920) who said '**wherever he went Pasteur was an initiator. Guided by an imagination so adventurous and at the same time so well controlled**'. Pasteur constantly was at the borders of interchange of biology, chemistry and medical sciences. These approaches laid the foundations of interdisciplinary understanding required in most scientific advancements. He 'saw and appreciated the significance of every observation and every opportunity which presented itself' during his career (Pearce, 1912).

As stated by Flexner (1911), 'remarkable achievements are never unique occurrences in nature. Even the greatest men rest on the shoulders of a large multitude of smaller ones who have preceded them, and epochal discoveries emerge out of a period of **intellectual restlessness** that affects many minds'. **So, what is the thinking process that lays the foundations of a prepared mind?** Moore et al. (1974) noted that there are two distinctly different kinds of thinking: '**creative thinking**' and '**critical thinking**' and '**effective thinking**' is both 'creative and critical'. They also noted that the **methods of modern science** are both **creative** and **critical.** During investigation of any phenomenon, 'a well-trained scientist first tries to formulate many tentative explanations' followed by rigorous tests to reveal possible explanations. During this process information is generated which allows the scientist to create and criticise all possible explanations 'until one that withstands rigorous testing is found' and 'create and criticise are the twin watchwords of an **effective thinker**' (Moore et al., 1974).

Prepared minds, in addition to the multidisciplinary understanding, must be effective thinkers. Higher Education Institutions (HEIs) can be significant contributors by providing guidance in shaping future '**prepared minds**' and facilitating the graduation of 'effective thinkers', who are ready to respond 'to the challenges of the 21st century with its complex environmental, social and economic pressures' (https://www.australiancurriculum.edu.au/f-10-curriculum/general-capabilities/critical-and-creative-thinking/).

Pasteur and the use-inspired research

> ***There is no such thing as a special category of science called applied science, there is science and its applications, which are related to one another as the fruit is related to the tree that has borne it***
>
> Louis Pasteur.

Stokes (1997) proposed that '**rigor and relevance are complementary notions** and when merged, they further the production, translation and implementation of instructional practices that are both rigorous (i.e., **evidence-based**) and relevant (i.e., **practice-based**)'. He also illustrated his concept in the **Pasteur's Quadrant** (PQ) (Fig. 1.1) in which three different approaches to the fundamental understanding of scientific problems versus their use for societal problems were illustrated. Stokes (1997) exemplified Pasteur as the one who bridges the gap between basic and applied research in contrast to Bohr (pure basic research) and Edison (applied research) (Smith et al., 2013).

Smith et al. (2013) noted that Stokes (1997) with the PQ communicated his 'vision to link research and practice, to improve the rigor and relevance of instructional practices and to connect researchers and practitioners in a meaningful and ongoing dialogue'. Such connections subsequently could enhance the field's ability to implement and sustain the use of practices that positively impact outcomes for learners. Moreover, they emphasised that educational design research and communities of practice can form frameworks through which the promise of PQ can be realised.

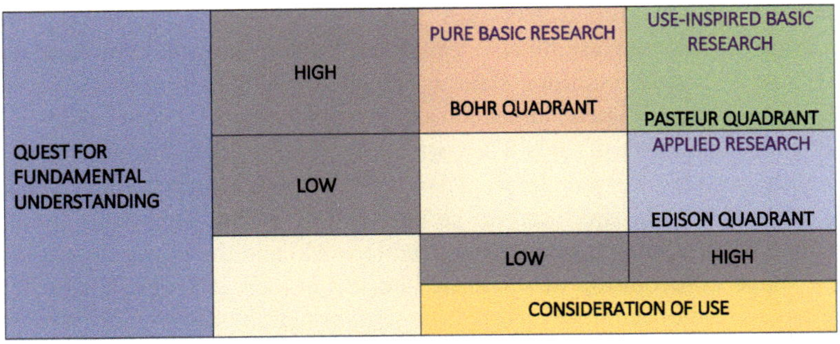

FIGURE 1.1 Quadrant model of scientific research. *Adapted from Stokes, D., 1997. Pasteur's Quadrant: Basic Science and Technological Innovation (p. 73). Brookings Institution Press, Washington, DC.*

Use-inspired research involves pursuing 'fundamental understanding but motivated by a question of use' (Stokes, 1997), and moreover, it enriches science that provides the crucial spark' (Krajewski and Chandawarkar, 2008).

> ***Oh, my goodness the mystery that has prompted my objective. My quality lies exclusively in my tirelessness. Do not promote what you can't explain, simplify, and prove early. Do not put forward anything that you cannot prove by experimentation***
>
> Louis Pasteur.

The way science is perceived, however, has changed in the 20th century and currently it might be interpreted as a commodity for public use and private sector utilisation (Godin and Schauz, 2016; Tijssen, 2018). It is demanded to be 'more aligned to pressing socioeconomic needs and practical problems, be it local communities, business interests or other user domains' (Sarewitz, 2016a; Tijssen, 2018). Accordingly, to capture 'the use-inspired identity of individual researchers within such conceptual change in terms of their concrete outputs and impacts' and within the realities of the current scientific climate, the PQ model is expanded by Tijssen (2018) into a **Pasteur's Cube (PC) model.** It comprises three, 'broadly defined constructs: a) knowledge production and skills creation (science and scientific research), b) technological development and artefacts production (engineering and technology) and c) end user engagement (commercialisation, entrepreneurship and innovation)'. In this model '**crossover researchers**' who can bridge science-oriented, user-oriented and application-oriented research are included.

Price and Behrens (2003) noted that if scientific 'research process proceeds in a linear fashion from a search for basic knowledge to application in the community context', the 'compelling insight offered by Stokes (1997) that is the drive for new knowledge and the pursuit of application (combined in a single effort)' can be missed and can become less capable of innovative and fail to contribute towards common good. 'Use-inspired knowledge prioritises practice, encourages translational research, fosters interdisciplinarity and dissolves rigid educational structures' (Tierney and Holley, 2008; Roy, 2018). Use-inspired research might also generate 'higher rates of "non-academic" outputs (e.g., policy recommendations, practice guidelines or prototype technologies) and associated impacts outside the scientific community' thus leading to generation of 'positive correlation with marketable outcomes and economic impact' (Tijssen, 2018). Tijssen (2018) highlighted that the '**Pasteur type scientists**', who are likely to be involved in use-inspired basic research, are often at interfaces between the academic world and the business sector, and hence shape co-evolution of scientific and technological development.

The term '**star scientist**' refers to most productive bio-scientists who 'had intellectual human capital of extraordinary scientific and pecuniary value' (Zucker and Darby, 1996). In the foundational years of biotechnology, the 'union of still-scarce-knowledge of new research techniques and the genius and vision to apply them in novel, valuable ways' coined the term (Zucker and Darby, 1996). Although productive in terms of breakthroughs and commercialisation of the products deriving from 'bioscientific revolution', the 'denial of knowledge-share by these scientists who were protective of their techniques, ideas and discoveries in the early years of the revolution with a tendency to collaborate more within their own institution resulted in slowed diffusion of advancements and knowledge to other scientists' (Zucker and Darby, 1996).

Kehoe and Tzabbar (2015) noted that 'whilst stars positively affect firms' productivity, their presence constrains the emergence of other innovative leaders in an organisation'. They also found that 'firm productivity and innovative leadership amongst non-stars in a firm are greatest when a star has broad expertise and collaborates frequently'. Baba et al. (2009) in their investigation related to the 'role of university—industry collaborations in shaping the innovative performances of universities and firms' concluded that 'engaging in research collaborations, measured as co-invention with "**Pasteur scientists**", increases firms' R&D productivity and the number of registered patents'. In contrast, they noted that 'firms' collaborations with star scientists exert little impact on their innovative output'.

Increasingly important role of regional engagement for timely delivery of the SDGs: 'globally competitive but locally engaged'

OECD (2007) recommended that to be competitive in a **globalising knowledge economy** its member countries needed to 'invest in their innovation systems at the national and regional levels'. As noted by the OECD (2007), dependency on access to new technologies, knowledge and skills at the local level in line with the parallel processes of globalisation and localisation turned OECD countries to knowledge-intensive products and services generation. They placed considerable emphasis 'on meeting regional development goals, by nurturing the unique assets and circumstances of each region, particularly in developing **knowledge-based industries**'. Accordingly, **HEIs** have become central to this process and had to change their strategic approaches as there was a lack of strategic focus on their contributions to the regions they belonged to. Moreover, such strategic contributions were not defined by the public policies either (OECD, 2007).

Long years of pursuit of knowledge unrelated to the surrounding environment or aligning with national goals had unintended detrimental effects in the long run 'in terms of providing opportunities for lifelong learning and contributing to the development of knowledge-intensive jobs thus enabling local workforce to remain within their communities'. It was time to change from sporadic approaches to systematic ones such as building of skills, entrepreneurialism and innovation within regions including indigenous development. To facilitate these changes, policy responses 'which initially focused on enhancing the capacity for technological innovation through technology transfer and interactions between HEIs and private industry were widened to include public services, social and organisational innovation, and to engage HEIs in the wider social fabric of which they are part' (OECD, 2007). OECD (2007) noted that HEIs as a regional 'key resource' can 'help serve regional development most obviously by contributing to a region's comparative advantage in knowledge-based industries and to its human capital base, but also for example by helping to generate new businesses, by contributing to tax revenues and by providing content and audience for local cultural programmes'.

However, 'in the past, neither public policy nor the HEIs themselves have tended to focus strategically on the contribution that they can make to the development of regions where they are located. Particularly for older, traditional HEIs, the emphasis has often been on serving national goals or on the pursuit of knowledge with little regard for the surrounding environment' (OECD, 2007). These facts set the scene and it has become clear that in order 'to be able to play their regional role, HEIs must do more than simply educate, and research, they must engage with others in their regions, provide opportunities for lifelong learning and contribute to the development of knowledge-intensive jobs which will enable graduates to find local employment and remain in their communities' (OECD, 2007).

Benneworth and Sanderson (2009) noted that for regional institutions located in 'areas of low demand' from external partners, the HEIs 'have to build up local demand for their knowledge'. However, risks related to 'long-term, costly and volatile' situations exist. Accordingly, with partnership and collaborative models of local capacity building, the HEIs might maximise benefits whilst minimising the risks. Academic staff preferably adopt and align with the engagement initiatives in universities as well as with the 'policy-makers who seek to foster stronger links between regional universities and their proximate spaces' Gunasekara (2006). Moreover, 'incorporating different ways of knowing in research and management has the potential to bring creativity to environmental problem-solving through integrating ways of knowing and innovation via co-producing knowledge' (Freitag, 2015).

In the above context, adapting the approaches recommended by the OECD and the UN and starting from the regional level knowledge-based sustainable practices and building in an incremental fashion towards global level targets can significantly reduce delays in reaching the SDGs (Fig. 1.2). Efforts can also be linked at similar ecosystem regions of the world; accordingly, initiatives, e.g., South to South Cooperation backed by Triangular Cooperation (https://www.unsouthsouth.org/about/about-sstc/) can be of significance in such incremental build-up. Some important aspects of such cooperation are as follows:

- 'increase and improve communications amongst developing countries, leading to a greater awareness of common problems and wider access to available knowledge and experience as well as the creation of new knowledge in tackling development problems;

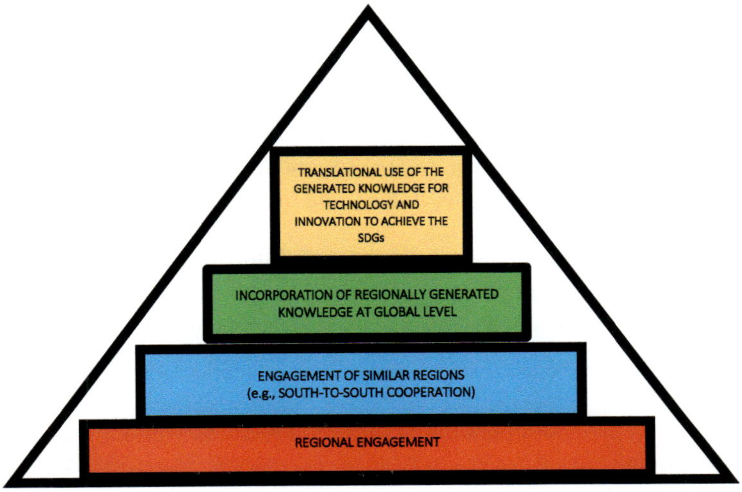

FIGURE 1.2 Incremental build-up of knowledge from regional to global level for timely delivery of the SDGs via implementation of effective and innovative technologies.

- recognise and respond to the problems and requirements of the least developed countries, land-locked developing countries, small island developing states and the countries most seriously affected by, for example, natural disasters and other crises and
- enable developing countries to achieve a greater degree of participation in international economic activities and to expand international cooperation for development' (https://www.unsouthsouth.org/about/about-sstc/).

Science is the highest personification of the nation because that nation will remain the first which carries the furthest the works of thought and intelligence. Science knows no country, because knowledge belongs to humanity, and is the torch which illuminates the world.

Louis Pasteur.

McNie et al. (2016) add that 'decision makers call upon and fund science to solve urgent problems, catalyse innovation and inform policy decisions. But the standard categories for describing, planning and assessing research, especially the persistence of "basic" and "applied," conceal much of the complexity and diversity of the contexts for conducting and using research, especially the role of knowledge users in the research process'. As a result, they propose a 'multi-dimensional research typology that divides research into three general activities: (1) knowledge production, (2) learning and engagement and (3) organisational and institutional processes, with a spectrum of value criteria ranging from strongly science-centric to strongly user-oriented. Their model enables consideration of the isolated knowledge value of science, the consideration and context of use and the engagement of intended users. Such approach they claim "can help inform and improve science-policy planning and decisions, aid in assessing the potential of existing projects, programs and institutions to achieve particular goals and yield insights about the strengths and weaknesses of completed projects"'.

Mori (2014) highlights that '**Autonomous Decentralised System** (ADS) concept and its chain of technologies and markets becomes a real-world example of concept-oriented R&D approach'. Technological advances and innovations including the **Smarter Planet initiative** (IBM100 - Smarter Planet) will subsequently apply these concepts into the needs of the 'sustainable society, economy and life'. Mori concludes that

- 'Technology without concept is reckless; concept without technology is arid,
- Technology without operation is illusion; operation without technology is time-killing'

Microbiology discipline-based teaching and Pasteurian pedagogies

My greatest desire, I repeat, is to find a way to perfect my teaching skills

Louis Pasteur.

National Research Council (2000) states that expertise may not be effectively translated into teaching within the area of expertise. 'Expert teachers know the kinds of difficulties that students are likely to face, and they know how to tap into their students' existing knowledge in order to make new information meaningful' whilst assessing their students' progress. Expert teachers thus must acquire pedagogical content knowledge and not just the content knowledge (Shulman, 1986, 1987; National Research Council, 2000).

Opinel (2008) noted that 'the structure and organisation of Pasteurian pedagogy mirrored Pasteur's conception of experimental science training, in which theory and practice were closely linked. Pasteur adopted the late 19th century trend of militant line to emphasise the fundamental importance of experiment' (Hulin, 2007; Opinel, 2008). 'The application of these principles in pedagogy made the Institut Pasteur a unique example of such instruction, a position that it kept for many years'. Courses offered at the Institut Pasteur still 'combine theoretical and practical teaching, as per the original pedagogical principles developed by Louis Pasteur and Emile Roux' (Opinel 2008).

Without theory, practice is but routine born of habit. Theory alone can bring forth and develop the spirit of invention

Louis Pasteur.

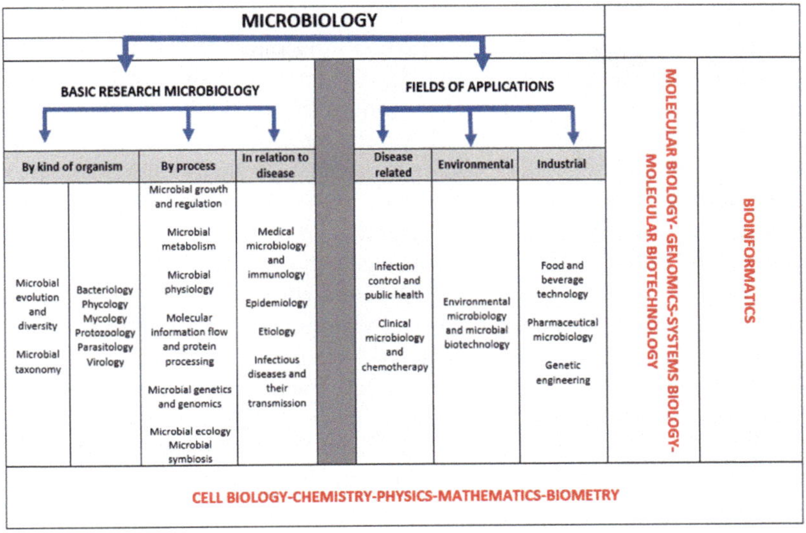

FIGURE 1.3 Minimum background requirement of foundational knowledge and multidisciplinary understanding in microbiology (built on the original structure by Black (2007) and including current teaching topics in microbiology by Madigan et al. (2021)).

Constructivism advocates teaching as a process that guides students to construct their own knowledge by building connections from what they already know to the information the teacher intends them to acquire. 'Students are more likely to learn and remember ideas that are reinforced by their own interests and experiences in the world around them'. Accordingly, 'classroom science teaching is recommended to make explicit links between scientific knowledge and everyday experiences' and the discipline of 'microbiology is particularly well suited for these recommendations because the effects of microorganisms are more evident in the world around us' (e.g., new infectious diseases, crop failures, food poisoning, antibiotic resistance ... ...) (Steffe and Gale, 1995; von Glaserfeld, 1995; Brooks and Brooks, 1999; Handelsman, 2002).

Microbiology discipline-based teaching (Kurtböke, 2008, 2016), thus, is as important as it has ever been for timely delivery of the SDGs for which involvement of many interdisciplinary prepared minds is required. Fig. 1.3 highlights the minimum background knowledge and multidisciplinary learning and understanding needs in microbiology on the way to possessing a prepared mind required to recognise the problem and implement effective and innovative solutions.

Global citizenship education

> ***The Greeks bequeathed to us one of the most beautiful words in our language – the word 'enthusiasm' – en theos – a God within. The grandeur of human actions is measured by the inspiration from which they spring. Happy is he who bears a god within, and who obeys it***
>
> Louis Pasteur.

It has been claimed that Plato (https://plato.stanford.edu/entries/plato-myths/) did not believe that 'most people were clever enough to look after their own and society's best interest, so the few clever people of the world needed to lead the rest of the flock' (Shaw, 2021). Such an approach thus resulted in the creation of an 'elite form of specialist and authoritative knowledge' (https://www.open.edu/openlearn/nature-environment/the-environment/nature-matters-systems-thinking-and-experts/content-section-2.3). 'In contemporary society, doctors and scientists have thus been considered the experts with dominant body of knowledge that is overall, inaccessible to the layman. Despite not understanding, public believed and highly valued the opinion of medical professionals or of scientific discoveries' (https://www.open.edu/openlearn/nature-environment/the-environment/nature-matters-systems-thinking-and-experts/content-section-2.3). Pasteur and scientists of his era greatly contributed to the establishment of such public trust and respect. Pasteur was instrumental in the establishment of the 'innate faith' placed in scientists to solve the problems of mankind (Henrick, 1991). Pasteur and his co-workers' efforts strengthened the views that 'in times of crisis, the governments and public in modern industrial societies automatically turn to scientists to solve society's problems'.

However, events taking place globally to the detriment of the environment in the 1980s such as the Bhopal gas disaster in India, Chernobyl nuclear reactor explosion and the oil spill from the tanker Exxon Valdez in Alaska 'have generated scepticism towards traditional expert support – particularly scientific or technical (technocratic) expertise, which is often viewed as being too closely aligned with business interests' (https://www.open.edu/openlearn/nature-environment/the-environment/nature-matters-systems-thinking-and-experts/content-section-2.3).

Most recently, as the world became more 'increasingly interconnected, human rights violations, inequality and poverty still threaten peace and sustainability'. The **Global Citizenship Education (GCED)** has been established as the 'UNESCO's response to these challenges' (https://mgiep.unesco.org/global-citizenship). It is designed to empower learners of all ages to understand global challenges as a whole, beyond their local issues and encourages them to be active promoters of more peaceful, tolerant, inclusive, secure and sustainable societies (https://mgiep.unesco.org/global-citizenship). There might be prime opportunity to utilise GCED's 'socially connected and emotionally engaged' platform with the inclusion of SDGs and an introductory level of literacy into the scientific disciplines involved in the design of required technologies to meet the SDGs.

Freitag et al. (2016) noted that although 'the success of citizen science in producing important and unique data is attracting interest from scientists and resource managers, the questions remain about the credibility of **citizen science data**'. Citizen science programs thus have 'to meet the same standards of credibility as academic science'. However, 'training and managing significant numbers of volunteers with limited resources' may not necessarily produce the desired outcomes.

Recently, Timmis et al. (2019) highlighted 'the urgent need for microbiology literacy in society'. As microorganisms play a key role in ecosystem functions, they noted that this fact 'necessitates their due consideration in decisions that are taken by individuals and families in everyday life as well as by individuals and responsible bodies at all levels and stages of community, national and planetary health assessment, planning and the formulation of pertinent policies'. They also added that unlike other subjects that have impact on our daily lives and there is widespread understanding, e.g., financial affairs, health and transportation, microbiology literacy is lacking in the general population. 'Choices involving microbial activity implications are often opaque, and the information available is sometimes biased and usually incomplete, and hence creates considerable uncertainty'. Therefore, they stress that microbiology literacy in society is indispensable for informed personal decisions, as well as for policy development in government and business, and for knowledgeable input of societal stakeholders in such policymaking (Timmis et al., 2019).

Timmis et al. (2019) also add that 'microbiology literacy needs to become part of the **world citizen job description**' and invite microbiologists, microbiological-learned societies and microbiology literate professionals, to participate in and contribute towards creation of microbiology literate societies for generation of required educational funds as well as 'convincing educators, policy makers, business leaders and relevant governmental and non-governmental agencies to support and promote this initiative'.

Understanding scholarship and expertise

In this section of the chapter, key definitions related to scholarship, expertise and beyond will be listed, which will form a background to the proceeding section. With rapidly advancing technology and availability of electronic tools, these key definitions must be visited frequently by novices to fully understand what is required to become another Pasteur.

Foundations of modern universities

Encyclopaedia Britannica's definition of modern universities highlights their evolution from the medieval schools known as ***studia generalia***, generally recognised places of study open to students from all parts of Europe. The earliest **studia** arose out of efforts to educate clerks and monks beyond the level of the cathedral and monastic schools. The inclusion of scholars from foreign countries then constituted the primary difference between the **studia** and the schools from which they grew (https://www.britannica.com/topic/university).

The first modern university in Halle, Europe (1694) set the scene 'to renounce religious orthodoxy of any kind in favour of rational and objective intellectual inquiry' (https://www.britannica.com/topic/university). 'In the later 18th and 19th centuries religion was gradually displaced as the dominant force as European universities became institutions of modern learning and research and were secularised in their curriculum and administration'. At the 'University of Berlin (1809) laboratory experimentation replaced conjecture; theological, philosophical and other traditional doctrines' that were from then on 'examined with a new rigour and objectivity; and modern standards of academic freedom were pioneered' (https://www.britannica.com/topic/university). 'The German model of the university as a complex of graduate schools performing advanced research and experimentation proved to have a worldwide influence' (https://www.britannica.com/topic/university).

Academic and academic freedom

Subsequent establishment of academic freedom allowed the freedom required by 'teachers and students to teach, study and pursue knowledge and research without unreasonable interference or restriction from law, institutional regulations or public pressure'. Such academic freedom, however defined, was not 'in the comfort or convenience of teachers and students but in the benefits to society, i.e., the long-term interests of a society are best served when the educational process leads to the advancement of knowledge, and knowledge is best advanced when inquiry is free from restraints by the state, by the church or other institutions or by special-interest groups' (https://www.britannica.com/topic/university). 'Protected by papal bulls and royal charters, the universities became legally self-governing corporations with the freedom to organise their own faculties, control admissions and establish standards for graduation' (https://www.britannica.com/topic/academic-freedom).

Academic is used to describe things that relate to the work done in schools, colleges and universities, especially work which involves **studying and reasoning** rather than practical or technical skills (https://www.collinsdictionary.com/dictionary/english/academic-expertise).

The word 'graduate' originates from Medieval Latin '**graduārī** "to take a degree". It ultimately derives from the Latin "**gradus**", meaning "a step"'. In each graduation, a step is taken to the next level of education (www.merriam-webster.com/dictionary/graduate). One of those highest levels of steps is the '**Doctor of Philosophy**' (PhD). The name derives from Latin 'philosophiae doctor or doctor philosophiae'. The word 'philosophy' in this context, however, is different to the field or academic discipline of 'philosophy' but is used in a broader sense in accordance with its original Greek meaning: '**lover of wisdom**'.

'**Sage**' or '**Sagdom**' in classical philosophy relates to someone who has attained wisdom and who lives 'according to an ideal which transcends the everyday', whereas 'the philosopher is not wise but possesses the self-awareness of lacking wisdom, and thus pursues it'. Wisdom involves an integration of knowledge, experience and deep understanding, as well as a tolerance for the uncertainties of life (https://www.psychologytoday.com/au/basics/wisdom).

Expert and expertise

Expertise, special skill or knowledge is acquired by training, study or practice which distinguish experts from **novices** and less experienced colleagues. An **expert** has a broad and deep competence in terms of knowledge, skills and experience through practice and education in a particular field with 'faculty for judging or deciding rightly, justly and wisely'. Such status is accorded by 'peers and public in a well-distinguished domain' (see https://www.nap.edu/read/9853/chapter/1).

Expertise characteristics, skills and knowledge of an expert or of a system distinguish experts from novices and less experienced people. In many domains, there are objective measures of performance capable of distinguishing experts from novices. 'Historically, an expert was referred to as a sage, with **profound thinking, distinguished for wisdom and sound judgement**'. 'Many accounts of the development of expertise emphasise that it comes about through long periods of deliberate practice' (Ericsson 2006; Ericsson et al., 2018).

Dreyfus and Dreyfus (2005) 'describe a five-stage phenomenological of skill acquisition': (1) novice, (2) advanced beginner, (3) competence, (4) proficiency and (5) expertise. 'Expertise is the highest stage' and they argue that contrary to the claims of 'knowledge engineers, expertise in general cannot be captured in rule-based expert systems'. It is because 'expertise is based on the making of immediate, unreflective situational responses; and the intuitive judgement is the hallmark of expertise'. They also add that 'deliberation is certainly used by experts, if time permits, but it is done for the purpose of improving intuition, not replacing it'. They also warn about the danger where expertise may be diminished through over-reliance on calculative rationality in bureaucratic societies.

Hartelius (2010, 2011) stated that what resources would non-experts have to evaluate claims put before them when faced with the competing claims of expertise. They also note that danger lies when encountering someone with autonomous expertise who may not possess the skill to persuade people to hold their points of view, where someone with merely attributed expertise may be persuasive but lack the actual knowledge pertaining to a given subject.

Scholarship and 'citizen scholar'

Scholarship relates to the systematised knowledge of a learned person, exhibiting accuracy, critical ability, thoroughness and erudition (https://www.collinsdictionary.com/dictionary/english/scholarship).

Hartelius and Cherwitz (2010) note that the 'engagement is the strategic leveraging of a scholar's intellectual capital in the interest of local, national and global communities and publics'. 'Engagement means being a **citizen scholar**'. It is the 'integration of theory, practice and production' with a purpose 'to provide fitting responses to social exigencies and the engineering of human action'.

Skilled memory, metacognition and Dunning–Kruger effect

'Experts knowledge is connected and organised around important concepts and it is "conditionalised" to specify the contexts in which it is applicable; it supports understanding and transfer (to other contexts) rather than only the ability to remember'. Moreover, as noted by Ericsson and Stasewski (1989) **skilled memory** enables experts to rapidly encode, store and retrieve information within the domain of their expertise and thereby circumvent the capacity limitations that typically constrain novice performance.

People's abilities to retrieve relevant knowledge can vary from being effortful to relatively effortless (fluent) to automatic (Schneider and Shiffrin, 1977, quoted in National Research Council, 2000). Automatic and fluent retrieval are important characteristics of expertise (National Research Council, 2000). Fluent retrieval does not mean that experts always perform a task faster than novices. Because experts attempt to understand problems rather than to jump immediately to solution strategies, they sometimes take more time than novices (Getzels and Csikszentmihalyi, 1976 quoted in National Research Council, 2000).

Chi et al. (1982) notes that novices sort problems into categories based upon surface features, whereas experts categorise them based upon their deep structures. Experts' ability to 'rapidly retrieve complex configurations from long-term memory' constitutes an important feature of expert performance. No number of 'rules and facts can capture the knowledge an expert has when the expert has stored experience of the actual outcomes of tens of thousands of situations' (Dreyfus and Dreyfus, 2005). National Research Council (2000) states that understanding expertise is important because it provides insight into the nature of thinking and problem solving. Research shows that it is not simply general abilities, such as memory or intelligence, nor the use of general strategies that differentiate experts from novices. Instead, experts have acquired extensive knowledge that affects what they notice and how they organise, represent and interpret information in their environment. This, in turn, affects their abilities to remember, reason and solve problems.

'**Metacognition** refers to the ability to monitor one's current level of understanding and decide when it is not adequate' (Brown, 1981; National Research Council, 2000). National Research Council (2000) states that 'the ability to monitor one's approach to problem solving to be metacognitive is an important aspect of the expert's competence. Experts step back from their first, oversimplistic interpretation of a problem or situation and question their knowledge that is relevant'.

Lifelong learning can be affected by one's perception of 'what it means to be an expert' (National Research Council, 2000). Models promoting the way of thinking that stresses 'accomplished novices are proud of their achievements but realise that there is much more to learn' rather than the models that assume 'experts know all the answers are the effective ones' (National Research Council, 2000). Adaptation of such models is important in the training of future experts. In addition, National Research Council (2000) highlights that 'it can be a mistake simply to expose novices to expert models and assume that the novices will learn effectively'; what they will learn thus will depend on how much they already know.

Moreover, 'hypothetical cognitive biases resulting in **illusory superiority**' (**Dunning–Kruger effect**) can pose a risk for novices. Kruger and Dunning's (1999) investigation on 'unskilled and unaware of it: how difficulties in recognising one's own competence can lead to inflated self-assessments', clearly highlights the problem of 'low ability' people's overestimation of their own ability compared to high ability people's underestimation of their own ability. They also add that without self-awareness of metacognition, people cannot evaluate their level of competence. Dunning et al. (2003) also add that failure to recognise one's own incompetence derives from incorrect self-assessment of competence stemming from their ignorance of a given activity's standards of performance. 'Poor performers', thus, 'are not in a position to recognise the shortcomings of their performance (Kruger and Dunning, 1999; Dunning, 2011; Dunning and Heltzer, 2014).

Future prospects and obstacles for a prepared mind

Washburn and Olbrys (2019) stated that 'today's workplace, and that of the future, requires individuals to be creative, creative thinkers and have strong communication and collaboration skills' (https://education-reimagined.org/resources/partnership-for-21st-century-learning/). Due to the tremendous amount of readily accessible information, a need to be digitally literate and tech savvy is also more important than ever before. In addition to these 'future ready' skill sets, a need exists to inspire and empower students to become active and engaged citizens who believe that their voice and societal contributions matter now and, in the future (Washburn and Olbrys, 2019).

Classroom teaching and graduate attributes

OECD defines the role of tertiary education as a contributor to social and economic development through four major missions (https://www.oecd.org/education/skills-beyond-school/41266690.pdf):

1. The formation of human capital (primarily through teaching),
2. The building of knowledge bases (primarily through research and knowledge development),

3. The dissemination and use of knowledge (primarily through interactions with knowledge users),
4. The maintenance of knowledge (inter-generational storage and transmission of knowledge)

HEIs are continuously looking into different delivery methods to contribute towards the above listed missions. Classroom teaching is now changing and focus is directed into '**student-centred teaching and learning**' (McCabe and O'Connor, 2014), including (i) inquiry-based learning (Inquiry-based learning - Department of Education, Skills and Employment, Australian Government (https://www.dese.gov.au/australian-curriculum/national-stem-education-resources-toolkit/i-want-know-about-stem-education/what-works-best-when-teaching-stem/inquiry-based-learning)) and (ii) problem-based learning (Problem-Based Learning | Center for Teaching Innovation (https://teaching.cornell.edu/teaching-resources/engaging-students/problem-based-learning)). Universities all around the world define key **graduate attributes** which are a combination of '**graduate qualities**' and '**generic skills**' and incorporate them into the program of enrolment to be able to deliver baseline needs of the future '**prepared minds**'. An Australian example is listed in Box 1.2.

BOX 1.2 Graduate attributes (example from the University of the Sunshine Coast in Australia (www.usc.edu.au)

Graduate qualities

1] **Creative and critical thinkers**: Independently solve problem and think critically; generate original ideas and concepts and appreciate innovation and entrepreneurship,
2] **Empowered**: Apply cognitive, technical and communication skills and knowledge to real work issues and contexts,
3] **Engaged:** Contribute positively to diverse communities through research, service and leadership,
4] **Ethical:** Understand ethical dilemmas, make ethical and professional choices and lead with integrity in intellectual, professional and community pursuits,
5] **Knowledgeable**: Build disciplinary and interdisciplinary knowledge through scholarly approach,
6] **Sustainability-focussed**: Recognise the interrelated nature of all systems and support a socially just and environmentally healthy future.

Generic skills

1] **Applying technologies**: Apply information and communication technologies and discipline specific technologies and equipment,
2] **Collaboration**: Work with other people to achieve a desired outcome, negotiating roles, responsibilities and contributions and accountability,
3] **Communication**: Share knowledge and understanding with diverse audiences effectively, respectfully and culturally appropriately using a variety of modes, techniques and media,

Box 1.2 Graduate attributes (example from the University of the Sunshine Coast in Australia (www.usc.edu.au)—cont'd

4] **Information literacy**: Know when and how to find relevant information using diverse sources and methods to make sense of information and construct new understandings,

5] **Organisation**: Systematically plan and set goals, manage time and information to successfully accomplish tasks

Bayh–Dole Act

In 1980, The Bayh-Dole Act (Public Law, 96–517, 12 December 1980) resulted in universities being able to retain property rights of federally funded research outcomes conducted by their employees and to exclusively license the results in return for royalties. Although this resulted in significant increases in university licensing activity, some argue that it drove 'faculty away from curiosity driven-basic research' (Washburn, 2008; Thursby and Thursby, 2010, 2011). Similarly, 'the kind of basic experimentation that leads to a greater understanding of how the world works has largely been set aside in favour of projects considered to have more immediate market potential' (Rae-Dupree, 2008).

> ***Do not share the opinion of those narrow minds who disdain everything in science which has not have immediate application … A theoretical discovery has out of merit of its existence: … it awakes hope, and that is all. But let it be cultivated, let it grow, and you will see what it will become***
>
> Louis Pasteur.

Eisenberg and Nelson (2002) noted that 'public science is more than a prelude to product development. At its best, it is a social commitment to disinterested investigation of the world by credible experts operating under the critical scrutiny of their peers'.

> ***I desire judgement and criticism upon all my contributions. Little tolerant of frivolous or prejudiced contradiction, contemptuous of that ignorant criticism which doubts on principle, I welcome with open arms the militant attack which has a method of doubting and whose rule of conduct has the motto 'more light'***
>
> Louis Pasteur.

Eisenberg and Nelson (2002) also highlighted that public science is a shared archive of an expanding knowledge base, a training ground for future researchers and from which future advances in human understanding will grow. Its social value does not depend on the ultimate profitability of the advances it spawns.

> ***Time is the best appraiser of scientific work, and I am aware that an industrial discovery rarely produces all its fruit in the hands of its first inventor***
>
> Louis Pasteur.

Publish or perish

It has been noted that the 'recurring criticism of our publishing principle is its emphasis on productivity, as a function of number and quality of publications. Any gains in efficiency are thought to come at the expense of innovation, boldness, heterogeneity and, ultimately, so-called scientific progress' (deRond and Miller, 2005; Sarewitz, 2016b). Rawat and Meena (2014), however, add that 'the emphasis on publishing has decreased the value of the resulting scholarship as scholar must spend time scrambling to publish whatever they can manage, rather than spend time developing significant research agenda'.

> ***When you believe you have found an important scientific fact, and are feverishly curious to publish it, constrain yourself for days, weeks, years sometimes, fight yourself, try and ruin your own experiments, and only proclaim your discovery after having exhausted all contrary hypotheses. But when, after so many efforts you have at last arrived at a certainty, your joy is one of the greatest which can be felt by a human soul***
>
> Louis Pasteur.

Rawat and Meena (2014) also point out that 'the pressure to publish-or-perish' can detract from the time and effort academics can devote to teaching undergraduate and post-graduates. As a result, this single-minded focus on the academics-as-researchers may cause faculty to neglect or be unable to perform some other key responsibilities (see Section Classroom teaching and graduate attributes). In addition, Neill (2008) highlights that 'under pressure to generate voluminous output, scientists often fall prey to double publishing, self-plagiarism and submitting the "minimal publishable unit" and asks, "are these ethical grey areas, or true transgressions?"'

Conclusions

This chapter lists key definitions (Section Understanding scholarship and expertise) and pressures academics face (Section Future prospects and obstacles for a prepared mind) whilst maintaining the focus and continuing to serve in line with the foundational principles of academia. Temponi (2005) states that 'the adoption of a **continuous improvement** (CI) approach in higher education requires not only upper administration commitment, but also uncovering the current underlying culture and examining the appropriateness of the objectives to adopt CI. A culture of a long-term commitment to CI implies engaging the administrative and academic systems and all the stakeholders of the institution. This was identified as a major roadblock for quality initiatives'.

The chapter also highlights that during such continuous improvements, academic core values (Section Understanding scholarship and expertise) must be visited and revisited by not only administrators but also by lifelong learners on the route to expertise. The chapter also notes the importance of maintaining target-directed focus all along, with particular emphasis placed on the Pasteurian pedagogies for the discipline of microbiology that value proof by experimentation. Technological advances are immense and information literacy is paramount as well as recognising quality amongst the vast ocean of information presented. This can be a difficult task for novices, and HEI's role is imperative in providing guidance.

The chapter also stresses the importance of keeping a Pasteurian spirit and adapting his dedication to scientific facts as well as valuing laboratory experimentations in the field of microbiology (Reich, 2008). The true meaning of laboratory experimentation must be revisited in the virtual laboratory skills learning environment (Lacey and Wall, 2021) to get the balance right for novices.

> ***The great art consists in devising decisive experiments, leaving no place to the imagination of the observer. Imagination is needed to give wings to thought at the beginning of experimental investigations on any given subject. When, however, the time has come to conclude and to interpret the facts derived from observations, imagination must submit to the factual results of the experiments***
>
> Louis Pasteur.

Whilst implementing Pasteurisation for sustainable futures, the key issue for the educators is how to install such Pasteurian spirit, and effective thinking to novices to be able to contribute towards the creation of future prepared minds who are crucially required for timely delivery of sustainable futures in current complex settings.

> ***I beseech you to take interest in these sacred domains so expressively called laboratories. Ask that there be more and that they be adorned for these are the temples of the future, wealth, and well-being. It is here that humanity will grow, strengthen, and improve. Here, humanity will learn to read progress and individual harmony in the works of nature, while humanity's own works are all too often those of barbarism, fanaticism and destruction***
>
> Louis Pasteur.

References

Baba, Y., Shichijo, N., Sedita, S.R., 2009. How do collaborations with universities affect firms' innovative performance? The role of "Pasteur scientists" in the advanced materials field. Res. Pol. 38 (5), 756–764.

Benneworth, P., Sanderson, A., 2009. The regional engagement of universities: building capacity in a sparse innovation environment. High Educ. Manag. Pol. 21 (1), 1–18.

Benyus, J.M., 2013. A biomimicry primer. In: Biomimicry Resource Handbook: A Seed Bank of Best Practices. Biomimicry 3.8. http://biomimicry.net/b38files/A_Biomimicry_Primer_Janine_Benyus.pdf.

Bhushan, B., 2009. Biomimetics: lessons from nature—an overview. Phil. Trans. Math. Phys. Eng. Sci. 367 (1893), 1445—1486.

Black, J.G., 2007. Microbiology: Principles and Explorations, seventh ed. Wiley, John Wiley and Sons Inc., p. 6 Figure 1.3.

Brooks, J.G., Brooks, M.G. (Eds.), 1999. Search of Understanding: The Case for Constructivist Classrooms. Association for Supervision and Curriculum Development, USA. ISBN 0871203588.

Brown, A.L., 1981. Metacognitive development and reading instruction. In: Shapiro, R. (Ed.), Theoretical Issues in Reading Comprehension. Erlbaum, Hillsdale, N.J.

Calkins, G.N., 1923. Debt of science to Pasteur. Sci. Mon. 17 (1), 5—16.

Cash, D.W., Belloy, P.G., 2020. Salience, credibility, and legitimacy in a rapidly shifting world of knowledge and action. Sustainability 12 (18), 7376.

Cash, D., Clark, W.C., Alcock, F., Dickson, N.M., Eckley, N., Jäger, J., 2002. Salience, credibility, legitimacy, and boundaries: linking research, assessment and decision making. SSRN Electr. J. https://doi.org/10.2139/ssrn.372280. KSG Working Papers Series.

Cash, D.W., Clark, W.C., Alcock, F., Dickson, N.M., Eckley, N., Guston, D.H., Jäger, J., Mitchell, R.B., 2003. Knowledge systems for sustainable development. Proc. Natl. Acad. Sci. U. S. A. 100 (14), 8086—8091.

Cavicchioli, R., Ripple, W.J., Timmis, K.N., Azam, F., Bakken, L.R., Baylis, M., Behrenfeld, M.J., Boetius, A., Boyd, P.W., Classen, A.T., Crowther, T.W., 2019. Scientists' warning to humanity: microorganisms and climate change. Nat. Rev. Microbiol. 17 (9), 569—586.

Chi, M.T.H., Glasser, R., Rees, E., 1982. Expertise in problem solving. In: Sternberg, R.J. (Ed.), Advances in the Psychology of Human Intelligence, vol. 1. Erlbaum, Hillsdale, NJ, pp. 7—75.

Choi, S.G., Chu, J., Brown, R.C., Wang, K., Wen, Z., 2017. Sustainable biocement production via microbially induced calcium carbonate precipitation: use of limestone and acetic acid derived from pyrolysis of lignocellulosic biomass. ACS Sustain. Chem. Eng. 5 (6), 5183—5190.

Chung, D., Cha, M., Guss, A.M., Westpheling, J., 2014. Direct conversion of plant biomass to ethanol by engineered *Caldicellulosiruptor bescii*. Proc. Natl. Acad. Sci. U. S. A. 111 (24), 8931—8936.

Clark, W.C., 2007. Sustainability science: a room of its own. *Proc. Natl. Acad. Sci.* U. S. A. 104 (6), 1737—1738.

Collado-Ruano, J., 2015. Biomimicry: a necessary eco-ethical dimension for a future human sustainability. Fut. Hum. Image (5), 23—57.

De Rond, M., Miller, A.N., 2005. Publish or perish: bane or boon of academic life? J. Manag. Inq. 14 (4), 321—329.

De Vrieze, J., De Mulder, T., Matassa, S., Zhou, J., Angenent, L.T., Boon, N., Verstraete, W., 2020. Stochasticity in microbiology: managing unpredictability to reach the sustainable development goals. Microbial Biotechnol. 13 (4), 829—843.

Demain, A.L., 1992. Microbial secondary metabolism: a new theoretical frontier for academia, a new opportunity for industry. In: Chadwick, D.J., Whelan, J. (Eds.), Secondary Metabolites: Their Function and Evolution, pp. 3—23. John Wiley and Sons, New York.

Dreyfus, H.L., Dreyfus, S.E., 2005. Peripheral vision: expertise in real world contexts. Organ. Stud. 26 (5), 779—792.

Duclaux, E., 1920. Pasteur, the History of a Mind. Saunders, Philadelphia PA.

Dunning, D., 2011. The Dunning-Kruger effect: on being ignorant of one's own ignorance. In: Olson, J., Zanna, M.P. (Eds.), Advances in Experimental Social Psychology, vol. 44. Elsevier, New York, NY, pp. 247–296.

Dunning, D., Helzer, E.G., 2014. Beyond the correlation coefficient in studies of self-assessment accuracy: commentary on Zell & Krizan (2014). Perspect. Psychol. Sci. 9 (2), 126–130.

Dunning, D., Johnson, K., Ehrlinger, J., Kruger, J., 2003. Why people fail to recognize their own incompetence. Curr. Dir. Psychol. Sci. 12 (3), 83–87.

Eisenberg, R.S., Nelson, R.R., 2002. Public vs. proprietary science: a fruitful tension? Daedalus 131 (2), 89–101.

Ericsson, K.A., 2006. The influence of experience and deliberate practice on the development of superior expert performance. In: The Cambridge Handbook of Expertise and Expert Performance, vol. 38(685–705), p. 2.

Ericsson, K.A., Stasewski, J.J., 1989. Chapter 9: skilled memory and expertise: mechanisms of exceptional performance. In: Klahr, D., Kotovsky, K. (Eds.), Complex Information Processing: The Impact of Herbert A. Simon. Lawrence Erlbaum Associates, Hillesdale N.J.

Ericsson, K.A., Hoffman, R.R., Kozbelt, A., Williams, A.M. (Eds.), 2018. The Cambridge Handbook of Expertise and Expert Performance. Cambridge University Press.

Fagunwa, O.E., Olanbiwoninu, 2020. Accelerating the sustainable development goals through microbiology: some efforts and opportunities. Access Microbiol. 2 (5).

Flexner, S., October 16, 1911. The Biological Basis of Specific Therapy. Ether Day address at the Massachusetts General Hospital.

Freitag, A., 2015. Collaborative research seed grants for integrating knowledges and creating new knowledge. J. Ext. 53 (5).

Freitag, A., Meyer, R., Whiteman, L., 2016. Strategies employed by citizen science programs to increase the credibility of their data. Citiz. Sci. Theory Pract. 1 (1), 1–11. https://doi.org/10.5334/cstp.6.

Getzels, J.W., Csikszentmihalyi, M., 1976. The Creativity Vision: A Longitudinal Study of Problem Finding in Art. Wiley, New York.

Ghosh, A., Mehta, A., Khan, A.M., 2019. Metagenomic analysis and its applications. Encycl. Bioinf. Comput. Biol. 3, 184–193.

Gieryn, T., 1999. Cultural Boundaries of Science: Credibility on the Line. Chicago University Press, Chicago.

Godin, B., Schauz, D., 2016. The changing identity of research: a cultural and conceptual history. Hist. Sci. 54 (3), 276–306.

Gross, B.L., Kellogg, E.A., Miller, A.J., 2014. Speaking of food: connecting basic and applied plant science. Am. J. Bot. 101 (10), 1597–1600.

Grove, J.M., Childers, D.L., Galvin, M., Hines, S., Muñoz-Erickson, T., Svendsen, E.S., 2016. Linking science and decision making to promote an ecology for the city: practices and opportunities. Ecosys. Health Sustain. 2 (9), e01239.

Gunasekara, C., 2006. Leading the horses to water: the dilemmas of academics and university managers in regional engagement. J. Sociol. 42 (2), 145–163.

Handelsman, J., 2002. Microbiology as a change agent in science education. ASM News 68, 163–167.

Harkness, J.M., 2002. In appreciation: a lifetime of connections: Otto Herbert Schmitt, 1913–1998. Phys. Perspect. 4 (4), 456–490.

Hartelius, E.J., 2010. The Rhetoric of Expertise. Lexington Books.

Hartelius, J., 2011. Rhetorics of expertise. Soc Epistemol 25 (3), 211–215.

Hartelius, E.J., Cherwitz, R.A., 2010. The Dorothy doctrine of engaged scholarship: the rhetorical discipline "had it all along". Q. J. Speech 96 (4), 436–442.

Hendrick, R., 1991. Biology, History & Louis Pasteur: a new approach to teaching science. Am. Biol. Teach. 53 (8), 467–478.

Henry, A.D., Vollan, B., 2014. Networks and the challenge of sustainable development. Annu. Rev. Environ. Resour. 39, 583–610.

Herndon, E.M., 2018. Permafrost slowly exhales methane. Nat. Clim. Change 8 (4), 273–274.

Hulin, N., 2007. L'enseignement secondaire scientifique en France d'un sie'cle a' l'autre, 1802e1980. Institut national de la recherche pe'dagogique, Paris, p. 63.

Kehoe, R.R., Tzabbar, D., 2015. Lighting the way or stealing the shine? An examination of the duality in star scientists' effects on firm innovative performance. Strat. Manag. J. 36 (5), 709–727.

Kennedy, E.B., Marting, T.A., 2016. Biomimicry: streamlining the front end of innovation for environmentally sustainable products: biomimicry can be a powerful design tool to support sustainability-driven product development in the front end of innovation. Res. Technol. Manag. 59 (4), 40–48.

Kennedy, E., Fecheyr-Lippens, D., Hsiung, B.K., Niewiarowski, P.H., Kolodziej, M., 2015. Biomimicry: a path to sustainable innovation. Des. Issues 31 (3), 66–73.

van Kerkhoff, L., Lebel, L., 2006. Linking knowledge and action for sustainable development. Annu. Rev. Environ. Resour. 31, 445–477.

Kirkewoog, S., 2017. The Influence of biomimicry on sustainability. Eng. Technol. Manag. Student Projects 87. http://archives.pdx.edu/ds/psu/21450.

Krajewski, A., Chandawarkar, R.Y., 2008. Pasteur's quadrant: preparing training programs for "use-inspired" surgical research. J. Surg. Educ. 65 (4), 283–288.

Kruger, J., Dunning, D., 1999. Unskilled and unaware of it: how difficulties in recognizing one's own incompetence lead to inflated self-assessments. J. Pers. Soc. Psychol. 77 (6), 1121.

Krüger, M., Frenzel, P., Conrad, R., 2001. Microbial processes influencing methane emission from rice fields. Global Change Biol. 7 (1), 49–63.

Kurtböke, D.İ., 2008. Microbial biotechnology education from a regional development perspective. Microbiol. Aust. 29 (2), 102–105.

Kurtböke, D.İ., 2016. Applied Microbiology and Biotechnology teaching tailored towards regional needs and graduate employment. Microbiol. Aust. 37 (2), 69–72.

Lacey, K., Wall, J.G., 2021. Video-based learning to enhance teaching of practical microbiology. FEMS Microbiol. Lett. 368 (2), fnaa203.

Mackelprang, R., Waldrop, M.P., DeAngelis, K.M., David, M.M., Chavarria, K.L., Blazewicz, S.J., Rubin, E.M., Jansson, J.K., 2011. Metagenomic analysis of a permafrost microbial community reveals a rapid response to thaw. Nature 480 (7377), 368–371.

Madigan, M.T., Bender, K.S., Buckley, D.H., Sattley, W.M., Stahl, D.A., 2021. Brock Biology of Microorganisms, 16th Global Edition. Pearson Education Ltd. ISBN 13: 9781292404790, ISBN 10: 1292404795.

McCabe, A., O'Connor, U., 2014. Student-centred learning: the role and responsibility of the lecturer. Teach. High. Educ. 19 (4), 350–359.

McNie, E.C., Parris, A., Sarewitz, D., 2016. Improving the public value of science: a typology to inform discussion, design and implementation of research. Res. Policy 45 (4), 884–895.

Moore, W.E., McCann, H., McCann, J., 1974. Creative and Critical Thinking, second ed. Houghton Mifflin Company, Boston.

Mori, K., 2014. In: Mori, K. (Ed.), Concept-oriented Research and Development in Information Technology, first ed. John Wiley & Sons, Inc.

Myrold, D.D., Zeglin, L.H., Jansson, J.K., 2013. The potential of metagenomic approaches for understanding soil microbial processes. Soil Sci. Soc. Am. J. 78 (1), 3–10.

Nagata, T., Tamburini, C., Arístegui, J., Baltar, F., Bochdansky, A.B., Fonda-Umani, S., Fukuda, H., Gogou, A., Hansell, D.A., Hansman, R.L., Herndl, G.J., 2010. Emerging concepts on microbial processes in the bathypelagic ocean–ecology, biogeochemistry, and genomics. Deep Sea Res. Part II Top. Stud. Oceanogr. 57 (16), 1519–1536.

National Research Council, 2000. How People Learn: Brain, Mind, Experience and School. National Academy Press, Washington D.C.

Neill, U.S., 2008. Publish or perish, but at what cost? J. Clin. Invest. 118 (7), 2368.

OECD, 2007. Higher Education and Regions-Globally Competitive, Locally Engaged. OECD Publishing.

Opinel, A., 2008. Teaching microbiology at the Institute Pasteur: the application of a pedagogical principle (1889–1939). Res. Microbiol. 159 (1), 36–39.

Pearce, R.M., 1912. Chance and the prepared mind. Science 35 (912), 941–956.

Price, R.H., Behrens, T., 2003. Working Pasteur's quadrant: harnessing science and action for community change. Am. J. Community Psychol. 31 (3–4), 219–223.

Rawat, S., Meena, S., 2014. Publish or perish: where are we heading? J. Res. Med. Sci. 19 (2), 87.

Rae-Dupree, J., 2008. When academia puts profit ahead of wonder. N. Y. Times 7.

Reich, J.W., 2008. Integrating science and practice: adopting the Pasteurian model. Rev. Gen. Psychol. 12 (4), 365–377.

Roy, R., 2018. Extending Pasteur's quadrant: university strategic groups and sources of heterogeneity in university technology transfer office activities. Int. J. Technol. Transf. Commer. 16 (1), 23–36.

Sarewitz, D., 2016a. Saving science. N. Atlantis 49 (1), 4–40.

Sarewitz, D., 2016b. The pressure to publish pushes down quality. Nature News 533 (7602), 147.

Scoones, I., 2007. Sustainability. Dev. Pract. 17 (4–5), 589–596.

Scott, R., 1991. The universal issue: information transfer. In: Hawksworth, D.L. (Ed.), The Biodiversity of Microorganisms and Invertebrates: Its Role in Sustainable Agriculture. CAB International, Wallingford, pp. 245–266 (CASAFA Report Series No. 4.).

Schneider, W., Shiffrin, R.M., 1977. Controlled and automatic human information processing: I. Detection, search, and attention. Psychol. Rev. 84 (1), 1.

Shaw, D., 2021. Plato's "Noble Lie" and the Management of Corporate Culture. Philosophy of Management, pp. 1–14.

Shulman, L., 1986. Those who understand: knowledge growth in teaching. Educ. Res. 15 (2), 4.

Shulman, L., 1987. Knowledge and teaching. Harv. Educ. Rev. 57 (1), 1–22.

Smith, G.J., Schmidt, M.M., Edelen-Smith, P.J., Cook, B.G., 2013. Pasteur's quadrant as the bridge linking rigor with relevance. Except. Child. 79 (2), 147–161.

Sorlini, C., 2020. Microbiology can make a great contribution to the new direction of global politics. Ann. Microbiol. 70, 64. https://doi.org/10.1186/s13213-020-01609-1.

Steffe, L., Gale, J. (Eds.), 1995. Constructivism in Education. Lawrence Erlbaum Associates, Inc., New Jersey.

Stein, L.Y., 2020. The long-term relationship between microbial metabolism and greenhouse gases. Trends Microbiol. 28 (6), 500–511.

Stokes, D., 1997. Pasteur's Quadrant: Basic Science and Technological Innovation. Brookings Institution Press, Washington, DC.

Temponi, C., 2005. Continuous improvement framework: implications for academia. *Qual. Assur. Educ.* 13 (1), 17–36.

Thursby, J., Thursby, M., 2010. University licensing: harnessing or tarnishing faculty research? Innovat. Pol. Econ. 10 (1), 159–189.

Thursby, J.G., Thursby, M.C., 2011. Has the Bayh-Dole act compromised basic research? Res. Pol. 40 (8), 1077–1083.

Tierney, W.G., Holley, K.A., 2008. Inside Pasteur's quadrant: knowledge production in a profession. Educ. Stud. 34 (4), 289–297.

Tijssen, R.J., 2018. Anatomy of use-inspired researchers: from Pasteur's Quadrant to Pasteur's Cube model. Res. Pol. 47 (9), 1626–1638.

Timmis, K., De Lorenzo, V., Verstraete, W., Ramos, J.L., Danchin, A., Brüssow, H., Singh, B.K., Timmis, J.K., 2017. The contribution of microbial biotechnology to economic growth and employment creation. Microbial Biotechnol. 10 (5), 1137–1144.

Timmis, K., Cavicchioli, R., Garcia, J.L., Nogales, B., Chavarría, M., Stein, L., McGenity, T.J., Webster, N., Singh, B.K., Handelsman, J., de Lorenzo, V., Pruzzo, C., Timmis, J., Martin, J.L.R., Verstraete, W., Jetten, M., Danchin, A., Huang, W., Gilbert, J., Lai, R., Santos, H., Lee, S.Y., Sessitsch, A., Bonfante, P., Gram, L., Lin, R.T.P., Ron, E., Karahan, Z.C., van der Meer, J.R., Artunkal, S., Jahn, Harper, L., 2019. The urgent need for microbiology literacy in society. Environ. Microbiol. 21 (5), 1513–1528.

von Glaserfeld, E., 1995. A constructivist approach to teaching. In: Steffe, L., Gale, J. (Eds.), Constructivism in Education. Lawrence Erlbaum Associates, Inc, New Jersey, pp. 3–16.

Washburn, J., 2008. University, Inc.: The Corporate Corruption of Higher Education. Basic Books, New York.

Washburn, E.K., Olbrys, S., 2019. Using student-centered teaching: inspiring and empowering teachers and learners. In: Young, N.D., et al. (Eds.), From Head to Heart: High Quality Teaching Practices in the Spotlight. Vernon Press, Delaware, USA, pp. 21–44 (Chapter 2).

WCED, 1987. Our Common Future (Report of the World Commission on Environment and Development). Oxford University Press, Oxford and New York.

Zucker, L.G., Darby, M.R., 1996. Star scientists and institutional transformation: patterns of invention and innovation in the formation of the biotechnology industry. Proc. Natl. Acad. Sci. U. S. A. 93 (23), 12709–12716.

Further reading

Guidance Note: Scholarship | Tertiary Education Quality and Standards Agency (teqsa.gov.au).

Partnership for 21st Century learning. https://education-reimagined.org/resources/partnership-for-21st-century-learning/.

Pasteur Quotes. https://todayinsci.com/P/Pasteur_Louis/PasteurLouis-Quotations.htmhttps://quotes.thefamouspeople.com/louis-pasteur-145.php.

Chapter 2

Public perception of microorganisms and microbiology education: a need for enhancing society's microbiology literacy

Graça S. Carvalho[1] **and Nelson Lima**[2]
[1]*CIEC (Research Centre on Child Studies), University of Minho, Braga, Portugal;* [2]*CEB (Centre of Biological Engineering), University of Minho, Braga, Portugal*

Chapter outline

Public perception: conceptions and social representations

The rapid spread of COVID-19 pandemic worldwide has forced governments to adopt strict measures to contain the infection. Different countries have been adopting different policies, but most of them intended to influence people's behaviours by keeping physical distancing to limit the SARS-CoV-2 virus transmission, such as imposing lockdown rules (WHO, 2020). Together with such legislative measures often following the World Health Organization (WHO), mixed conflicting messages from different agencies, media and social platforms have been causing unnecessary scares and inappropriate people's risk perceptions (Faour-Klingbeil et al., 2021). Therefore, understanding the

Importance of Microbiology Teaching and Microbial Resource Management for Sustainable Futures
https://doi.org/10.1016/B978-0-12-818272-7.00006-7

public perception of health risk and trust in information sources is a matter of great relevance to policymakers to take appropriate measures for better and more adequate public information.

The example of SARS-CoV-2 and COVID-19 pandemic social messages influencing public perception is used in this chapter to understand the notion of conceptions and social representations and its implications in education, particularly in microbiology education.

The history of science has shown that the difference between scientific thinking and common-sense thinking is not that simple. Indeed, scientists, like lay people, are also susceptible to historically situated circumstances and to inaccuracies that can be later amended and updated. Likewise, lay people's common sense can also be nourished by scientific knowledge. Traditionally, common sense can be understood in individual and collective dimension. The former is seen as what each person interprets and constructs of what surrounds him/her during his/her life, according to what he/she perceives and does; the latter is seen as what a group of people share in interpreting the world, beyond their singular view (Canguilhem, 1965, 1977).

This has been a matter of intense research intending to understand and define people's conceptions and people's representations. A '**situated conception**' (Clément, 2010) is created when someone interprets and reacts to a specific phenomenon or a question. The person's interpretation is firstly dependent on the way the phenomenon is presented to him/her or as the questions is formulated (i.e., a situated question) and, secondly, dependent on the person's previous conceptions that he/she mobilises to answer and interpret the phenomenon. Therefore, to move up from 'situated conceptions' to '**conceptions**' about a specific topic, one has to be asked with different questions, either redundant or very diverse (Clément, 2010). For example, to know someone's conceptions about the topic 'how the SARS-CoV-2 virus infects a person', several questions must be used in a questionnaire, interview or by crossing different information from different data collecting tools.

As said above, one's previous conceptions are mobilised to build newly reconstructed conceptions (Popp, 2018). For example, asking a microbiologist 'how the SARS-CoV-2 virus infects a person?' He/she may answer 'the virus spike protein attaches onto the ACE2 receptor in lung cells and injects its RNA that will be replicated inside the cell to produce new virus particles that will infect other cells causing disease'; a non-expert person may simply say 'the person breaths in the virus, and it causes disease'. They both mobilised their (scientific) knowledge and their previous conceptions to answer the question. The mechanism for interpreting the question may be the same for a scientist or a lay person, but the level of interpretation and understanding can achieve different quality levels. The more advanced conceptions one has, the more elaborated conceptualisation he/she can reach. This is why health literacy, scientific literacy and microbiology literacy, amongst other literacies, are so relevant nowadays.

When individual conceptions about a given topic are shared by a group of people, they were initially called 'collective representations' (Durkheim, 1968). Subsequently, the sociologist Serge Moscovici (1984) renamed them as '**social representations**'. Sociologists define a social group as a set of individuals united by at least one common characteristic (e.g., historical, cultural, religious, ethnic), having a feeling of belonging to the group, and recognised by the society (Durkheim, 1968). Therefore, social groups can be defined according to gender, religion, socio-professional categories, amongst others. In the socio-professional category, several social groups can be identified, e.g., civil servants, athletes, artists, teachers. Similarly, scientists are a social group, and share certain social representations, for example, science values, science research methods or even criteria of their scientific career.

Even within the scientific world, there are diverse social groups with specific specialisations and share different sets of science understanding and routines. For example, in the health sector, several social groups of scientific researchers and medical practitioners have been working intensively on the COVID-19 pandemic, such as microbiologists (more specifically virologists), immunologists, clinicians (more specifically, pulmonologists or intensive care specialists) and epidemiologists. Each of these social groups has its specific social representations, but they all contribute to developing science and welfare. The knowledge produced by this scientific community is in a continuous construction of new conceptions about the SARS-CoV-2 and COVID-19 pandemic. Therefore, the scientific community is sharing these new conceptions and are renewing their social representations about the SARS-CoV-2 and COVID-19 pandemic.

Whereas these scientific community's social representations are mostly based on science data, the general society is invaded by conflicting messages from a variety of sources that transmit scientifically correct information and scientifically erroneous information (or even fake news). All this information, either scientifically correct or incorrect, can participate in peoples' construction of their conceptions about the SARS-CoV-2 and COVID-19 pandemic. When a group of people, a community, shares the same conceptions (being they scientifically correct or not), they construct their social representation of the SARS-CoV-2 and COVID-19 pandemic.

Public perception on this topic varies largely, depending on what people accept as real or false, and consequently construct their conceptions. Therefore, diverse citizen's groups can emerge with their specific social representations on SARS-CoV-2 and COVID-19 pandemic. Since people's conceptions are the bases of general public perceptions, it is relevant to understand how conceptions are constructed and reconstructed. For this matter, Pierre Clément (2004) has created the KVP model that assumes person's conceptions result from Knowledge (K), Values (V) and Practices (P) interactions. This model has been applied to teachers and other education actors and pedagogical resources such as textbooks (Clément, 2010, 2019; Clément and Carvalho, 2017; McComas et al., 2018; Silva et al., 2021).

Microbiology education: educational actors' conceptions and didactic transposition

Although conceptions can be analysed in the context of other conceptual frameworks, the **KVP model** (Clément, 2004) has shown to be a very useful tool in the analysis of teachers conceptions (Carvalho and Clément, 2007; Clément, 2010; Clément and Carvalho, 2017; Magalhães-Júnior et al., 2020; Silva et al., 2017; Silva et al., 2019a,b) and syllabus and textbook conceptions (Franzolin et al., 2020; Tracana and Carvalho, 2014) of several countries.

In this KVP model, Knowledge (K) refers to the scientific community's information, referring to new scientific and technological knowledge, accepted for publication after analysed by pair reviewers; it is a continuous knowledge construction built upon previously published knowledge. Values (V) are assumed in a broad sense of the term, including ideologies, beliefs, fashion and lifestyles; for example, sexism, racism or xenophobia are considered as values, as well as the search for truth by science and the 'scientific ideologies' defined by the epistemologist (Canguilhem, 1977) to characterise the trends within the biological sciences, such as reductionism, anatomisation or absolute genetic determinism. Social Practices (P) refer to teachers' teaching practices, school textbooks authors publishers' practices, including their conceptions related to students' current and future social practices to whom textbooks are addressed to in the view of their professional future and their responsibility as present and future citizens (Carvalho and Clément, 2007).

This KVP model can be adjusted to anyone's conceptions about scientific knowledge or even to non-scientific knowledge. Therefore, anyone's conception about the SARS-CoV-2 and COVID-19 pandemic can be interpreted using this KVP model. Indeed, people's conceptions on this pandemic are the result of the interactions between these three poles: their previously scientific knowledge (K) on the topic, their cultural and personal values (V) and their social and personal practices (P), as exemplified in Fig. 2.1.

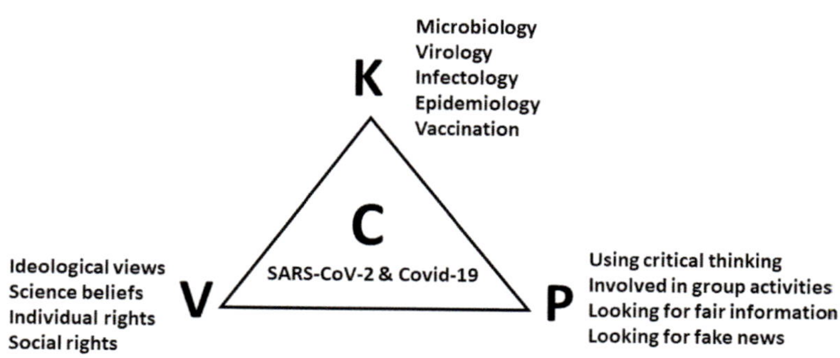

FIGURE 2.1 One's conceptions (C) on SARS-CoV-2 and COVID-19 pandemic can be analysed as interactions between Knowledge (K), Values (V) and Practices (P).

Science education is interested in the processes of teachers' transmission and students' appropriation of scientific knowledge, necessarily using the epistemological and historical approaches to scientific contents and social objectives. In the 1960s, Durkheim (1968) clearly expressed that:

'Education is the influence exercised by its adult generations on those not yet ripe for social life. It has as its object the awakening and development amongst each child certain number of physical, intellectual and moral states which are demanded of her by the political society in which she finds herself and the special groups for which she is particularly destined'.

In this sense, the school is an instrument to persons' training to acquire knowledge, skills, capacities, attitudes, habits and values that must constitute education programmes background network (Develay, 1992). Therefore, the school constitutes a privileged setting for learning international scientific knowledge and local socio-economic and political context knowledge (Carvalho, 2009).

Selecting contents to be learned in school is a permanent tension matter of great social and political relevance. The problem arises in deciding what knowledge is considered relevant for specific socio-cultural populations (Tracana and Carvalho, 2014). The sociologist Michel Verret (1975) gave great attention to this matter of selecting what should be taught in school and, subsequently, the mathematics educator Yves Chevallard (1985) developed the concept of '**Didactic transposition**'. This concept includes two phases:

- External didactic transposition (EDT), which analyses why some contents, and not others, are selected to be taught; such selected contents constitute school curricula and national programmes.
- Internal didactic transposition (IDT), which is concerned with how the contents are transposed in classroom teaching-learning processes; it is the phenomenon of 'didacticization' expressed by Develay (1992).

These authors (Chevallard, 1985; Verret, 1975) focussed mainly on knowledge contents; however, social values and social practices are also implicitly or explicitly involved in the selection process of contents to be taught (Clément, 2004, 2010). The K pole represents the scientific knowledge based on scientific publications that can be far or close to the learners' knowledge and therefore hard or easy for children and young people to understand. In addition, their ability to acquire knowledge takes place in the interaction of the other two poles, P and V as Clément (2004) says:

'The K pole represents scientific knowledge. [...] The personal knowledge assimilation occurs according to the two other poles, P and V. On the one hand, one uses one's previous knowledge that allows assimilating, retaining and reshaping everything useful for his/her practices: professional, personal and/or social - pole P. On the other hand, the attention that each one gives to knowledge, the importance that he/she gives to it, often depends on the interaction between this knowledge and his/her system of values - pole V'.

For example, in the decision to select the SARS-CoV-2 and COVID-19 pandemic topic to be included in the school curriculum, several tensions related to knowledge, values and practices (KVP) will arise. Topic contents to be selected from the international scientific community knowledge must be adjusted to children and young people's ages and their previous learning about microbiology (K), to the school community values (V) and to the teaching-learning practices (P), which vary all over the world and also within regions and countries.

There are three levels of didactic transposition (Chevallard, 1985): (i) the 'international scientific knowledge', (ii) the 'knowledge to be taught' in school (selected by decision-makers) (iii) and the 'knowledge learned' at school (in the process of teaching and learning). Between the first and second levels occurs 'External didactic transposition'; between the second and third levels the 'Internal didactic transposition'.

Fig. 2.2 exemplifies the didactic transposition in the SARS-CoV-2 and COVID-19 pandemic case, with the three didactic transposition levels. Scientists (Fig. 2.2: right column), through their successive newly formed conceptions (resulting of KVP interactions), develop 'international scientific knowledge' (Fig. 2.2: middle column) on specific topics (Fig. 2.2: left column), in this particular case on SARS-CoV-2 and COVID-19 pandemic.

In the 'external didactic transposition' process, policymakers select the 'knowledge to be taught' in schools from the 'international scientific knowledge'. These decisions are subject to influences of several sectors of the society, such as politicians, agencies, educators and other stakeholders. They all claim, based upon their conceptions (resulting of KVP interactions), about the needs they emphasise as being relevant for SARS-CoV-2 and COVID-19 pandemic education.

In the 'internal didactic transposition' process, teachers, using textbooks and other pedagogical resources, teach their students about SARS-CoV-2 and COVID-19 pandemic. In the teaching and learning process, teachers and

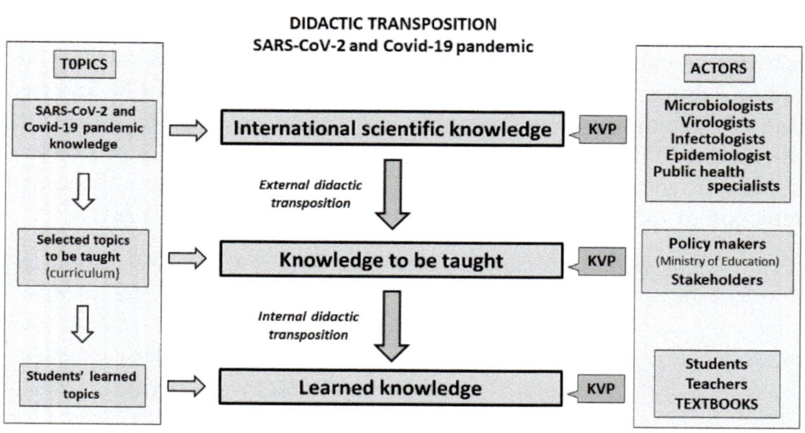

FIGURE 2.2 Didactic transposition of the SARS-CoV-2 and COVID-19 pandemic topic emphasising the actors' conceptions as resulting from KVP interactions.

textbook authors' conceptions (resulting from KVP interactions) are transmitted to students, who acquire them by imbibing in their previous conceptions, also constructed by KVP interactions (Fig. 2.2).

Pupils' perception of microorganisms and microbiology education

Children show interest in nature and natural phenomena since their early years, and the family and media are the primary sources of information before they attend formal education (Byrne and Grace, 2010; Dominguez et al., 2018). In the constructivist view of learning, children develop their ideas in trying to make sense of everyday situations and events (Byrne and Grace, 2010) to construct and reconstruct their conceptions about life, nature and natural phenomena from various sources (e.g., family, media). Consequently, it is expected that children's conceptions about many phenomena differ from the scientific view and that these alternative conceptions may influence further learning, often creating learning obstacles (Carvalho et al., 2004).

Children's conceptions about microorganisms are often scientifically incorrect since the notion of 'being microscopic' is difficult to them because it is 'something' not visible with the naked eye (Mafra et al., 2015) and they often have alternative ideas regarding microorganisms' size and morphology (Byrne, 2011). Many studies have shown that children, young people and adults tend to conceive microorganisms in an anthropocentric perspective, considering them as being unpleasant or harmful by associating them with spoiled food and garbage bins (Simonneaux, 2002) and to causing illness in humans (Byrne, 2011; Byrne and Grace, 2010; Mafra et al., 2020; Ruiz-Gallardo and Paños, 2018); 'but the mechanism of infection and subsequent recovery is poorly understood' (Byrne, 2011). In addition, it has been shown that children's conceptions on microorganisms' technological applications (Byrne and Sharp, 2006; Simonneaux, 2002) and their role in environmental balance (Leach et al., 1996) are scientifically incorrect or incomplete.

It has been argued that microorganisms can be explored in initial years of schooling (Byrne, 2011; Byrne and Sharp, 2006; Carey, 1985; Mafra et al., 2020; Ruiz-Gallardo and Paños, 2018) and that working with their previous conceptions is a relevant aspect on the constructivist approach to learning (Byrne and Grace, 2010; Driver et al., 1985; Osborne and Freyberg, 1985). However, even after formal education, some children continue to hold alternative and contradictory conceptions to the scientific ones, which seem to endure over time because many of these conceptions are rooted in their imagination and fantasy, usually acquired in the family, and are not based in scientific evidence, making learning difficult (Ballesteros de la Cruz et al., 2018; Byrne, 2011; Byrne and Sharp, 2006; Mafra et al., 2015). Therefore, alternative conceptions can constitute epistemological learning obstacles (Bachelard, 1938; Carvalho et al., 2004, 2007; Mafra et al., 2020). This suggests the need for a new type of approach to microbiology teaching and learning, preferably as early as possible in schooling (Byrne, 2011).

Indeed, the teacher's role in the classroom and the choice of appropriate teaching methodologies are essential to motivate students to develop scientific thinking, instead of just telling them what to know, think and why (Ruiz-Gallardo and Paños, 2018). **Experimental activities** to address microorganisms in primary school have shown to be an efficient approach (Faccio et al., 2013; Mafra et al., 2015, 2020; Patrick et al., 2008; Ruiz-Gallardo and Paños, 2018). Indeed, they capture the children's innate curiosity, make them involved in activities that make sense and encourage them to learn in this active process of cognitive development, which leads to a better understanding of the physical world (Myers, 2010; Ruiz-Gallardo and Paños, 2018).

Besides the teaching and learning processes, the **curriculum** must be adapted to the modern society and include from the early schooling the study of microorganisms and their influence on the daily children's life, to help them to make appropriate decisions now and in the future (Byrne, 2011). The curriculum for young people should also emphasise relationships between microbiology and medicine, agriculture, biotechnology or environmental protection (Ballesteros de la Cruz et al., 2018; Byrne, 2011). Despite this need of innovation, the research on the curricula and textbooks of various countries addressing microorganisms in primary school has shown to be inadequate or even non-existent in various countries, for example in England (Byrne, 2011), Germany (Hilge and Kattmann, 2003), India (Vijapurkar and Konde, 2014), Italy (Bandiera, 2007), Portugal (Mafra et al., 2015, 2016) or Spain (Ballesteros de la Cruz et al., 2018; López-Luengo et al., 2021).

It is expected that the SARS-CoV-2 and COVID-19 pandemic topic is added soon to the national curricula and textbooks of most countries because it is a matter of great social interest. On other words, its EDT (see Fig. 2.2) is likely to be quick. There is a period between the emergence of a scientific concept and its appearance in the national curriculum called '**Didactic Transposition Delay**' (DTD) (Quessada and Clément, 2007). It is dependent on two processes 'The delay of acceptance of a new scientific finding by the scientific community [...] The delay taken by the education system to decide to introduce new scientific conceptions in the syllabus, at a precise level of the curriculum' (Quessada and Clément, 2007). DTD has been analysed in different countries to understand why specific biology knowledge topics are integrated into national curricula quickly, take a long time or never appear at all (Clément, 2019; Quessada and Clément, 2007).

A DTD between a new emerging scientific concept and the corresponding appearance in school textbooks also occurs. Typically, the textbook's DTD is longer than the curriculum's DTD (but not always) because the topics in textbooks usually appear after being stated in the curriculum (Clément, 2019). However, we believe the SARS-CoV-2 and COVID-19 pandemic textbook's DTD will be shorter than the curriculum's DTD due to its severe social impact. Indeed, textbook authors and publishers will make this topic appear in textbooks before the countries' national curricula are updated, which involves heavier socio-political and administrative actions. This shortcut is shown in Fig. 2.3, where the SARS-CoV-2 and COVID-19 pandemic topic ('International scientific knowledge') surpasses the curriculum step ('Knowledge to be taught') to be at once taught in school ('Learned knowledge').

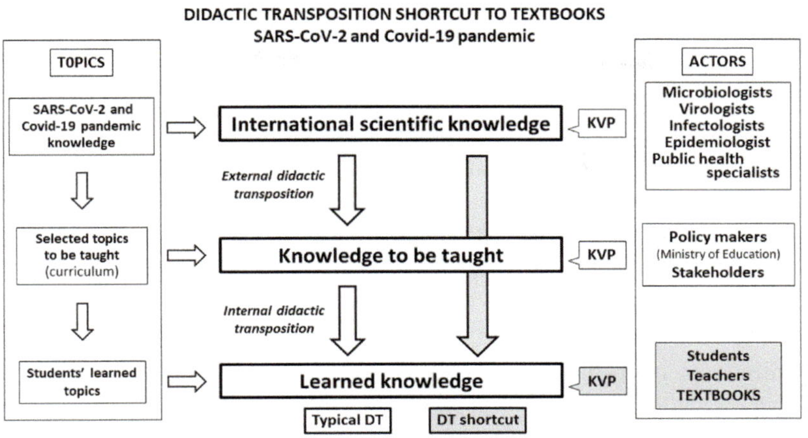

FIGURE 2.3 SARS-CoV-2 and COVID-19 pandemic topic shortcut in the didactic transposition (DT) that surpasses the 'Knowledge to be taught' step (curriculum).

The SARS-CoV-2 and COVID-19 pandemic topic that has emerged in late 2019 and had a massive social impact on the whole planet in 2020 is already (early 2021) a matter of classroom discussions between teachers and students ('Learned knowledge'), who will have textbooks soon to support the teaching and learning process. In these school discussions, teachers and students argue with their conceptions resulting from their KVP on the SARS-CoV-2 and COVID-19 pandemic; textbooks will also carry their publishers' and authors' KVP-derived conceptions (Fig. 2.3).

This discussion about microbiology education in early school years was an example of how scientific knowledge is introduced in curricula and how it takes time to be taught in the classroom. It was also referred to the particular focus on microorganisms' adverse or pathological effects on humans, which we believe may become even more stressed in the coming days due to the emerging SARS-CoV-2 and COVID-19 pandemic. However, it must be noticed that the school should not neglect the other microorganisms' important aspects such as their distribution in all Earth ecosystems and their relevance in medicine, agriculture, biotechnology or environmental protection. This understanding of key microbial activities is of great relevance not only for school children and young people but also for adults enhancing their microbiology literacy.

Enhancing society's microbiology literacy

As mentioned above, there is a general tendency to assume that microorganisms are harmful to humans. Therefore, the microorganisms' crucial role in all life on Earth must be highlighted (Connor et al., 2020) as they are essential in food and medicine production and in thriving the human ecosystem (Homburger et al., 2015) and take part in the carbon, nitrogen, phosphate and sulphur biogeochemical cycles (Tortora et al., 2013), which are processes vital for life to exist and persist on Earth (Falkowski et al., 2008). Microorganisms ubiquity, their diverse activities and influence on all living beings on Earth require an understanding of essential microbial processes for informed, evidence-based and decision-making at all levels of society, which is attained by enhancing scientific literacy, more specifically, microbiology literacy in society (Timmis et al., 2019; McGenity et al., 2020).

Scientific literacy has been defined as:

> *'An individual's scientific knowledge and use of that knowledge to identify questions, to acquire new knowledge, to explain scientific phenomena and to draw evidence-based conclusions about science-related issues, understanding of the characteristic features of science as a form of human knowledge and enquiry, awareness of how science and technology shape our material, intellectual, and cultural environments, and willingness to engage in science-related issues and with the ideas of science, as a reflective citizen'*
>
> OCDE (2009).

Similarly, **microbiology literacy** is related to knowledge of relevant microbial activities that are indispensable for informed personal decisions and central to society's empowerment to improve everyday life (Timmis et al., 2019). These authors exhort microbiology experts to collaborate to enhance microbiology literacy in society by delivering adequate knowledge of microbial activities to those professionals with significant societal impacts such as teachers, politicians, industry leaders and heads of international agencies. They strengthen the need to introduce microbiology core elements in all levels of education, from kindergarten to high school, because:

> *'It is essential that knowledge of microbes in society is raised to dispel the harmful belief and practice of microbiophobia. This is one central message of the microbiology literacy aid and will be at the forefront of its use in school curricula'*
>
> Timmis et al. (2019).

Being so relevant to teach microbiology in schools nowadays, microbiologists (with their 'international scientific knowledge') must develop their efforts to influence policymakers to transpose to national curricula ('knowledge to be taught') so that teachers and pedagogical resources (e.g., textbooks) help students to learn adequate microbiology in school ('Learned knowledge') (see Fig. 2.2). This is what didactic transposition is all about.

Several teaching and learning methods have been developed to provide students' positive experiences in microbiology learning, in several microbiology topics such as food, agriculture, health, medicine, environment and biotechnology, often including hands-on and experimental activities (Connor et al., 2020; Faccio et al., 2013; Mafra et al., 2015, 2020; Patrick et al., 2008; Ruiz-Gallardo and Paños, 2018).

Besides emphasising the need to improve school microbiology teaching, Kenneth Timmis et al. (2019) have claimed for a public service to promote adults' microbiology updating and to help those who have not studied microbiology at school, to learn the basics to be able to enhance their microbiology literacy. In this way, people will better understand 'the difference between what is rather certain, what is probable and what is unknown' (Timmis et al., 2019) and will be empowered to make evidence-based risk assessments and make informed decisions on their behalf.

Therefore, society's microbiology literacy is of utmost importance at the current SARS-CoV-2 and COVID-19 pandemic so that well-informed politicians can take adequate policies and actions to fight the pandemics and empowered people can understand them and make appropriate personal decisions. In other words, in a microbiology literate society, public perception (involving KVP interactions) of microorganisms and SARS-CoV-2 and COVID-19 pandemic is of higher level and, most probably, the society as a whole can react in a more effective way to overcome it.

Although recently it has been claimed strong efforts to enhance microbiology literacy in society (e.g., McGenity et al., 2020; Timmis et al., 2019), we must recognise that for a long time now, the international microbiology scientific community has shown increasing concern to bring the latest scientific developments closer to lay citizens. A pioneering approach goes back to the origins of microbiology with Frantisek Král, the founder of the first microbial culture collection in the world named 'Král'sche Sammlung von Mikroorganismen' around 1890. Working with Prof. Isidor Soyka of the Institute of Hygiene of the German University of Prague on bacteria cultivation methods in solid media, Král created a museum of permanent microorganism culture preparations for teaching purposes (Kocur, 1990).

Currently, several microbiology museums are exposing permanent or temporary exhibitions, such as the 'Microbe museum Micropia'[1] in Amsterdam, the Netherlands; 'Microbiology Museum of Butantan Institute'[2] in São Paulo, Brazil; 'Meet your microbes!'[3] at Melbourne Museum, Australia. These are museum examples that illustrate the continuous efforts to disseminate knowledge about microorganisms.

Furthermore, microbiology scientific societies (e.g., Microbiology Society[4]) or scientific journals (e.g., The *Journal of Microbiology & Biology Education* – JMBE,[5] published by the American Society for Microbiology; the Scientific Video Journal - JoVE[6]) promote microbiology activities and provide educational materials. Moreover, the 'International Microorganism Day' (IμD),[7] launched by the Portuguese Society of Microbiology in 2017, celebrates each year worldwide on the 17th of September everything that has to do with microbes. In addition, the 'World Microbiome Day'[8] focusses on microbe diversity, microbiome habitats, diets to feed and fuel microbiomes.

1. www.micropia.nl.
2. https://butantan.gov.br/atracoes/museu-de-microbiologia.
3. https://museumsvictoria.com.au/media-releases/meet-your-microbes/.
4. https://microbiologysociety.org/.
5. https://www.asmscience.org/content/journal/jmbe.
6. https://www.jove.com/education/84/microbiology.
7. https://www.internationalmicroorganismday.org/.
8. https://worldmicrobiomeday.com/.

All these enthusiastic activities allow us to be optimistic in the sense that formal and informal microbiology literacy has had a long path and has the potential to grow and reach wider audiences.

Acknowledgements

This research had partial financial support of Portuguese national funds through the FCT (Foundation for Science and Technology) within the framework of the CIEC project UIDB/00317/2020 and CEB project UIDB/04469/2020. It also had the partial support of the European Union's Horizon 2020 research and innovation programme under grant agreement N° 871129 - IS_MIRRI21 Project.

References

Bachelard, G., 1938. La Formation de L'Esprit Scientifique. Vrin.

Ballesteros de la Cruz, M., Paños, E., Reyes Ruiz-Gallardo, J., 2018. Los microorganismos en la educación primaria: ideas de los alumnos de 8 a 11 años e influencia de los libros de texto. In: Enseñanza de Las Ciencias: Revista de Investigación y Experiencias Didácticas, vol. 1, pp. 79–98. https://dialnet.unirioja.es/servlet/articulo?codigo=6373916.

Bandiera, M., 2007. Micro-organisms: everyday knowledge predates and contrasts with school knowledge. In: Pintó, R., Couso, D. (Eds.), Contributions from Science Education Research. Springer, pp. 213–224.

Byrne, J., 2011. Models of Micro-organisms: children's knowledge and understanding of micro-organisms from 7 to 14 years old. Int. J. Sci. Educ. 33 (14), 1927–1961. https://doi.org/10.1080/09500693.2010.536999.

Byrne, J., Grace, M., 2010. Using a Concept Mapping Tool with a Photograph Association Technique (CoMPAT) to elicit children's ideas about microbial activity. Int. J. Sci. Educ. 32 (4), 479–500. https://doi.org/10.1080/09500690802688071.

Byrne, J., Sharp, J., 2006. Children's ideas about micro-organisms. Sch. Sci. Rev. 88 (322), 71–79.

Canguilhem, G., 1965. La connaissance de la vie. Librairie philosophique J. Vrin.

Canguilhem, G., 1977. Idéologie et rationalité dans l'histoire des sciences de la vie. Librairie philosophique J. Vrin.

Carey, S., 1985. The human body. In: Press, M.I.T. (Ed.), Conceptual Change in Childhood. MIT Press, pp. 41–71.

Carvalho, G.S., 2009. A transposição didáctica e o ensino da biologia. In: Caldeira, A.M.A., Araújo, E.S.N.N. (Eds.), Introdução à Didática da Biologia. Escrituras, pp. 34–57.

Carvalho, G.S., Clément, P., 2007. Projecto "Educação em biologia, educação para a saúde e educação ambiental para uma melhor cidadania": análise de manuais escolares e concepções de professores de 19 países (europeus, africanos e do próximo oriente). In: Revista Brasileira de Pesquisa Em Educação Em Ciências, vol. 7, pp. 1–21 (2). http://hdl.handle.net/1822/7775.

Carvalho, G.S., Silva, R., Clément, P., 2007. Historical analysis of Portuguese primary school textbooks (1920–2005) on the topic of digestion. Int. J. Sci. Educ. 29 (2), 173–193. https://doi.org/10.1080/09500690600739340.

Carvalho, G.S., Silva, R., Lima, N., Coquet, E., Clément, P., 2004. Portuguese primary school children's conceptions about digestion: identification of learning obstacles. Int. J. Sci. Educ. https://doi.org/10.1080/0950069042000177235.

Chevallard, Y., 1985. La transposition didactique: du savoir savant au savoir enseigné. La pensée sauvage.

Clément, P., 2004. Science et idéologie: exemples en didactique et épistémologie de la biologie. In: Lang, P. (Ed.), Colloque Science - Médias — Société, pp. 53—69. http://science.societe.free.fr/documents/pdf/Sciences_medias_societe_2004/Clement.pdf.

Clément, P., 2010. Conceptions, représentations sociales et modèle KVP. Skholê 16, 55—70.

Clément, P., 2019. Umwelt, KVP & DTD: interactions entre Connaissances (K), Valeurs (V), Pratiques (P) & Délai de Transposition Didactique. Educ. Matem. Pesq. 21 (4), 18—35. https://doi.org/10.23925/1983-3156.2019v21i4p018-035. São Paulo.

Clément, P., Carvalho, G.S., 2017. Les conceptions d'enseignants de 31 pays sur la santé, entre approches biomédicale et socio-écologique. RDST 16, 33—56. http://journals.openedition.org/rdst/1545.

Connor, C.R., Watkins, M., Walter, B., Harper, J.D.I., 2020. Food for thought: bringing primary school microbiology to life. Teach. Sci. J. Aust. Sci. Teachers Assoc. 66 (1), 20—28. https://eds.a.ebscohost.com/eds/pdfviewer/pdfviewer?vid=4&sid=a7d32f33-f752-4497-ad06-06cd72e7ae9a%40sdc-v-sessmgr01.

Develay, M., 1992. De l'apprentissage à l'enseignement. ESF.

Dominguez, C.R.C., Leporo, N., Tino De Franco, M., Inglez, G.C., Gonçalves, V.M., Bizerra, A.F., 2018. Learning about microorganisms in childhood: four- to six-year-old children's voice in kindergartens and museums. Revista Brasileira de Pesqui. Em Educ. Em Ciências 18 (1), 1—25. https://doi.org/10.28976/1984-2686rbpec20181811.

Driver, R., Guesne, E., Tiberghien, A., 1985. Children's Ideas in Science. Open University Press.

Durkheim, É., 1968. Éducation et sociology. PUF.

Faccio, E., Costa, N., Losasso, C., Cappa, V., Mantovani, C., Cibin, V., Andrighetto, I., Ricci, A., 2013. What programs work to promote health for children? Exploring beliefs on microorganisms and on food safety control behavior in primary schools. Food Contr. 33 (2), 320—329. https://doi.org/10.1016/j.foodcont.2013.03.005.

Falkowski, P.G., Fenchel, T., Delong, E.F., 2008. The microbial engines that drive earth's biogeochemical cycles. Science 320 (5879), 1034—1039. https://doi.org/10.1126/science.1153213.

Faour-Klingbeil, D., Osaili, T.M., Al-Nabulsi, A.A., Jemni, M., Todd, E.C.D., 2021. The public perception of food and non-food related risks of infection and trust in the risk communication during COVID-19 crisis: a study on selected countries from the Arab region. Food Contr. 121, 107617. https://doi.org/10.1016/j.foodcont.2020.107617.

Franzolin, F., Prochazka, L.D.S., De Pietri, A.P.Z.S., Carvalho, G.S., 2020. Complexidade genética e a expressão da cor da pele, cor dos olhos e estatura humana: transposição didática (Genetic complexity and the expression of human skin color, eye color, and height: didactic transposition). Invest. Em Ensino de Ciências 25 (1), 239—261. https://doi.org/10.22600/1518-8795.ienci2020v25n1p239.

Hilge, C., Kattmann, U., 2003. The significance of microbes for biology teaching — a study of scientific students' conception. In: 4th ESERA Conference.

Homburger, S.A., Drits-Esser, D., Malone, M., Stark, L.A., 2015. Microbes as friends, not foes. Shifting the focus from pathogenesis to symbiosis. Am. Biol. Teach. 77 (9), 659—668.

Kocur, M., 1990. History of the Král collection. In: Sly, L.I., Iijima, T., Kirsop, B. (Eds.), 100 Years of Culture Collections. Proceedings of the Král Symposium to Celebrate the Centenary of the Establishment of the First Recorded Service Culture Collection. Institute for Fermentation, pp. 4—12.

Leach, J., Driver, R., Scott, P., Wood-Robinson, C., 1996. Children's ideas about ecology 3: ideas found in children aged 5–16 about the interdependency of organisms. Int. J. Sci. Educ. 18 (2), 129–141. https://doi.org/10.1080/0950069960180201.

López-Luengo, M.A., Díaz, E.G., Paños, E., Ruiz-Gallardo, J.-R., 2021. Microorganismos y hábitos de higiene. ¿Se aprende más en la Educación Infantil mediante fichas? Rev. Eureka Sobre Enseñanza Divulg. Ciencias 18 (2), 2302. https://doi.org/10.25267/Rev_Eureka_ensen_divulg_cienc.2021.v18.i2.2302.

Mafra, P., Lima, N., Carvalho, G.S., 2015. Experimental activities in primary school to learn about microbes in an oral health education context. J. Biol. Educ. 49 (2), 190–203. https://doi.org/10.1080/00219266.2014.923485.

Mafra, P., Carvalho, G.S., Lima, N., 2016. Os microrganismos nos programas e manuais escolares do 1.º e 2.º ciclo do ensino básico português. Gaia Scient. 10 (2), 52–59.

Mafra, P., Lima, N., Carvalho, G.S., 2020. Primary school health education: how children can learn about microbes and hand hygiene. In: Carvalho, G.S., Palhares, P., Azevedo, F., Parente, M.C. (Eds.), Improving Children's Learning and Well-Being. CIEC, Institute of Education, University of Minho, pp. 52–63. https://www.amazon.com.br/Improving-Childrens-Learning-Well-Being-Carvalho/dp/9728952635/ref=sr_1_1?__mk_pt_BR=ÅMÅŽÕÑ&dchild=1&keywords=Improving+children%27s+learning+and+well-being&qid=1590921126&sr=8-1.

Magalhães-Júnior, C.A., Corazza, Maria Júlia Kiouranis, N.M.M., Pansera-de-Aaraújo, M.C., Oliveira, A.L., Maricato, F.E., Schneider, E.M., Ortiz, A.J., Carmo, T. do, Carvalho, J. L. de, Araújo, J.L.D., RangelL, M., Gimenes, E.R., Carvalho, G.S., 2020. Concepções e representações sociais de professores sobre a sua formação inicial: construção e validação de um questionário. Ensaio Pesqui. Em Educa. Em Ciências (Belo Horizonte) 22, 1–35. https://doi.org/10.1590/21172020210110.

McComas, W.F., Reiss, M.J., Dempster, E., Lee, Y.C., Olander, C., Clément, P., Boerwinkel, D.J., Waarlo, A.J., 2018. Considering grand challenges in biology education: rationales and proposals for future investigations to guide instruction and enhance student understanding in the life sciences. Am. Biol. Teach. 80 (7), 483–492. https://doi.org/10.1525/abt.2018.80.7.483.

McGenity, T.J., Gessesse, A., Hallsworth, J.E., Garcia Cela, E., Verheecke-Vaessen, C., Wang, F., Chavarría, M., Haggblom, M.M., Molin, S., Danchin, A., Smid, E.J., Lood, C., Cockell, C.S., Whitby, C., Liu, S., Keller, N.P., Stein, L.Y., Bordenstein, S.R., Lal, R., et al., 2020. Visualizing the invisible: class excursions to ignite children's enthusiasm for microbes. Microb. Biotechnol. 1–44. https://doi.org/10.1111/1751-7915.13576.

Moscovici, S., 1984. The phenomenon of social representations. In: Farr, R., Moscovici, S. (Eds.), Social Representations. Cambridge University Press, pp. 3–69.

Myers, D., 2010. Psicología [Psychology]. Editorial Médica Panamericana.

OCDE, 2009. PISA 2009 Assessment Framework. Key competencies in reading, mathematics and science. In: OECD Report. http://www.oecd-ilibrary.org/education/pisa-2009-assessment-framework_9789264062658-en.

Osborne, R.H., Freyberg, P., 1985. Learning Science: The Implication of Children's Science. Heinemann.

Patrick, H., Mantzicopoulos, P., Samarapungavan, A., French, B.F., 2008. Patterns of young children's motivation for science and teacher-child relationships. J. Exp. Educ. 76 (2), 121–144. https://doi.org/10.3200/JEXE.76.2.121-144.

Popp, J.A., 2018. John Dewey's reconstructed conception of growth. Contemp. Pragmat. 15 (1), 165–178. https://doi.org/10.1163/18758185-01501008.

Quessada, M.P., Clément, P., 2007. An epistemological approach to French syllabi on human origins during the 19th and 20th centuries. Sci. Educ. 16 (9–10), 991–1006. https://doi.org/10.1007/s11191-006-9051-9.

Ruiz-Gallardo, J.R., Paños, E., 2018. Primary school students' conceptions about microorganisms. Influence of theoretical and practical methodologies on learning. Res. Sci. Technol. Educ. 36 (2), 165–184. https://doi.org/10.1080/02635143.2017.1386646.

Silva, H.M., Clément, P., Leão, I.M.S., Garros, T.V., Carvalho, G.S., 2017. Biology teachers' conceptions about the origin of life in Brazil, Argentina, and Uruguay: a comparative study. Zygon 52 (4), 943–961. https://doi.org/10.1111/zygo.12371.

Silva, H.M., Mortimer, E.F., Diaz, M.A., Tomasco, I.H., Gibram, D.E., Carvalho, G.S., 2019a. The human species origin: views of biology teachers from three Latin American countries. Theol. Sci. 17 (2), 257–272. https://doi.org/10.1080/14746700.2019.1596343.

Silva, H.M., Oliveira, A.W., Belloso, G.V., Díaz, M.A., Carvalho, G.S., 2021. Biology teachers' conceptions of Humankind Origin across secular and religious countries: an international comparison. Evol. Educ. & Outr. 14 (2), 1–12. https://doi.org/10.1186/s12052-020-00141-9.

Silva, H.M., Peñaloza, G., Tomasco, I.H., Carvalho, G.S., 2019b. Chimpanzee included in the genus Homo? How biology teachers from three Latin American countries conceive it. J. Biol. Educ. 53 (5), 506–515. https://doi.org/10.1080/00219266.2018.1501408.

Simonneaux, L., 2002. Analysis of classroom debating strategies in the field of biotechnology. J. Biol. Educ. 37 (1), 9–12. https://doi.org/10.1080/00219266.2002.9655839.

Timmis, K., Cavicchioli, R., Garcia, J.L., Nogales, B., Chavarría, M., Stein, L., McGenity, T.J., Webster, N., Singh, B.K., Handelsman, J., de Lorenzo, V., Pruzzo, C., Timmis, J., Martín, J.L.R., Verstraete, W., Jetten, M., Danchin, A., Huang, W., Gilbert, J., et al., 2019. The urgent need for microbiology literacy in society. Environ. Microbiol. 21 (5), 1513–1528. https://doi.org/10.1111/1462-2920.14611.

Tortora, G.J., Funke, B.R., Case, C.L., 2013. Microbiology: An Introduction. Pearson Education Limited.

Tracana, R., Carvalho, G., 2014. Poluição em livros didáticos portugueses: identificação de abordagens para resolver problemas de poluição e estratégias de prevenção. Revista Amazônica 13 (1), 8–21.

Verret, M., 1975. Le temps des études. Librairie Honoré Champion.

Vijapurkar, J., Konde, P., 2014. "Omne vivum ex vivo"? A study of middle school students' explanations of the seemingly sudden appearance of some life forms. Res. Sci. Educ. 44 (6), 885–902. https://doi.org/10.1007/s11165-014-9406-1.

WHO, 2020. COVID-19 and Food Safety: Guidance for Food Businesses. World Health Organization, pp. 1–6.

Chapter 3

Skills and education for engineering biology

Jim Philp
Organisation for Economic Cooperation and Development (OECD), Paris, France

Chapter outline

Importance of Microbiology Teaching and Microbial Resource Management for Sustainable Futures
https://doi.org/10.1016/B978-0-12-818272-7.00005-5

Introduction

The bioeconomy paradigm calls for a transformation of the economy away from fossil resources, gradually to be replaced by biological feedstocks for industrial production (e.g., Asada and Stern, 2018; OECD, 2018; Priefer et al., 2017). At least three different narratives for the bioeconomy are apparent in national strategies (Bracco et al., 2018): one which focuses on replacement of fossil carbon, another that is driven by biotechnology and a third that seeks to optimise the use of available biomass in an ecologically sustainable manner. It can be argued that biotechnology plays roles in the other two narratives.

These different narratives are played out in the way different countries write their national bioeconomy strategies. Those of the United Kingdom (HM Government, 2018) and the United States (US White House, 2012) have a very clear focus on synthetic biology as a driver of the bioeconomy. However, the revised bioeconomy strategy of the European Commission (2018) plays down the roles of biotechnology compared to its original bioeconomy strategy of 2012 (European Commission, 2012). Canada may adopt the functional definition of the bioeconomy of the European Commission but is also likely to 'rely on biotechnology as a competitive advantage' (Bioindustrial Innovation Canada, 2019). Japan's national bioeconomy strategy (Council for Integrated Innovation Strategy, 2020) is new and also has a strong focus on biotechnologies. This could change the nature of biology and biotechnology education non-uniformly, with different needs in those countries following an industrial bioeconomy agenda.

Traditional undergraduate and postgraduate curricula in biology and various engineering disciplines display some very marked differences (Waters and Sarin, 2011). Perhaps this is because a natural fit for biology graduates has been public research, whilst engineers are more likely to occupy jobs in industry (Delebecque and Philp, 2018). This chapter will examine why the education of biologists must change for them to fulfil key roles in this biotechnology-driven manufacturing paradigm. The essence can be summed up in the terms 'engineering biology' and 'synthetic biology'. In this chapter both terms are used, without attempting to distinguish them by definitions. Sometimes synthetic biology is used as it is the term used in particular publications.

The chapter is not about industrial biotechnology or biorefining, although these are also part of the new paradigm. Readers will find an emphasis here on design and prototyping rather than scale-up (see Fig. 3.2A). This is a hallmark of modern manufacturing – the separation of design from manufacturing (OECD, 2014b). Due to advances in information technologies, the two can be highly separated geographically and the communications done mostly electronically.

Three pillars of context: engineering biology and sustainability

In at least the last 5 years, evidence and media concentration on the climate emergency have made it clear to society that a transition to a more sustainable future is a necessity (e.g., Welch and Southerton, 2019). What is much less clear to society is that there is such an option as a bioeconomy, and that the bioeconomy can be driven by biotechnology (e.g., VTT, 2017). Most bioeconomy strategies stress that the goals are threefold: economic, environmental and social (e.g., Majer et al., 2018). This section looks at some aspects of all three of these pillars of sustainability and gives some context to the roles of engineering or synthetic biology.

Economic

> *'A confluence of advances in biological sciences — decades in the making — with the accelerating development of computing, automation and artificial intelligence (AI), is fuelling a new wave of innovation that could have significant impact on economies and societies, from health and agriculture to consumer goods and energy'.*
>
> McKinsey Global Institute (2020).

Many publications emphasise the environmental context of sustainability. To have a long-term future, sustainability must make sense economically. For biotechnologies to take a major place in this future, there has to be evidence of economic relevance. Publications addressing this have started to emerge. Most recently a study by the McKinsey Global Institute (2020) has examined the economic contributions of the 'bio-revolution' in some detail. Fig. 3.1 is an overarching chart from the report.

The figures for the impact of biotechnologies on the economy are notoriously difficult to quantify. Carlson (2016) discovered that in the United States there is no official mechanism to distinguish between products made through biology and products manufactured through other technologies. Data collection methodologies are not optimal for a number of reasons. This is due in part to the use of new bio-based pathways and value chains to create products previously manufactured in sectors completely dissociated from biology. Nevertheless, Carlson estimated that US revenues generated by biotechnology in 2012 were at least $ 324 billion, or the equivalent of over 2% of US GDP.

Environmental

Weiss et al. (2012) compared cradle-to-grave greenhouse gas emissions associated with conventional and bio-based chemicals, based on 44 life cycle

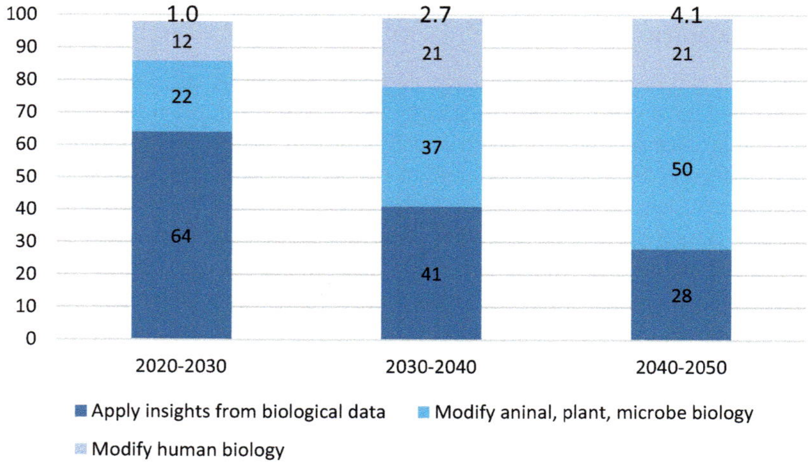

FIGURE 3.1 Estimated economic impact of the bio-revolution in the near, medium and long term. Sized applications by technology platform (partial estimate). The figure at the top of each column represents $ trillion. *Adapted from McKinsey Global Institute, May 2020. The Bio Revolution: Innovations Transforming Economies, Societies, and Our Lives.*

analysis studies covering approximately 60 individual bio-based materials and 350 different life cycle scenarios. They found that the bio-based materials save, on average, 55 ± 34 MJ non-renewable energy and 3 ± 1 kg CO_2 per kg material compared to their fossil-based counterparts.

Whilst liquid biofuels have been supported in public policy (quotas and mandates) to help governments meet climate targets (e.g., OECD, 2014a), the impacts of bio-based chemicals and materials has been largely ignored (Philp, 2015). Quite apart from the impacts microorganisms can have on climate-friendly industrial production, the key roles they have in carbon and nutrient cycling, animal (including human) and plant health, agriculture and the global food web are ignored at the political level (Cavicchioli et al., 2019). It could be that the next generation of engineering biologists will also bear the burden of moving microorganisms up the political agenda.

Social

> *'Advances in synthetic biology have the potential to develop new products, materials and services that could contribute to the 2030 United Nations Sustainable Development Goals. Support for synthetic biology initiatives in developing countries is needed to ensure that these benefits are open to all'.*
>
> French (2019).

The United Nations Sustainable Development Goals (SDGs)[1] cover the economic, environmental and social challenges of now and the future (e.g., Kroll et al., 2019). The social challenges are more difficult to characterise and quantify (Miola and Schiltz, 2019). For example, compliance with international conventions, such as the UN Convention on Biological Diversity (Lai et al., 2019) and attention to access and benefits sharing (ABS), are challenging political issues for which evidence gathering is fraught.

By contrast, job creation is relatively easy to measure. A claim for engineering biology is that it can create many high-value, knowledge-based jobs. Zymergen of the United States describe themselves as 'biofacturers'.[2] On 08 September 2020,[3] Zymergen announced $ 300 million in new investment to accelerate the delivery of novel, high performance materials using a bio-based approach to overcome fossil dependence. It is a rapidly changing company with around 700–800[4] employees, having been founded only in 2013.

At a larger scale, Scotland provides an interesting national example. For many decades, cornerstones of the Scottish economy were coal, steelmaking and shipbuilding. Deep coal mining is now extinct in Scotland, and steelmaking and shipbuilding are a fraction of their former size compared to even the 1980s. Meanwhile, the country employs over 41,000 people across some 770 organisations in the life sciences. Companies in the sector contribute in excess of £ 6.5 billion in turnover and about £ 2.4 billion in gross value added to the Scottish economy. The Scottish life sciences sector has grown at 9.6% per annum since 2010.[5]

What does the public know of these developments?

There is a need for learning on the part of both scientists and the public generally. To biologists, it is intuitive how microorganisms and sunlight form the life support system for the biosphere. The public is familiar with other arenas that have large impact on humans and humanity, such as finance, health, energy and transport and even nowadays climate change. But how microbial activities impact society – what Timmis et al. (2019) referred to as 'microbiology literacy' – is lacking.

1. https://sdgs.un.org/goals Last accessed: 10 December 2020.
2. https://www.zymergen.com/Last accessed: 10 December 2020.
3. https://markets.businessinsider.com/news/stocks/zymergen-takes-aim-at-3-trillion-chemical-materials-industry-with-300-million-in-new-funding- 1029571230#:~:text=Zymergen%20Takes%20Aim%20at%20%243,in%20New%20Funding%20%7C%20Markets%20Insider Last accessed: 10 December 2020.
4. One source quotes 732 employees for 2020: https://pitchbook.com/profiles/company/97130-35#overview Last accessed: 10 December 2020.
5. https://www.lifesciencesscotland.com/advantages/vibrant-diverse-growing-company-base Last accessed: 10 December 2020.

The acceptance of a new era of biomanufacturing based on engineering biology would be easier with a greater knowledge of these fundamentals in society at large. Enabling this acceptance is far from easy and requires that the scientists and engineers themselves improve communication of the messages in accessible language (Aguilar and Patermann, 2020). About 70% of the total potential impact of the bio-revolution could hinge on consumer, societal and regulatory acceptance (McKinsey Global Institute, 2020). Along with the formal science and engineering skills of the engineering biologist, communication at various levels will also be a necessary skill.

Bio-foundries and the design-build-test-learn cycle as a new manufacturing paradigm

The future education of the bio-based workforce can be discussed with a focus on the bio-foundry, no longer a concept but a reality (Hillson et al., 2019). The bio-foundry can be seen as a missing link in the biomanufacturing value chain (Kitney et al., 2019). The hope for the bio-foundry concept is that it will bring the reliability and reproducibility to biotechnology that has always been absent (Jessop-Fabre and Sonnenschein, 2019). In pre-clinical research alone, the lack of reproducibility has an estimated cost of $ 28 billion per annum in the United States (Freedman et al., 2015). Automation and robotics offer ways to overcome the challenges by removing lumpen human variation and 'craft', by allowing a much greater level of rational design, high throughput in experiments, improving statistical analysis and allowing programming with standardised protocols.

Burgard et al. (2016) suggested that what is needed is 'an integrated technology platform encompassing metabolic modelling, high-throughput (HT) pathway and strain construction, quantitative small-scale screening and systems biology, all of which are intimately linked to fermentation and process engineering'. This aligns with the view of Lee and Kim (2015) that 'researchers often fail to consider a fully integrated industrial bioprocess when developing microbial strains with new activities'.

The bio-foundry is the DBTL part of the fully integrated industrial bioprocess (Fig. 3.2A), and the part that this chapter concentrates on. Bio-foundries are highly automated facilities that comprise the extensive and coordinated use of laboratory robots (Fig. 3.2B) that are programmed to perform specific tasks according to a workflow. Typically, different platforms within the bio-foundries perform different tasks, for example, liquid handling, genetic assembly and characterisation functions. Bio-foundries are based on information infrastructures that allow the robots and other equipment within the bio-foundries to be programmed to follow detailed, complex workflows (Chao et al., 2017). The combination of bio design tools (BioCAD) and bio-foundries is rapidly producing a new type of biology - digital biology (Morris et al., 2005). Eventually, the automated generation of robotic build

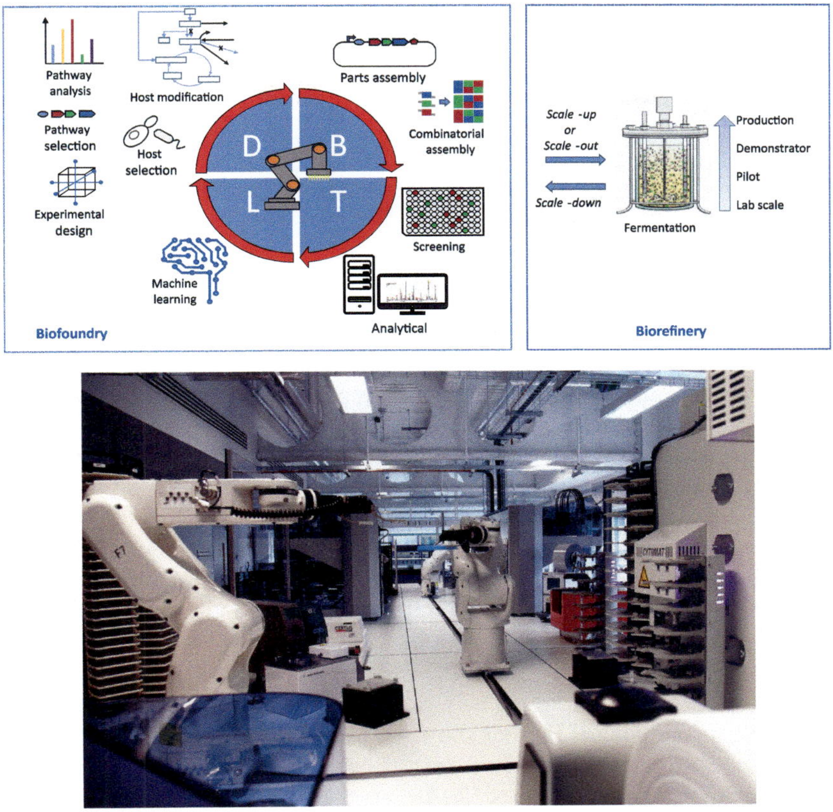

FIGURE 3.2 Publicly funded bio-foundries as key infrastructure. (A) The business relationship between the bio-foundry and biorefineries. Engineering biology seeks to increase reproducibility to enable the quantitative precision required for modern manufacturing. Standards, automation and machine learning are key to the success of this approach. Scale-down refers to acquiring data at production scale and transferring the information back to the laboratory via scale-down simulators. The production strain can be re-engineered as new information is back-translated to the fermentation operation for its fine-tuning. This is consistent with modern manufacturing in that the site of the design (the bio-foundry) can be totally separated from the site of manufacturing (typically the biorefinery). (B) The Edinburgh Genome Foundry. *(B) Courtesy of the Edinburgh Genome Foundry, University of Edinburgh.*

worklists and the integration of machine learning algorithms will collectively allow fully automated smart biomanufacturing (Carbonell et al., 2020).

Clearly someone whose education has been based on pipetting in wet laboratories will be unsuited to working in this environment without retraining. As the bio-revolution advances, then it will become increasingly necessary to educate the workforce for the DBTL approach, and there will be much less of an emphasis on pipetting and the one-factor-at-a-time (OFAT) approach (Czitrom, 1999). The DBTL approach will allow highly qualified scientists to spend more time on design, rather than slavishly pipetting. With reference to the DeepMind[6] AI leap in solving protein structures announced in late November (Callaway, 2020), 'It's going to require more thinking and less pipetting'.

Combining bio-foundries, biological resource centres and automated synthetic chemistry

Biological resource centres (BRCs) have been defined as follows: 'They consist of service providers and repositories of the living cells, genomes of organism and information relating to heredity and the functions of biological systems. BRCs contain collections of culturable organisms (e.g., microorganisms, plant, animal and human cells), replicable parts of these (e.g., genomes, plasmids, viruses, cDNAs), viable but not yet culturable organisms, cells and tissues, as well as databases containing molecular, physiological and structural information relevant to these collections and related bioinformatics' (OECD, 2001).

Subsequent to earlier OECD work, the Microbial Resources Research Infrastructure was accepted as a European research infrastructure within the European Strategy Forum for Research Infrastructures, and preparations began in 2012 (Smith et al., 2014).

It can be said that whilst the BRC is the curator of the known biology, the bio-foundry is the creator of the new biology. In combination, the possibilities for cooperation are huge. Fig. 3.3 shows how the two types of facilities may be related. The closer interaction of the various stakeholders and facilities, as called for by Clarke and Kitney (2020), can be envisaged in this diagram. It seems inevitable that at some point in the future they may merge to become a single, all-inclusive facility. One reason for this is projects like the Earth BioGenome Project,[7] which will vastly increase the volume of DNA sequence available as it intends to obtain the sequences of all known eukaryotes (Lewin et al., 2018).

Overcoming some of the limitations of biology in industrial processes will require workflows to determine how to split a process between biochemical conversion and chemical catalysis (Whitehead et al., 2020). Given that synthetic chemistry is also being automated (Coley et al., 2019), here is one more aspect of convergence that needs to be incorporated into the education of the engineering biologist. Perhaps this is the triple nexus of future manufacturing: **the BRC, the automated bio-foundry and automated synthetic chemistry**.

6. https://deepmind.com/Last accessed: 10 December 2020.
7. https://www.earthbiogenome.org/Last accessed: 10 December 2020.

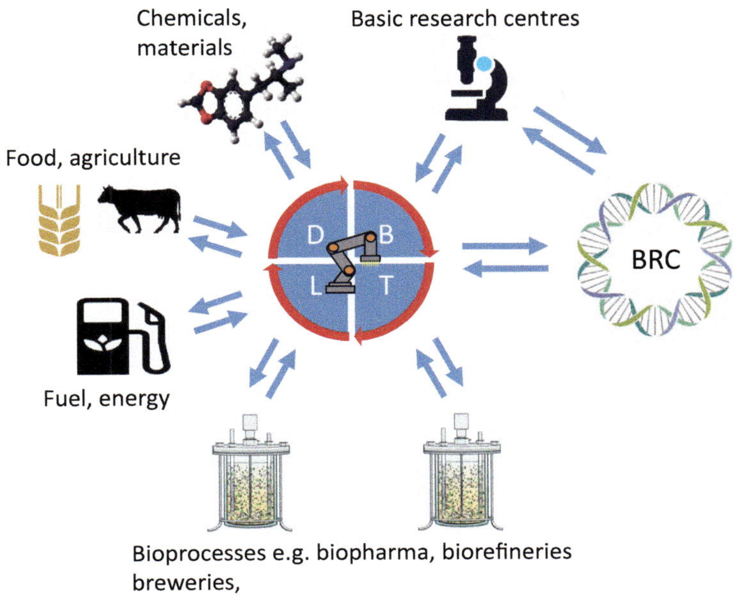

FIGURE 3.3 The discovery of new biological resources, their curation in BRCs, and optimisation in bio-foundries is a future infrastructure platform for both basic research and industrial production.

Three examples of engineering biology and sustainability

As synthetic biology is a platform technology that cuts across many economic sectors, the number of applications and potential products is very large: one bio-foundry, over the course of 85 days, was able to produce 17 material monomers and key intermediates (Robinson et al., 2020). For the purposes of this chapter, three examples are chosen for different reasons. The common denominator is sustainability (Bieber-Freudenberger et al., 2018). Indeed, Marvik and Philp (2020) state that the fundamental justification for public policy intervention in the bioeconomy is increased sustainability.

Vaccines and pandemics

The established vaccines industry has been troubled for decades. Fixed and operating costs are high and the technology has only changed in detail. Until the COVID-19 crisis, mRNA vaccines were only potentially useful, but the crisis has produced effective candidate vaccines at breakneck speed. One of the reasons for this speed is that there is no need for the complicated manufacturing processes that rely on cells or eggs. The potential of the approach was demonstrated by Crone et al. (2020) who showed that an automated SARS-CoV-2 clinical diagnostics platform designed and developed in a bio-foundry can be quickly deployed and scaled.

The mRNA vaccines are prime candidates for design by synthetic biology (Kitney et al., 2021). A simplified production process lends itself to small-scale manufacturing close to the point-of-care, for example, in hospitals, small companies on medical campuses and even in mobile production plants. The German company CureVac and Tesla, the electric vehicle manufacturer, have a patent application on a technology to 'print' RNA in 'micro-factories' (WO 2020002598, 2020).

Until COVID-19, relating vaccines to sustainability would have been an uphill political struggle. As a result of the virus, the global economy is expected to contract by a staggering 4.3% in 2020 (much worse than during the financial crisis in 2009), with 130 million more people entering extreme poverty (UNCTAD, 2020). Clearly, pandemics are not economically or socially sustainable and vaccines – and engineering biology – have a crucial role in their elimination.

Spider silk: the microbial resource is the production factory

Spider silks are amongst the toughest known biological materials. Dragline silk is the strongest, with superior mechanical properties to steel and Kevlar. Combined with flexibility, the range of applications is large. They are lightweight and virtually invisible to the human immune system, giving them 'revolutionary potential' for medicine and industry (Babb et al., 2017).

Tailor-making different spider silks for different materials and applications can be imagined, a perfect fit for engineering biology: there are at least seven silk types with a specialised role in elements of web construction, prey immobilisation or reproduction. Amongst the newer applications of spider silk being considered are microphones in hearing aids and cell phones. The German company AMSilk has entered into an agreement with Airbus to develop structural materials for aircraft using synthetic spider silk.[8] Silk has high-value applications in cosmetics, and Givaudan has acquired the cosmetics business of AMSilk.

Technical challenges remain that are attractive to engineering biologists. Synthetic biology, metabolic engineering and powerful genome-editing tools have created a selection of host systems capable of producing recombinant silks that both match and outperform the impressive properties of natural spider silk (Whittall et al., 2020).

In relation to sustainability, the stated applications, whilst by no means comprehensive, demonstrate that spider silks represent totally new materials, not drop-in replacements for existing fossil-derived ones. Their manufacture

---

8. https://www.airbus.com/newsroom/stories/strong-as-a-spider-web.html Last accessed: 10 December 2020.

is dislocated from the environmentally damaging mining and fossil industries.

Gas fermentation

The microbial fermentation of waste industrial gases such as CO, produced in large volumes in steelmaking,[9] to useful products such as ethanol (Liew et al., 2016), bioplastics (WO/2020/186173, 2020) and animal feed (Pikaar et al., 2018) is becoming reality. Synthetic biology and metabolic engineering approaches will play essential roles in expanding the product spectrum beyond ethanol to other fuels and commodity chemicals (Heijstra et al., 2017).

There are various ways that fermenting waste gases to products contributes to sustainability. It is one of the emerging technologies of carbon capture and utilisation (Zhu, 2019), effectively a value-adding strategy compared to carbon capture and storage. Perhaps its greatest contribution will be that it relieves pressure on land use, thus ameliorating concerns over indirect land use change (Searchinger et al., 2008).

Multidisciplinary education

The challenge of multi- and interdisciplinary education

Interdisciplinarity implies a problem-solving function not necessarily implicit to multidisciplinarity. A useful distinction was made by Boix Mansilla et al. (2000) who framed interdisciplinarity as the capacity to integrate knowledge and modes of thinking in two or more disciplines. To equip engineering biology students with problem-solving capability implies the need for both. This, in fact, challenges the teachers as much as the students: traditional scientific education and training has remained divided by disciplines such as microbiology, chemistry and computing. The challenges to higher education remain on many levels (MacLeod, 2018).

The long-standing conundrum of multi- and interdisciplinary education is the need for both breadth and depth (Collura et al., 2004). Training in sustainability itself needs multi- and interdisciplinarity as some of the depth skills needed are systems thinking, strategic planning, and evaluating environmental, social and economic performance. This educational conundrum for sustainability (Mascarelli, 2013) is the same for industrial biotechnology and engineering biology: how to make the interdisciplinary approach not only substantive, but also practical for early career scientists. Multiple agency funding for projects that span disciplines is needed (National Academies of Sciences, Engineering, and Medicine, 2020).

---

9. https://www.britannica.com/technology/steel/Primary-steelmaking Last accessed: 10 December 2020.

Life sciences industry-wide issues

Life sciences PhD-level education remains focused on training for academic careers (American Society for Microbiology, 2013). In the *Nature* PhD survey of 2017, nearly 75% of (general PhD student) respondents said that they would like a job in academia as an option upon graduation (Woolston, 2017). The number of US post-doctoral researchers far exceeds the growth in the number of tenure-track job openings (National Academy of Engineering, and Institute of Medicine, 2014). There are simply too many PhD students and too few senior posts (Nature Editorial, 2016).

Some of the life sciences-wide industry issues are clearly crystallised in an American report (Coalition of State Bioscience Institutes, 2013). Employer interviews identified several industry-wide gaps in the capabilities and talent of the workforce pool:

- The life science industry has a decreased need for deeply trained senior scientists. There is an over-specialisation surplus, whereas employers are looking for a workforce with greater breadth and more soft skills.
- Academic programmes are training students by discipline and not by problem solving, which typically requires cross-disciplinary skills and capabilities.
- Apprenticeships and long-term training programmes are lacking.

The UK biopharmaceuticals industry recently highlighted major skills gaps in mathematical and computational areas. These have emerged due to the rapid development of new disciplines such as systems biology and health informatics (ABPI, 2015).

Synthetic biology presents unique challenges

Given the multiple disciplines involved in synthetic biology (Fig. 3.4), it is improbable that a standard curriculum can deliver skills and education adequately. This calls for multidisciplinarity that flies in the face of how much of higher education is delivered (Hannon et al., 2018). Its acceptance by academia will depend on the teachers recognising that they are teaching for careers in manufacturing and engineering, with less emphasis on research careers.

The critical workforce gaps in bio-based manufacturing

'The need for a specially trained workforce for the synthetic biology sector is considered by many in industry to be a key pinch point for the industrial development of the area. The industrial view, regarding personnel, is that for every engineer, mathematician and computer scientist there are at least 30 PhDs in biology and biochemistry available. What is crucial for the development of the

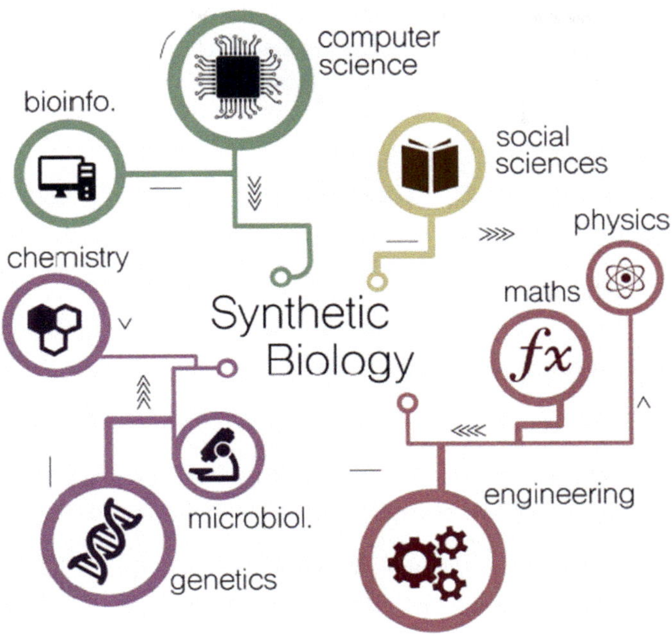

FIGURE 3.4 Engineering biology challenges educators as never before. *Reproduced under a Creative Commons Attribution Licence (http://creativecommons.org/licenses/by/3.0/) from Hallinan, J., Wipat, A., Kitney, R., Woods, S., Taylor, K., Goñi-Moreno, Á., 2019. Future-proofing synthetic biology: educating the next generation. Eng. Biol. 3, 25–31.*

> *synthetic biology industry is the training of many more engineers, mathematicians and computer scientists with an understanding of biology'.*
>
> Kitney et al. (2021).

Industrial biotechnology is the fastest growing subsector of biotechnology ahead of biologics and GM crops (Carlson, 2016). But this burgeoning industry is faced with skills gaps. Finding biologists is not the most difficult task for employers. Automation engineers specialising in high-throughput strain production critical to engineering biology-based manufacturing are rare. Managing automated systems will have to be a skill set for graduates in biology and chemistry in the future (Extance, 2016). For a long time, it has also been difficult to find fermentation staff: this is the province of the biochemical engineer, who combines the mathematics of cell growth with bioreactor and bioprocess design and yet biomanufacturing is the common operation that links together all the different market sectors of the world's biotechnology industry.

Perhaps hardest to find of all are employees well versed in experimental design and statistics, especially now that large data sets are becoming more common (Philp, 2018). Big data is creating an imperative for more complex design that enables fewer experiments and trials. Scientific irreproducibility – the inability to repeat others' experiments and reach the same conclusion – is a growing concern. Yet few early career researchers receive formal instruction on topics like experimental design and flaws in statistical analysis (Baker, 2016).

This diverse group of employees is essential for a functional engineering biology-based production plant but remains rare as this business sector is still a small niche. As sector growth is difficult to forecast, it challenges governments to predict how to invest in and reform higher education to create a workforce that matches the growth dynamics of the sector.

Bioinformatics may be a major roadblock

The bottleneck for the growing industrial biotechnology industry is shifting to bioinformatics and data mining. Data mining tools akin to the ones revolutionising social sciences and linguistics will become essential (Misirli et al., 2016). As high-throughput sequencing is increasingly deployed, the acquisition of genomic information is also rapidly increasing. DNA synthesis costs have dropped during this century but need to drop further. These price decreases, combined with advances in next-generation sequencing, are increasing the need for and role of advanced software design tools. 'Dry lab' skills have traditionally been isolated from 'wet lab' ones. Nevertheless, bioinformatics requires deep knowledge of biology theory and mathematics/computing: the BLAST algorithm speaks loudly to this (Lobo, 2008). These fields are usually not taught in depth in the same programme in higher education.

Beyond science and engineering

Given the history of the genetic modification (GM) debate (National Academies of Sciences, Engineering, and Medicine, 2016b), such matters as public perception will also shadow industrial biotechnology and engineering biology. There is already evidence that political and economic pressures will guide development of engineering biology (Rai and Boyle, 2007). Kuldell (2007) argued that educational efforts that fail to equip students for these aspects of the emerging discipline are unsound. Public engagement is necessary for the acceptance of engineering biology more broadly, but it is weakened by a lack of a standard approach (National Academies of Sciences, Engineering, and Medicine, 2016a). Policy makers could include social scientists and ethicists in strategies for developing and encouraging the uptake of bio-based products and have this embedded in education. On the other hand, public engagement should not become a 'mode of governance' of research (Kuntz, 2016).

Bio-based production creates regulatory challenges across boundaries as well. The use of wastes as feedstocks in biorefineries has regulatory mountains to climb (Lin et al., 2013). The metabolically engineered microbes (i.e., process) are subject to GMO regulation. At the same time, bio-based chemicals and fuels (i.e., products), often being drop-in substitutes for fossil-derived materials, are subject to chemical regulations. These include the Toxic Substances Control Act (TSCA)[10] in the United States and Registration, Evaluation, Authorisation and Restriction of Chemicals (REACH)[11] in Europe. More mundane, perhaps, but ever-present are regulations associated with health and safety, packaging and traceability. Another huge challenge has been the intellectual property issues around synthetic biology (Kahl et al., 2018).

The limits of multidisciplinarity

It should be asked if there are limits to which multi-disciplinarity can be taught for at least two reasons:

1. In the quest for breadth, depth may become too diluted. This could certainly be cause for concern in engineering degrees if the certification bodies such as the Institution of Chemical Engineers[12] found too little depth to certify engineers and courses.
2. The burden of learning required to become familiar, if not expert, in various disciplines may be too great, which could be especially true within the time limitations of Master programmes. The flexibility is greater within undergraduate degrees; nevertheless, careful attention would have to be given to curriculum design to prevent overload, with the attendant problem of student drop-out.

Regarding this latter point, it is worth remembering that the drop-out rates in STEM subjects (science, technology, engineering and mathematics) are already alarming (Chen et al., 2018): some 50% of American college students who major in STEM drop out (Lowery, 2010). Rates with biologists could be even higher unless amelioration measures are taken (Hamm et al., 2020).

Some approaches to engineering biology education and training

As with more traditional engineering, the education of engineering biologists should start in high school in future. It will carry all the way through higher

10. https://www.epa.gov/laws-regulations/summary-toxic-substances-control-act Last accessed: 10 December 2020.
11. https://echa.europa.eu/regulations/reach/understanding-reach Last accessed: 10 December 2020.
12. https://www.icheme.org/education/find-a-degree-programme/Last accessed: 13 December 2020.

education to PhD level, whether the PhD is aimed at a career in research or a passport to industry. However, the latter PhD might differ substantially in its structure from the former. The workforce, of course, does not need to be trained to PhD level. Rather, practical skills will be needed, just as they are in, say, pharmaceutical manufacturing, e.g., quality assurance and quality control. Unconventional teaching methods could assume increasing importance due to the emerging, rapidly changing nature of engineering biology. The international genetic engineering machines (iGEM) competition is a remarkable exemplar of the unconventional approach.

High school

There are multiple advantages to engaging high school students and their teachers in synthetic biology projects (Dubé et al., 2017). School pupils are able to explore the scientific as well as ethical, legal and social implications of the field. However, significant challenges to establishing synthetic biology at high school level exist, particularly infrastructure, specialised equipment, mentorship and funds.

A key barrier is access to equipment at high schools. To this end, Stark et al. (2018) have developed solutions. First is a synthetic biology kit that produces a visually engaging endpoint: a shelf-stable, just-add-water kit enabled by expression of fluorescent proteins in freeze-dried, cell-free reactions. Second, they have developed inexpensive incubators and imagers, resulting in a comprehensive kit costing less than $100 for a 30-person classroom.

A high school international Genetically Engineered Machine (iGEM) competition was established in 2011, then integrated into the main competition in 2015 as a dedicated stream. Box 3.1 shows the quality of what can be achieved at high school level.

Undergraduate courses: preparing the way

It is probably too early for entire undergraduate degrees to train biologists in industrial manufacturing. In Canada, for example, at the time of writing, there are only three universities that formally teach synthetic biology programmes, two of which are at the undergraduate level (Diep et al., 2021). However, relevant science undergraduate degrees could be re-designed to serve as a platform for post-graduate study.

For example, microbiology curricula are dominated by medical microbiology. A re-orientation of microbiology undergraduate education could include, apart from the core microbiological skills, quantitative skills that are important for success in industry. Students so equipped with skills in calculus, linear algebra, statistics, large dataset management and programming (American Society for Microbiology, 2013) would be better prepared for future education in, and a career based on, engineering biology.

BOX 3.1 iGEM 2020 High School Grand Prize winner: Taipei American School (TAS), Taipei City, Taiwan

Seasonal flu and pandemics require rapid and reliable detection mechanisms to implement preventive and therapeutic measures. Current detection methods of viral infections have limitations in speed, accuracy, accessibility and usability. This project presented a novel, widely applicable viral diagnostic test that uses a modified version of rolling circle amplification to be sensitive, specific, direct RNA targeted, colorimetric and operable at room temperature. The project was to specifically detect the following high-impact viruses: SARS-CoV-2, Influenza A (H1N1pdm09) and Influenza B (Victoria Lineage), although the test can be adapted to any viral infection. Results using synthetic viral DNA and RNA sequences showed that the diagnostic test takes approximately 1 h, detects femtomolar concentrations of RNA strands and differentiates between virus strains.

Indeed, Bishop and Eley (2001) wondered why microbiology students are scared of mathematics and why, at least in the UK, they were 'completely incapable of handling numbers properly'. Given that microbiology is associated with achieving SDGs addressing food security, health and wellbeing, clean energy, environmental degradation and climate change (Fagunwa and Olanbiwoninu, 2020), then there is a direct link between microbiology and engineering biology beyond the obvious technical links.

The University of Cambridge (UK) has a particular specialisation in plant synthetic biology and offers a few undergraduate courses: Synthetic Biology and Plant Biotechnology, Introduction to Molecular Bioengineering; Synthetic and Executable Biology. A module in Data Acquisition and Handling[13] is particularly appropriate in the genomics age due to the widespread availability of large datasets. The module presents the techniques used to acquire data in the various -omics approaches (transcriptomics, proteomics and metabolomics), as well as in high-throughput genetics. Because of their size and experimental limitations, the handling of these datasets presents unique challenges. Therefore, the module emphasises the practical aspects of dealing with these types of data in a set of 18 lectures and 6 computer-based practical sessions.

The international Genetically Engineered Machine student competition

'The competition is based on a simple idea: synthetic biology engineering principles of standardisation, abstraction and modularity can be applied to

13. https://www.sysbiol.cam.ac.uk/Part%20III/Structure Last accessed: 10 December 2020.

biotech to make engineering new functions in life systems less intimidating, more accessible and more predictable'.

Vilanova and Porcar (2014).

In these early days of the development of synthetic biology, it is essential to build communities of common interest (Ces and Elani, 2019). The iGEM competition[14] takes place within a time frame of less than 1 year, in such a way that even undergraduate students without prior training in biology, but with reasonable technical and theoretical support, can participate. It culminates in the iGEM Jamboree held at the Massachusetts Institute of Technology, at which students present their research projects to peers and policy experts and a mixed audience from academia, industry and social sciences (Smolke, 2009). The iGEM Jamboree is now the largest synthetic biology event in the world.

The competition is a shining example of what can be done when dedicated tutors unleash the potential of 20-year-olds. It started in 2004 with only five teams. Since then, there have been over 2000 teams from about 60 countries (Millett et al., 2019). By the end of the 2020 competition, there had been some 40,000+ participants.

Kelwick et al. (2015) consider synthetic biology a challenging area for the training of microbiologists and suggested that the iGEM competition provides a useful platform to develop this training. Fig. 3.5 shows how participation in the competition can develop diverse and interdisciplinary skills.

Taught and research masters

The University of Newcastle-upon-Tyne (UK) offers a 1-year taught MSc in synthetic biology that aims to provide a 'balance of molecular biology, engineering, computing and modelling skills necessary for a career in synthetic biology'.[15] Fig. 3.6 shows an overview of this MSc. One of its compulsory modules is Numeric Skills for Digital Biology. It blends fundamental statistical and mathematical concepts and techniques of importance in bioinformatics, synthetic biology and neuroinformatics (collectively referred to as digital biology). The syllabus contains: introduction to data analysis; probability axioms; combinatorics; conditional probability; discrete probability models; continuous probability models; properties of estimators; sampling distributions; central limit theorem; statistical inference and maximum likelihood estimation.

Synthetic biology lends itself well to a Masters by research degree, emphasising practice-led research combined with relatively few taught modules compared with other graduate degrees. This sort of degree is often

14. https://igem.org/Main_Page Last accessed: 10 December 2020.
15. https://www.ncl.ac.uk/postgraduate/courses/degrees/synthetic-biology-msc/#profile Last accessed: 10 December 2020.

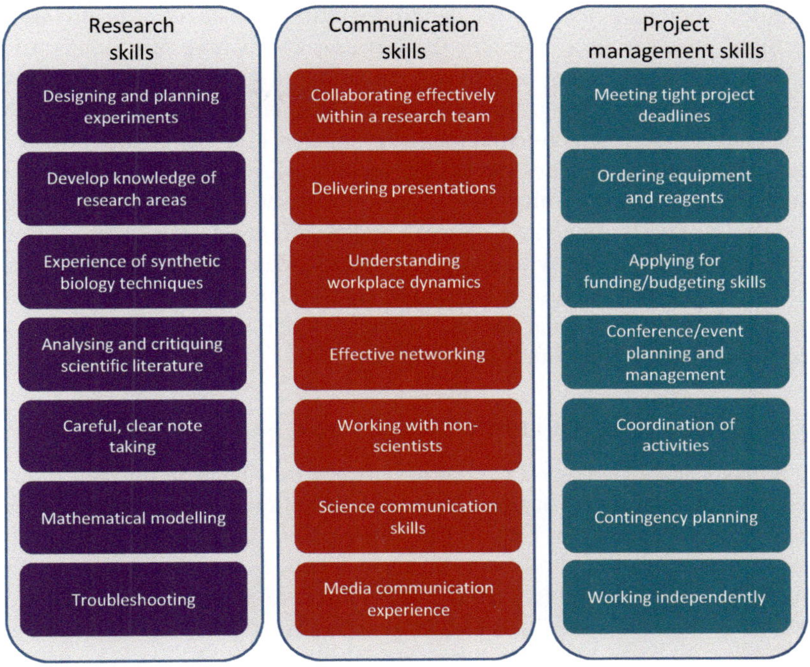

FIGURE 3.5 iGEM students perceive that the competition helps develop and improve research-focused skills, and transferable skills including communication and project management skills. *Adapted from Kelwick, R., Bowater, L., Yeoman, K.H., Bowater, R.P., 2015. Promoting microbiology education through the iGEM synthetic biology competition. FEMS Microbiol. Lett. 362, fnv129, https://doi.org/10.1093/femsle/fnv129.*

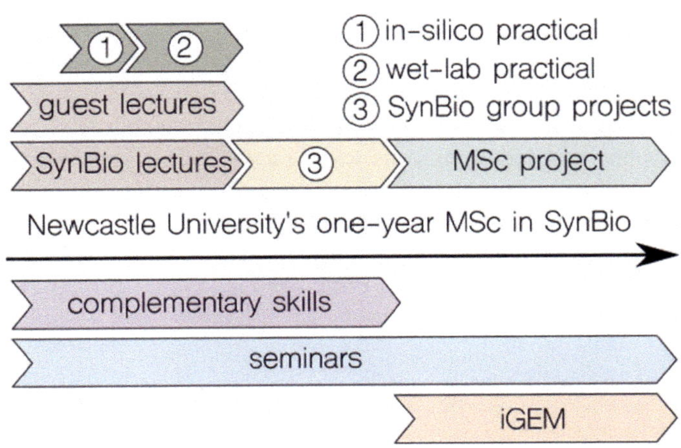

FIGURE 3.6 Overview of the University of Newcastle-upon-Tyne (UK) 1-year MSc in synthetic biology. *Reproduced under a Creative Commons Attribution Licence (http://creativecommons.org/licenses/by/3.0/) from Hallinan, J., Wipat, A., Kitney, R., Woods, S., Taylor, K., Goñi-Moreno, Á., 2019. Future-proofing synthetic biology: educating the next generation. Eng. Biol. 3, 25–31.*

designed to prepare students for doctoral research. However, it is also useful for those considering a career in the private sector where research is a key focus, but a PhD is not specifically required (Masood, 1998). University College London (UCL) offers a 1-year MRes in synthetic biology[16] that demonstrates the research-led emphasis compared to a taught MSc. Students undertake modules to the value of 180 credits. An extended research project constitutes 120 credits, whilst the balance of 60 is satisfied by three core taught modules. Similarly, Imperial College, London, offers a 1-year MRes degree[17] in which students begin with compulsory core lecture courses and practicals. Subsequently, they choose a topic for an 8-month long multidisciplinary, theoretical or experimental research project, supervised by at least two supervisors with different kinds of expertise.

Training PhD students in interdisciplinarity

As far back as 1995, a report expressed the need for change in the education of scientists and engineers (Institute of Medicine, National Academy of Sciences, and National Academy of Engineering, 1995). This report was concerned that the United States was producing too many PhDs. It said that industry often complained that graduates were too specialised to accomplish the range of tasks they would be confronted with. Also, scientists from small biotechnology companies are often placed in a managerial role in which they may have no training or know-how (Corolleur et al., 2004). This has all brought about a call for a new type of PhD, one that offers much more breadth and flexibility.

The United States National Science Foundation (NSF) had a flagship interdisciplinary training programme at North Carolina State University – the Integrative Graduate Education and Research Traineeship (IGERT), educating PhD scientists and engineers by building on the foundations of their disciplinary knowledge with interdisciplinary training. Since 1998, the IGERT programme has made 278 awards and has provided funding for approximately 6500 graduate students. Research topics were selected specifically to transcend traditional disciplinary boundaries and so encourage teamwork. Diversity in the background of selected students was a factor in their preparation to solve large and complex research problems of significant scientific and societal importance – typical of the so-called grand challenges that engineering biology is supposed to be able to address. Insights from IGERT are provided by a personal communication from Professor Fred Gould (Box 3.2).

16. https://www.ucl.ac.uk/prospective-students/graduate/taught-degrees/synthetic-biology-mres Last accessed: 10 December 2020.
17. https://www.imperial.ac.uk/life-sciences/postgraduate/masters-courses/mres-in-systems-and-synthetic-biology/Last accessed: 10 December 2020.

BOX 3.2 Impact of interdisciplinary training on disciplinary depth at NSF - Integrative Graduate Education and Research Traineeship (IGERT)

'Our NSF-IGERT graduate training programme was titled "Genetic Engineering and Society: The case of transgenic pests." Our vision for the programme was development of "a group of interactive students, trained in technologies used for manipulation of pest genomes as well as methods needed to assess the environmental and socio-cultural appropriateness of specific products of these manipulations"'.

When we first started the programme, some faculty were reluctant to have their students participate because they felt that such broadly trained PhD students would not have the depth in a single field that would be attractive to academic and industry employers. The Co-PIs on the grant were well aware of this criticism and understood their origins. As the programme progressed, it became apparent from our students' interactions with industry representatives that the private sector was searching for PhD-level employees who had greater breadth than those that they were generally encountering.

We recently polled some of our now advanced molecular biology graduate students about their perspectives on the depth of their training. They were all confident that they could match their peers in genetics and entomology in terms of depth. They did say that they had taken a couple fewer disciplinary courses than their peers in those departments, but they had learned the important subject matter on their own. Some felt that it would take them one extra year to complete their degrees because of the interdisciplinary work.

Each of the cohorts did a group project that resulted in a publication or website. Students in our third (last) cohort have published two peer reviewed interdisciplinary articles as a group.

Courtesy of Professor Fred Gould, North Carolina State University.

Massive Open Online Courses

The traditional on-campus experience could be radically changed by Massive Open Online Courses (MOOCs), which will enhance classroom and laboratory work via a combination of videos, lecture notes, problems and user forums. An important feature is that they can reach large audiences electronically in any number of locations. Unlike the traditional lecture, typically each lesson is a 10-minute video on a single concept followed by self-assessment tools. This makes MOOCs easily scalable and adaptable (Hill, 2012), crucial to an emerging subject like synthetic biology. Synthetic biology capability is increasing exponentially (Opgenorth et al., 2019). As a result, educational materials lose their freshness, if not their relevance, quickly. When the hard foundational work of creating an MOOC is done, software and screencasts can replace or upgrade course content in a matter of minutes. The evidence for the impact of MOOCs is still embryonic (Buhl and Andreasen, 2018), but there is evidence that they improve information retention rates and motivation (Barak et al., 2016).

A number of MOOC platforms, such as Coursera[18] and edX,[19] now propose a wide array of classes spanning engineering to molecular biology and all the building blocks in between that can provide the basic toolset to start practising engineering biology. Recently, a dedicated MOOC in synthetic biology has been offered (Box 3.3).

BOX 3.3 The Principles of Synthetic Biology (PoSB) MOOC (Anderson et al., 2019)

PoSB is structured around five fundamentals (Fig. 3.7). It is meant to appeal to both students of biology and engineering. Students with a background in biology should benefit from the insights into model design, digital logic and design-build-test approaches. Those with engineering backgrounds can learn how to model gene expression, how biological 'machines' can implement familiar functions, and how biological systems have fundamental limitations.

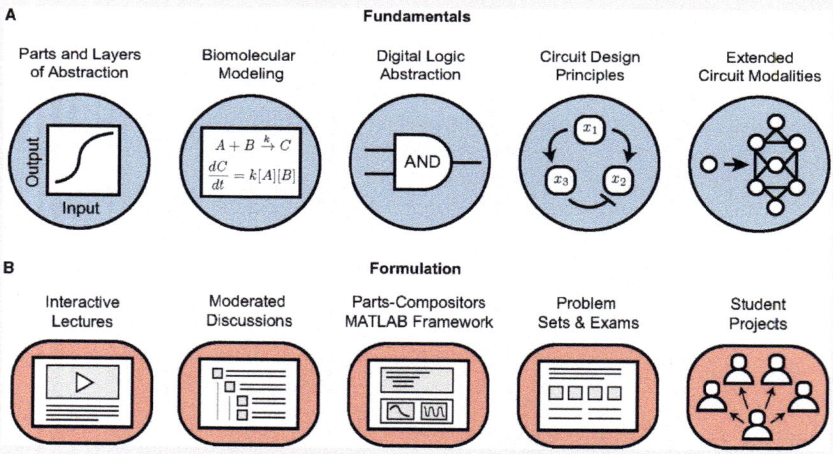

FIGURE 3.7 Course fundamentals and formulation. (A) The fundamental methodologies and abstractions used to design the course curriculum. (B) The many different modes that students use to engage with the course. *Reproduced under the terms of the Creative Commons Attribution Non-Commercial License (http://creativecommons.org/licenses/by-nc/4.0/) from Anderson, D.A., Jones, R.D., Arkin, A.P., Weiss, R., 2019. Principles of synthetic biology: a MOOC for an emerging field. Synt. Biol. 4, ysz010. https://doi.org/10.1093/synbio/ysz010.*

Over 2 years, PoSB managed to reach over 10,000 students from more than 150 countries, a scale not possible through traditional teaching methods.

18. https://www.coursera.org/Last accessed: 10 December 2020.
19. https://www.edx.org/Last accessed: 10 December 2020.

Inevitably hybrid approaches that have elements of the MOOC approach will emerge. For example, SynBio4All, being developed at the *Centre de Recherches Interdisciplinaires* in Paris, can be described as part MOOC, part open laboratory notebook and part crowdsourcing discussion platform on synthetic biology (SynBio4All[20]).

A role for intermediate research organisations and laboratories

Intermediate research organisations, typified by the Fraunhofer Society of Germany,[21] can enable work in selected fields without the conflicting pressures of publishing and teaching explicit in academic research (Gauvreau et al., 2018). The concept seems enshrined in the UK Catapult[22] model, which encourages both regional and open innovation (Kerry and Danson, 2016). Further, such a model has great potential to fill skill gaps (more on the apprentice, hands-on model rather than on the academic student model).

Roles for engineering societies and academies

Whilst at a distance from universities, professional engineering societies can influence undergraduate education in various ways as they serve as a bridge between employers and schools of engineering (National Academy of Engineering, 2017). They set and maintain professional standards within their fields. Most importantly they can help clarify the knowledge and skill base needed by those practicing in the field, and thereby influence the undergraduate curriculum. These activities all help to keep curricula up to date and ensure that graduates emerge with the skills needed for practice.

Business management education and training for the engineering biology industry

One solution to a shortage of experienced managers in the biotechnology industry has been to create a specific stream for biotechnology within the normally generic MBA programme (OECD, 2005). Theories of business administration have their roots in commerce, which has in the past been focused on non-technological issues (Lambert, 2004). Therefore, the typical MBA programme is not particularly well suited to engineering biology business management. Given the pressures on small companies active in engineering biology, much shorter courses that focus on specific knowledge gaps may be more appropriate.

---

20. https://scistarter.org/synbio4all Last accessed: 10 December 2020.
21. https://www.fraunhofer.de/en.html Last accessed: 10 December 2020.
22. https://catapult.org.uk/Last accessed: 10 December 2020.

SynbiCITE developed a 4-day More Business Acumen (MBA)[23] interactive course (Imperial College) that covered the main strategies required to establish, build and manage a biotechnology company built around engineering biology. It has focused on the early stages of setting up a company, getting funding and understanding the wider reaches of intellectual property. It is open to both industry staff and students.[24]

Training technicians: providing the biomanufacturing workforce

Technicians are workforce employees, not researchers. As they are responsible for day-to-day work in biomanufacturing, technicians will be required in higher numbers than researchers and need a broader range of skills. Industrial biotechnology and engineering biology should be part of their training, not all of it. Some foreseeable functions will be routine maintenance of metabolically engineered strains; embedding engineering biology with GMP guidance; regulatory and compliance training, e.g., bio-banking, transportation of live biologicals and document management; standard operating procedures to deal with accidental spills and releases. They should also be trained in matters such as scale-up, knowledge of packaging and labelling protocols (Wallman et al., 2013). Scale-up is a massive technical barrier in the bio-based industry, especially at the scale of transportation fuels (Westfall and Gardner, 2011), stretching the skills of both strain and fermentation engineers.

Manufacturing does not fit well into normal boundaries of university degree programmes. As a result, it is often marginalised (Glaser, 2013). An approach that creates a vocational workforce locally and separately from the universities – in technical and community colleges,[25] for example – would take pressure off universities. It would also create more jobs and investment in local or community colleges. This aligns well with thinking that envisages the creation of biorefineries and bioeconomy clusters in rural and semi-rural environments (Hassan et al., 2019).

One approach is an apprenticeship training route to keep day-release apprentices from companies close to the training centres. However, a 'tyranny of small numbers' has been identified in the UK (Lewis, 2016), meaning that there is not a critical mass of apprentices in one geographical area to make it worthwhile for colleges to offer specialised industrial biotechnology courses. This relates to industrial biotechnology and engineering biology being emerging sectors.

23. http://www.synbicite.com/knowledge/biotechnology-mba/Last accessed: 10 December 2020.
24. https://www.alliancembs.manchester.ac.uk/news/building-a-synthetic-biology-rich-biotech-business-from-scratch-/Last accessed: 13 December 2020.
25. https://www.capitalcc.edu/course/synthetic-biology/Last accessed: 10 December 2020.

The Industrial Biotechnology Innovation Centre[26] in Scotland has developed a range of skills and education programmes with its collaborative partners to meet the need of the biotechnology industry (Archer, 2016). Jointly with the Forth Valley College and Glasgow Kelvin College, the Higher National Diploma is specifically designed to create a cadre of technical staff versed in biology, chemistry and engineering.

Revisit mechatronics to shape the education of the future engineering biologist?

The French standard NF E 01−010 (Norme Française, 2008) gives the following definition of mechatronics, translated from the French as 'an approach aiming at the synergistic integration of mechanics, electronics, control theory and computer science within product design and manufacturing, in order to improve and/or optimise its functionality'.

Much of what has been discussed in this chapter can be seen as an integration of biotechnology into mechatronics, albeit with less emphasis on, but not a trivialisation of, the mechanical aspects of manufacturing. The rise of the mechatronics engineer depended on uniting the principles of mechanics, electronics and computing to generate simpler, more economical and reliable systems (Schiehlen, 2019). Education was refined over decades to optimise the undergraduate curriculum to create the mechatronics engineers that have revolutionised manufacturing. This education has necessitated multi- and interdisciplinarity in critical fields such as mechanical, electrical, electronic, computer and control engineering. The integration of these fields has resulted in problem-solving design and manufacture (Acar and Parker, 1996). Several sections of this chapter have discussed integration of various fields for engineering biology to take up a major role in manufacturing. There are surely lessons for engineering biology to learn from the evolution of mechatronics education.

Concluding remarks

The transition from research orientation to production orientation will require a paradigm shift in biotechnology education (Delebecque and Philp, 2018). Universities will need to be able to attract into biotechnology school-leavers with a more mathematical background, students who will graduate equally capable in DNA engineering and computation. The shift calls for bold reforms in education. If Descartes reintroduced reductionism to Western science (Beresford, 2010), then the approach to biotechnology and biomanufacturing described here is quite the opposite. This approach requires experimental

26. https://www.ibioic.com/Last accessed: 10 December 2020.

design that involves changing several variables simultaneously, questioning the relevance of OFAT (Sadowski et al., 2016).

The approach relies on high levels of computing power and literacy. With astonishing foresight, Alan Turing, whilst recognising the complexity of biological phenomena, posited that some biological patterns could be treated 'in detail with the aid of a digital computer' as far back as 1952 (Turing, 1952). Almost 70 years later, this vision remains to be fulfilled, but it is now clear that it will be. This heralds a transition in biology from a descriptive to a quantitative science: it is only this century, in the age of genomics, that biology has become data rich.

Microbiology as an independent discipline is threatened with subsumption into molecular biology and the -omics: many university microbiology departments have shrunk or disappeared (American Society for Microbiology, 2007). Ironically, there is reason to believe that the bioeconomy concept has been mainly driven by advances in microbiology (Wesseler and von Braun, 2017), and that microbiology is central to the development of engineering biology (Huang and Nikel, 2019). The glorious metabolic and catabolic diversity of microorganisms represents huge opportunities for microbiology to be the source of new biocatalysts and chassis strains for industrial production (Chi et al., 2019).

Microbiology should be seen as a key discipline for the future engineering biologist in an education that levels the playing field of the wet laboratory with the quantitative skills of the design engineer. This education has to start in high school to create the manufacturing workforce of the future.

Disclaimer statement

The views expressed are those of the author and not necessarily those of the OECD or of the governments of its member countries.

References

ABPI, 2015. Bridging the Skills Gap in the Biopharmaceutical Industry: Maintaining the UK's Leading Position in Life Sciences. Association of the British Pharmaceutical Industry, London.

Acar, M., Parker, R.M., 1996. Engineering education for mechatronics. IEEE Trans. Ind. Electron. 43, 106–112.

Aguilar, A., Patermann, C., 2020. Biodiplomacy, the new frontier for bioeconomy. N. Biotech. 59, 20–25.

American Society for Microbiology, 2007. Basic Research on Bacteria. The Essential Frontier. American Society for Microbiology, Washington, DC. https://asm.org/ASM/media/docs/NIHASMBacteriaReport.pdf. (Accessed 11 December 2020).

American Society for Microbiology, 2013. Microbe-powered Jobs: How Microbiologists Can Help Build the Bioeconomy. American Academy of Microbiology, Washington, DC.

Anderson, D.A., Jones, R.D., Arkin, A.P., Weiss, R., 2019. Principles of synthetic biology: a MOOC for an emerging field. Synt. Biol. 4, ysz010. https://doi.org/10.1093/synbio/ysz010.

Archer, I., 2016. How Scotland is tackling the IB knowledge gap. Ind. Biotechnol. 12, 268–271.

Asada, R., Stern, T., 2018. Competitive bioeconomy? Comparing bio-based and non-bio-based primary sectors of the world. Ecol. Econ. 149, 120–128.

Babb, P.L., Lahens, N.F., Correa-Garhwal, S.M., Nicholson, D.N., Kim, E.J., Hogenesch, J.B., Kuntner, M., Higgins, L., Hayashi, C.Y., Agnarsson, I., Voight, B.F., 2017. The *Nephila clavipes* genome highlights the diversity of spider silk genes and their complex expression. Nat. Genet. 49, 895–903.

Baker, M., 2016. Seek out stronger science. Nature 537, 703–704.

Barak, M., Watted, A., Haick, H., 2016. Motivation to learn in massive open online courses: examining aspects of language and social engagement. Comput. Educ. 94, 49–60.

Beresford, M.J., 2010. Medical reductionism: lessons from the great philosophers. QJM: Int. J. Med. 103, 721–724.

Biber-Freudenberger, L., Basukala, A.K., Bruckner, M., Börner, J., 2018. Sustainability performance of national bio-economies. Sustainability 10, 2705. https://doi.org/10.3390/su10082705. www.mdpi.com/journal/sustainability.

Bioindustrial Innovation Canada, 2019. Canada's Bioeconomy Strategy: Leveraging Our Strengths for a Sustainable Future. https://www.bincanada.ca/biodesign. (Accessed 25 November 2020).

Bishop, R., Eley, A., 2001. Microbiologists and maths. Microbiol. Today 28, 62–63. The Society for General Microbiology, ISSN: 1464-0570.

Boix Mansilla, V., Miller, W.C., Gardner, H., 2000. On disciplinary lenses and interdisciplinary work. In: Wineburg, S., Grossman, P. (Eds.), Interdisciplinary Curriculum: Challenges of Implementation. Teachers College Press, New York.

Bracco, S., Calicioglu, O., San Juan, M.G., Flammini, A., 2018. Assessing the contribution of bioeconomy to the total economy: a review of national frameworks. Sustainability 10, 1698. https://doi.org/10.3390/su10061698.

Buhl, M., Andreasen, L.B., 2018. Learning potentials and educational challenges of massive open online courses (MOOCs) in lifelong learning. Int. Rev. Educ. 64, 151–160.

Burgard, A., Burk, M.J., Osterhout, R., Van Dien, S., Yim, H., 2016. Development of a commercial scale process for production of 1,4-butanediol from sugar. Curr. Opin. Biotechnol. 42, 118–125.

Callaway, E., November 30, 2020. 'It will change everything': DeepMind's AI makes gigantic leap in solving protein structures. Nat. News 588, 203–204. https://www.nature.com/articles/d41586-020-03348-4.

Carbonell, P., Le Feuvre, R., Takano, E., Scrutton, N.S., 2020. In silico design and automated learning to boost next-generation smart biomanufacturing. Synt. Biol. ysaa020. https://doi.org/10.1093/synbio/ysaa020.

Carlson, R., 2016. Estimating the biotech sector's contribution to the US economy. Nat. Biotechnol. 34, 247–255.

Cavicchioli, R., Ripple, W.J., Timmis, K.N., Azam, F., Bakken, L.R., Baylis, M., Behrenfeld, M.J., Boetius, A., Boyd, P.W., Classen, A.T., Crowther, T.W., Danovaro, R., Foreman, C.M., Huisman, J., Hutchins, D.A., Jansson, J.K., Karl, D.M., Koskella, B., Welch, D.B.M., Martiny, J.B.H., Moran, M.A., Orphan, V.J., Reay, D.S., Remais, J.V., Rich, V.I., Singh, B.K., Stein, L.Y., Stewart, F.J., Sullivan, M.B., van Oppen, M.J.H., Weaver, S.C., Webb, E.A., Webster, N.S., 2019. Scientists' warning to humanity: microorganisms and climate change. Nat. Rev. Microbiol. 17, 569–586.

Ces, O., Elani, Y., 2019. Community building in synthetic biology. Exp. Biol. Med. 244, 281–282.

Chao, R., Mishra, S., Si, T., Zhao, H., 2017. Engineering biological systems using automated biofoundries. Metab. Eng. 42, 98–108.

Chen, Y., Johri, A., Rangwala, H., 2018. Running out of STEM: a comparative study across STEM majors of college students at-risk of dropping out early. In: Proceedings of the International Conference on Learning Analytics and Knowledge, Sydney, Australia, March 2018 (LAK'18). https://dl.acm.org/doi/10.1145/3170358.3170410. (Accessed 15 December 2020).

Chi, H., Wang, X., Shao, Y., Qin, Y., Deng, Z., Wang, L., Chen, S., 2019. Engineering and modification of microbial chassis for systems and synthetic biology. Synt. & Syst. Biotechnol. 4, 25–33.

Clarke, L., Kitney, R., 2020. Developing synthetic biology for industrial biotechnology applications. Biochem. Soc. Trans. 48, 113–122.

Coalition of State Bioscience Institutes (CBSI), 2013. Demand for Talent: Current and Projected Workforce Trends in the Life Science Industry. A joint report of CBSI and Booz & Co.

Coley, C.W., Thomas III, D.A., Lummiss, J.A.M., Jaworski, J.N., Breen, C.P., Hart, V.T., Fishman, J.S., Rogers, L., Gao, H., Hicklin, R.W., Plehiers, P.P., Byington, J., Piotti, J.S., Green, W.H., Hart, A.J., Jamison, T.F., Jensen, K.F., 2019. A robotic platform for flow synthesis of organic compounds informed by AI planning. Science 365, eaax1566. https://doi.org/10.1126/science.aax1566.

Collura, M.A., Aliane, B., Daniels, S., Nocito-Gobel, J., 2004. Development of a multidisciplinary engineering foundation spiral. In: Proceedings of the 2004 American Society for Engineering Education Annual Conference and Exposition. Educational Research & Methods - Session 2630, pp. 9.439.2–439.439.14.

Corolleur, C.D.F., Carrere, M., Mangematin, V., 2004. Turning scientific and technological human capital into economic capital: the experience of biotech start-ups in France. Res. Pol. 33, 631–642.

Council for Integrated Innovation Strategy, 2020. Bio-Strategy 2020 (Basic Measures). https://www.dwih-tokyo.org/files/2020/10/bio2020_honbun_en_rev-1.pdf. (Accessed 12 December 2020).

Crone, M.A., Priestman, M., Ciechonska, M., Jensen, K., Sharp, D.J., Anand, A., Randell, P., Storch, M., Freemont, P.S., 2020. A role for biofoundries in rapid development and validation of automated SARS-CoV-2 clinical diagnostics. Nat. Commun. 11, 4464.

Czitrom, V., 1999. One-factor-at-a-time versus designed experiments. Am. Statistician 53, 126–131.

Delebecque, C.J., Philp, J., 2018. Education and training for industrial biotechnology and engineering biology. Eng. Biol. 3, 6–11.

Diep, P., Boucinha, A., Yeung, B.A., Kell, B., Chen, X., Tsyplenkov, D., Serra, D., Escobar, A., Gnanapragasam, A., Emond, C.A., Sajtovich, V.A., Mahadevan, R., Kilkenny, D.M., Gini-Newman, G., Kaern, M., Ingalls, B., 2021. Advancing undergraduate synthetic biology education: insights from a Canadian iGEM student perspective. Can. J. Microbiol. 67, 749–770. https://doi.org/10.1139/cjm-2020-0549.

Dubé, S., Orr, D., Dempsey, B., Wieden, H.-J., 2017. A synthetic biology approach to integrative high school STEM training. Nat. Biotechnol. 35, 591–595.

European Commission, 2012. Innovating for Sustainable Growth: a Bioeconomy for Europe. Available at: https://op.europa.eu/en/publication-detail/-/publication/1f0d8515-8dc0-4435-ba53-9570e47dbd51. (Accessed 26 November 2020).

European Commission, 2018. A Sustainable Bioeconomy for Europe: Strengthening the Connection between Economy, Society and the Environment. COM(2018) 673 Final. European Commission, Brussels. https://knowledge4policy.ec.europa.eu/publication/sustainable-bioeconomy-europe-strengthening-connection-between-economy-society_en. (Accessed 15 December 2020).

Extance, A., 2016. The Human Element. Chemistry World careers Blog. www.chemistryworld.com/careers/what-skills-will-you-need-in-2026/1017241.article. (Accessed 11 December 2020).

Fagunwa, O.E., Olanbiwoninu, A.A., 2020. Accelerating the sustainable development goals through microbiology: some efforts and opportunities. Access Microbiol. https://doi.org/10.1099/acmi.0.000112.

Freedman, L.P., Cockburn, I.M., Simcoe, T.S., 2015. The economics of reproducibility in pre-clinical research. PLoS Biol. 13, e1002165.

French, K.E., 2019. Harnessing synthetic biology for sustainable development. Nat. Sustain. 2, 250–252.

Gauvreau, D., Winickoff, D., Philp, J., 2018. Engineering biology and the grand challenges: do we need a new R&D&I model? Eng. Biol. 2, 2–6.

Glaser, V., 2013. Overview: meeting current needs and assessing future opportunities to drive the global bioeconomy. Ind. Biotechnol. 9, 271–274.

Hallinan, J., Wipat, A., Kitney, R., Woods, S., Taylor, K., Goñi-Moreno, Á., 2019. Future-proofing synthetic biology: educating the next generation. Eng. Biol. 3, 25–31.

Hamm, J.M., Perry, R.P., Chipperfield, J.G., Hladkyj, S., Parker, P.C., Weiner, B., 2020. Reframing achievement setbacks: a motivation intervention to improve 8-year graduation rates for students in science, technology, engineering, and mathematics (STEM) fields. Psychol. Sci. 31 (6), 623–633. https://doi.org/10.1177/0956797620904451.

Hannon, J., Hocking, C., Legge, K., Lugg, A., 2018. Sustaining interdisciplinary education: developing boundary crossing governance. High Educ. Res. Dev. 37, 1–15.

Hassan, S.S., Williams, G.A., Jaiswal, A.K., 2019. Lignocellulosic biorefineries in Europe: current state and prospects. Trends Biotechnol. 37, 231–234.

Heijstra, B.D., Leang, C., Juminaga, A., 2017. Gas fermentation: cellular engineering possibilities and scale up. Microb. Cell Factories 16, 60. https://doi.org/10.1186/s12934-017-0676-y.

Hill, P., 2012. Online educational delivery models a descriptive view. Educ. Rev. 47, 84–86.

Hillson, N., Caddick, M., Cai, Y., Carrasco, J.A., Chang, M.W., Curach, N.C., Bell, D.J., Le Feuvre, R., Friedman, D.C., Fu, X., Gold, N.D., Herrgård, M.J., Holowko, M.B., Johnson, J.R., Johnson, R.A., Keasling, J.D., Kitney, R.I., Kondo, A., Liu, C., Martin, V.J.J., Menolascina, F., Ogino, C., Patron, N.J., Pavan, M., Poh, C.L., Pretorius, I.S., Rosser, S.J., Scrutton, N.S., Storch, M., Tekotte, H., Travnik, E., Vickers, C.E., Yew, W.S., Yuan, Y., Zhao, H., Freemont, P.S., 2019. Building a global alliance of biofoundries. Nat. Commun. 10, 2040.

HM Government, 2018. Growing the bioeconomy. In: Improving Lives and Strengthening Our Economy: A National Bioeconomy Strategy to 2030. Department for Business, Energy and Industrial Strategy (BEIS), London.

Huang, W.E., Nikel, P.I., 2019. The Synthetic Microbiology Caucus: from abstract ideas to turning microbes into cellular machines and back. Microbial. Biotechnol. 12, 5–7.

Institute of Medicine, National Academy of Sciences, and National Academy of Engineering, 1995. Reshaping the Graduate Education of Scientists and Engineers. The National Academies Press, Washington, DC.

Jessop-Fabre, M.M., Sonnenschein, N., 2019. Improving reproducibility in synthetic biology. Front. Bioeng. Biotechnol. 7, 18. https://doi.org/10.3389/fbioe.2019.00018.

Kahl, L., Molloy, J., Patron, N., Matthewman, C., Haseloff, J., Grewal, D., Johnson, R., Endy, D., 2018. Opening options for material transfer. Nat. Biotechnol. 36, 923–927.

Kelwick, R., Bowater, L., Yeoman, K.H., Bowater, R.P., 2015. Promoting microbiology education through the iGEM synthetic biology competition. FEMS Microbiol. Lett. 362, fnv129. https://doi.org/10.1093/femsle/fnv129.

Kerry, C., Danson, M., 2016. Open innovation, triple helix and regional innovation systems: exploring CATAPULT centres in the UK. Ind. High. Educ. 30, 67–78.

Kitney, R., Adeogun, M., Fujishima, Y., Goñi-Moreno, Á., Johnson, R., Maxon, M., Steedman, S., Ward, S., Winickoff, D., Philp, J., 2019. Enabling the advanced bioeconomy with engineering biology. Trends Biotechnol. 37, 917–920.

Kitney, R., Bell, J., Philp, J., 2021. Build a sustainable vaccines industry with synthetic biology. Trends Biotechnol. (in press).

Kroll, C., Warchold, A., Pradhan, P., 2019. Sustainable Development Goals (SDGs): are we successful in turning trade-offs into synergies? Palgrave Commun. 5, 140. https://doi.org/10.1057/s41599-019-0335-5 | www.nature.com/palcomms.

Kuldell, N., 2007. Authentic teaching and learning through synthetic biology. J. Biol. Eng. 1. https://doi.org/10.1186/1754-1611-1-8.

Kuntz, M., 2016. Scientists should oppose the drive of postmodern ideology. Trends Biotechnol. 34, 943–945.

Lai, H.-E., Canavan, C., Cameron, L., Moore, S., Danchenko, M., Kuiken, T., Sekeyová, Z., Freemont, P.S., 2019. Synthetic biology and the united Nations. Trends Biotechnol. 37, 1146–1151.

Lambert, R., 2004. Dip into the Reservoir of Innovation. Cambridge Manufacturing Review, Spring, pp. 2–5.

Lee, S.Y., Kim, H.U., 2015. Systems strategies for developing industrial microbial strains. Nat. Biotechnol. 33, 1061–1072.

Lewin, H.A., Robinson, G.E., Kress, W.J., Baker, W.J., Coddington, J., Crandall, K.A., Durbin, R., Edwards, S.V., Forest, F., Gilbert, M.T.P., Goldstein, M.M., Grigoriev, I.V., Hackett, K.J., Haussler, D., Jarvis, E.D., Johnson, W.E., Patrinos, A., Richards, S., Castilla-Rubio, J.C., van Sluys, M.-A., Soltis, P.S., Xu, X., Yang, H., Zhang, G., 2018. Earth Bio-Genome project: sequencing life for the future of life. Proc. Natl. Acad. Sci. USA 115, 4325–4333.

Lewis, P., 2016. Technician Roles, Skills, and Training in Industrial Biotechnology: An Analysis. The Gatsby Charitable Foundation, London.

Liew, F., Martin, M.E., Tappel, R.C., Heijstra, B.D., Mihalcea, C., Köpke, M., 2016. Gas fermentation – a flexible platform for commercial scale production of low-carbon-fuels and chemicals from waste and renewable feedstocks. Front. Microbiol. 7, 694. https://doi.org/10.3389/fmicb.2016.00694.

Lin, C.S.K., Pfaltzgraff, L.A., Herrero-Davila, L., Mubofu, E.B., Abderrahim, S., Clark, J.H., Koutinas, A.A., Kopsahelis, N., Stamatelatou, K., Dickson, F., Thankappan, S., Mohamed, Z., Brocklesby, R., Luque, R., 2013. Energy Environ. Sci. 6, 426–464.

Lobo, I., 2008. Basic local alignment search tool (BLAST). Nat. Educ. 1, 215.

Lowery, G., April 02, 2010. Tougher Grading Is One Reason for High STEM Dropout Rate. Cornell Chronicle.

MacLeod, M., 2018. What makes interdisciplinarity difficult? Some consequences of domain specificity in interdisciplinary practice. Synthese 195, 697–720.

Majer, S., Thrän, D., Wurster, S., Ladu, L., Sumfleth, B., Moosmann, D., 2018. Gaps and research demand for sustainability certification and standardisation in a sustainable bio-based economy in the EU. Sustainability 10, 2455. https://doi.org/10.3390/su10072455.

Marvik, O.J., Philp, J., 2020. The systemic challenge of the bioeconomy: a policy framework for transitioning towards a sustainable carbon cycle economy. EMBO Rep. 21, e51478.

Mascarelli, A., 2013. Environmental puzzle solvers. Sustainability training is on the rise, and institutions are working out how best to translate it into marketable skills. Nature 494, 507–509.

Masood, E., 1998. A degree of success for MRes courses. Nature 391, 729.

McKinsey Global Institute, May 2020. The Bio Revolution: Innovations Transforming Economies, Societies, and Our Lives.

Millett, P., Binz, T., Evans, S.W., Kuiken, T., Oye, K., Palmer, M.J., van der Vlugt, C., Yambao, K., Yu, S., 2019. Developing a comprehensive, adaptive, and international biosafety and biosecurity program for advanced biotechnology: the iGEM experience. Appl. Biosaf. 24, 64–71.

Miola, A., Schiltz, F., 2019. Measuring sustainable development goals performance: how to monitor policy action in the 2030 Agenda implementation? Ecol. Econ. 164, 106373.

Mısırlı, G., Hallinan, J., Pocock, M., Lord, P., McLaughlin, J.A., Sauro, H., Wipat, A., 2016. Data integration and mining for synthetic biology design. ACS Synth. Biol. 5, 1086–1097.

Morris, R.W., Bean, C.A., Farber, G.K., Gallahan, D., Jakobsson, E., Liu, Y., Lyster, P.M., Peng, G.C.Y., Roberts, F.S., Twery, M., Whitmarsh, J., Skinner, K., 2005. Digital biology: an emerging and promising discipline. Trends Biotechnol. 23, 113–117.

National Academy of Engineering, 2017. Engineering Societies and Undergraduate Engineering Education: Proceedings of a Workshop. The National Academies Press, Washington, DC.

National Academy of Engineering, and Institute of Medicine, 2014. The Postdoctoral Experience Revisited. The National Academies Press, Washington, DC.

National Academies of Sciences, Engineering, and Medicine, 2016a. Gene Drives on the Horizon: Advancing Science, Navigating Uncertainty, and Aligning Research with Public Values. National Academies Press, Washington, DC.

National Academies of Sciences, Engineering, and Medicine, 2016b. Genetically Engineered Crops: Experiences and Prospects. The National Academies Press, Washington, DC.

National Academies of Sciences, Engineering, and Medicine, 2020. Reopening U.S. Research Universities: Confronting Long-Standing Challenges and Imagining Novel Solutions: Proceedings of a Workshop in Brief. The National Academies Press, Washington, DC.

Nature Editorial, 2016. Life after academia. Nat. Ed. 537, 585.

Norme Française, 2008. Mécatronique. Vocabulaire. NF E 01-010, ISSN 0335-3931.

OECD, 2001. Biological Resource Centres: Underpinning the Future of Life Sciences and Biotechnology. OECD Publishing, Paris.

OECD, 2005. Supply and Demand Chain Inhibitors in Industrial Biotechnology. OECD Publishing, Paris.

OECD, 2014a. Biobased chemicals and plastics. Finding the right policy balance. In: OECD Science, Technology and Industry Policy Papers No. 17. OECD Publishing, Paris.

OECD, 2014b. Emerging Policy Issues in Synthetic Biology. OECD Publishing, Paris.

OECD, 2018. Meeting Policy Challenges for a Sustainable Bioeconomy. OECD Publishing, Paris.

Opgenorth, P., Costello, Z., Okada, T., Goyal, G., Chen, Y., Gin, J., Benites, V., de Raad, M., Northen, T.R., Deng, K., Deutsch, S., Baidoo, E.E.K., Petzold, C.J., Hillson, N.J., Martin, H.G., Beller, H.R., 2019. Lessons from two design–build–test–learn cycles of dodecanol production in *Escherichia coli* aided by machine learning. ACS Synth. Biol. 8, 1337–1351.

Philp, J.C., 2015. Balancing the bioeconomy: supporting biofuels and bio-based materials in public policy. Energy Environ. Sci. 8, 3063–3068.

Philp, J., 2018. The bioeconomy, the challenge of the century for policy makers. N. Biotech. 40, 11–19.

Pikaar, I., Matassa, S., Bodirsky, B.L., Weindl, I., Humpenöder, F., Rabaey, K., Boon, N., Bruschi, M., Yuan, Z., van Zanten, H., Herrero, M., Verstraete, W., Popp, A., 2018. Decoupling livestock from land use through industrial feed production pathways. Environ. Sci. Technol. 52, 7351–7359.

Priefer, C., Jörissen, J., Frör, O., 2017. Pathways to shape the bioeconomy. Resources 6, 10. https://doi.org/10.3390/resources6010010.

Rai, A., Boyle, J., 2007. Synthetic biology: caught between property rights, the public domain, and the commons. PLoS Biol. 5, 0389–0393.

Robinson, C.J., Carbonell, P., Jervis, A.J., Yan, C., Hollywood, K.A., Dunstan, M.S., Currin, A., Swainston, N., Spiess, R., Taylor, S., Mulherin, P., Parker, S., Rowe, W., Matthews, N.E., Malone, K.J., Le Feuvre, R., Shapira, P., Barran, P., Turner, N.J., Micklefield, J., Breitling, R., Takano, E., Scrutton, N.S., 2020. Rapid prototyping of microbial production strains for the biomanufacture of potential materials monomers. Metab. Eng. 60, 168–182.

Sadowski, M.I., Grant, C., Fell, T.S., 2016. Harnessing QbD, programming languages, and automation for reproducible biology. Trends Biotechnol. 34, 214–227.

Schiehlen, W., 2019. From cybernetics to mechatronics: seven decades of interdisciplinary engineering. Proc. Appl. Math. Mech. 19, e201900108.

Searchinger, T., Heimlich, R., Houghton, R.A., Dong, F., Elobeid, A., Fabiosa, J., Tokgoz, S., Hayes, D., Yu, T.-H., 2008. Use of US croplands for biofuels increases greenhouse gases through emissions from land-use change. Science 319, 1238–1240.

Smith, D., McCluskey, K., Stackebrandt, E., 2014. Investment into the future of microbial resources: culture collection funding models and BRC business plans for biological resource centres. SpringerPlus 3, 81.

Smolke, C.D., 2009. Building outside of the box: iGEM and the BioBricks foundation. Nat. Biotechnol. 27, 1099–1102.

Stark, J.C., Huang, A., Nguyen, P.Q., Dubner, R.S., Hsu, K.J., Ferrante, T.C., Anderson, M., Kanapskyte, A., Mucha, Q., Packett, J.S., Patel, P.K., Patel, R., Qaq, D.A., Zondor, T., Burke, J., Martinez, T.S., Miller-Berry, A., Puppala, A., Reichert, K., Schmid, M., Brand, L., Hill, L.R., Chellaswamy, J.F., Faheem, N., Fetherling, S., Gong, E.A., Gonzalzles, E.M., Granito, T., Koritsaris, J., Nguyen, B., Ottman, S., Palffy, C., Patel, A., Skweres, S., Slaton, A.E., Woods, T., Donghia, N., Pardee, K., Collins, J.J., Jewett, M.C., 2018. BioBits Bright: a fluorescent synthetic biology education kit. Sci. Adv. 4, eaat5107. https://doi.org/10.1126/sciadv.aat5107.

Timmis, K., Cavicchioli, R., Garcia, J.L., Nogales, B., Chavarría, M., Stein, L., McGenity, T.J., Webster, N., Singh, B.K., Handelsman, J., de Lorenzo, V., Pruzzo, C., Timmis, J., Ramos, J.L., Verstraete, M.W., Jetten, M., Danchin, A., Huang, W., Gilbert, J., Lal, R., Santos, H., Lee, S.Y., Sessitsch, A., Bonfante, P., Gram, L., Lin, R.T.P., Ron, E., Karahan, Z.C., van der Meer, J.R., Artunkal, S., Jahn, D., Harper, L., 2019. The urgent need for microbiology literacy in society. Environ. Microbiol. 21, 1513–1528.

Turing, A.M., 1952. The chemical basis of morphogenesis. Phil. Trans. Roy. Soc. Lond. B Biol. Sci. 237, 37–72.

UNCTAD, 2020. Impact of the COVID-19 Pandemic on Trade and Development. Transitioning to a New Normal. UNCTAD/OSG/2020/1. United Nations Publications, New York.

US White House, 2012. National Bioeconomy Blueprint. https://obamawhitehouse.archives.gov/sites/default/files/microsites/ostp/national_bioeconomy_blbueprint_april_2012.pdf. (Accessed 25 November 2020).

Vilanova, C., Porcar, A., 2014. iGEM 2.0 - refoundations for engineering biology. Nat. Biotechnol. 32, 420–424.

VTT, 2017. Synthetic Biology as an Enabler of Sustainable Bioeconomy - A Roadmap for Finland. VTT Technical Research Centre of Finland.

Wallman, S., Coleman, M., Fino, M., Kainer, D., King, K., Kuehnle, A., Nedosa, I., 2013. Roundtable discussion. Training technicians to support the bioeconomy: defining the need; designing and implementing innovative solutions. Ind. Biotechnol. 9, 275–281.

Waters, C., Sarin, S., 2011. AC 2011–715: Integrating Biology and Engineering. American Society for Engineering Education. https://www.asee.org/file_server/papers/attachment/file/0001/1019/Integrating_Biology_and_Engineering__final.pdf. (Accessed 11 December 2020).

Weiss, M., Haufe, J., Carus, M., Brandão, M., Bringezu, M.S., Hermann, B., Patel, M.K., 2012. A review of the environmental impacts of bio-based materials. J. Ind. Ecol. 16, S169–S181.

Welch, D., Southerton, D., 2019. After Paris: transitions for sustainable consumption. Sustain. Sci. Pract. Pol. 15, 31–44.

Wesseler, J., von Braun, J., 2017. Measuring the bioeconomy: economics and policies. Ann. Rev. Resour. Econom. 9, 275–298.

Westfall, P.J., Gardner, T.S., 2011. Industrial fermentation of renewable diesel fuels. Curr. Opin. Biotechnol. 22, 344–350.

Whitehead, T.A., Banta, S., Bentley, W.E., Betenbaugh, M.J., Chan, C., Clark, D.S., Hoesli, C.A., Jewett, M.C., Junker, B., Koffas, M., Kshirsagar, R., Lewis, A., Li, C.-T., Maranas, C., Papoutsakis, E.T., Prather, K.L.J., Schaffer, S., Segatori, L., Wheeldon, I., 2020. The importance and future of biochemical engineering. Biotechnol. Bioeng. 117, 2305–2318.

Whittall, D.R., Baker, K.V., Breitling, R., Takano, E., 2020. Host systems for the production of recombinant spider silk. Trends Biotechnol. https://doi.org/10.1016/j.tibtech.2020.09.007 (in press).

WO/2020/186173, 2020. Gas Fermentation for the Production of Protein-Based Bioplastics. International Application No. PCT/US2020/022659.

WO2020002598, 2020. Bioreactor for RNA in Vitro Transcription. International Application No. PCT/EP2019/067323.

Woolston, C., 2017. A love-hurt relationship. Nature 550, 549–552.

Zhu, Q., 2019. Developments on CO_2-utilization technologies. Clean Energ. 3, 85–100.

Chapter 4

Multimedia in microbiology education: a critical sustainable resource

Danilla Grando
School of Science, RMIT University, Melbourne, VIC, Australia

Chapter outline

If I look back on what made the most impact on my learning of microbiology, it was the story telling. Great teachers painted a vivid picture in my mind of how case studies unfolded. The impact on the patient and my role as a diagnostic microbiologist cemented my passion to be the best health professional I could possibly be. The stories became real to me and became part of my lived experience.

Importance of Microbiology Teaching and Microbial Resource Management for Sustainable Futures
https://doi.org/10.1016/B978-0-12-818272-7.00011-0

We know that learning is best done when it relates to a student's own experiences and involves some active component of interaction. But oral passing of knowledge has been shown to be even more effective when combined with visuals, hence the value of picture books in our youth. Textbooks that have numerous pictures are preferred by students and the visual aspects help to engage students. When we combine picture and text, this is called multimedia learning. Today multimedia learning also makes great use of video. In this chapter, I will examine how my multimedia teaching has developed over the years and how multimedia can be used to design teaching packages that are effective learning tools. It is important to understand that the choice of multimedia needs to be thoughtful, and the attributes of effective multimedia use are informed by research.

Of important note will be that these learning tools can be developed to be sustainable and able to be repurposed. This was never more important than during the teaching year of 2020 when I no longer had access to microscopes or practical class materials. I needed to replace my practical classes with meaningful alternatives, notwithstanding that this approach is not a replacement for the skill development required for diagnostic microbiologists. My lectorials were live streamed instead of in the classroom. Centralised exams were cancelled and so I decided to do away with the final exam. I decided to rewrite all my lectorials so that my delivery was in modules, with end of module online tests rather than the traditional end of semester formative assessment. Like the various lockdowns experienced in my community, my teaching aimed to tell students 'We can do this, let's get through one step at a time'. In this chapter, I will examine the successes of the 2020 teaching year resulting from a time-honoured curated approach to multimedia instruction and how this year has forced a major shift in teaching delivery that has meant we can move forward in more sustainable ways for the future.

Using multimedia to create active learning opportunities

In 2011, the American Association for the Advancement of Science (AAAS, 2011) reported on the need for change to the delivery of biology education. They drew attention to the fact that too many students were disengaged from the traditional models of teaching that imparted information and required the rote memorisation of science concepts. Instead, students need to be actively engaged in reflecting on science concepts and activities that enable them to assess their understanding and application of the core knowledge. These are said to be 'Active Learning exercises' (Michael, 2006). Michael and Modell (2003) describe examples of active learning opportunities and include Technology-Enhanced Learning as part of their description. Today it is simple to create multimedia through mobile devices such as mobile phones and laptop computers and there are now many freely available tools on the internet to enable the creation of rich multimedia learning including easy to use photo and video editing tools.

In the past I have often used the term 'Learning Object' which can be defined as any visual object that engages the student's attention (Wiley, 2000). However, it is debatable whether simply grabbing the student's attention can always be a learning opportunity. Other definitions require degrees of interactivity with the object. In the past, we have used Flash software to enable interactivity; however, the phasing out of Flash for the Web in late 2020 shows us the dangers of being too reliant on the use of particular software. In this chapter, I will focus on the ways we can use images and the importance of careful use of those images to maximise learning opportunities.

Types of graphics

It is common to see graphics used in lessons that purportedly may be interesting to students but ultimately distract students. If a graphic does not serve to inform or to help students assimilate new knowledge, then it should not be part of the lesson. There are some graphics that serve to amuse, and these can be powerful tools when used during lectures as it helps to create a more relaxed atmosphere and to lighten the seriousness of the matter under exploration. Bjorngard-Basayne et al. (2018) give worked examples of how the use of memes and GIFS can help break down barriers between students and students and staff. They also make important comments about avoiding cultural appropriation such as the concept of 'digital blackface', explained in Stark (2018). The term digital blackface embodies the use of reaction GIFs and is illustrated by the following post (https://www.teenvogue.com/story/digital-blackface-reaction-gifs) with examples of its use. It is often seen when non-black people use animated GIFs of black people to embody a reaction that they want to emphasise. Interestingly at the time of writing, many of the twitter accounts of examples used in this article had been suspended, supporting that its use is unacceptable. We have all sat through presentations where the presenter has sought to amuse by using images that have stereotypic representations, without regard to whether marginalised groups find this an uncomfortable reminder of inflicted prejudices.

Graphics, including animations and videos, help students explore concepts and when used with assessment of learning, are engaging. If used during lectures or when audio is supplied to accompany the graphic, use of captions is not required. It is useful though to supply students with a PowerPoint version of the lesson where captions can be supplied in the notes pane of the normal view. Thus, students can refer to the notes pane if they have trouble recalling the points and they cannot be otherwise bothered to go through the lecture recording to find the explanation. The caption can also guide the student into making notes of the important aspects of the graphic. An example created by the author is given in Fig. 4.1.

There are many ways that images can be used to enhance the learning experience. To break up my lecture time, I have a classroom activity where

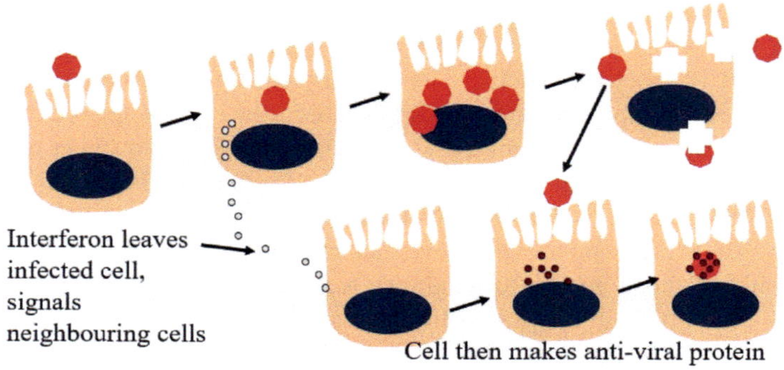

FIGURE 4.1 A chemical produced by host cells during viral attack: Interferon. Top left: a virus such as influenza attaches to a respiratory cell. Moving across the sequence at the top, the viral genetic material is used with the host machinery to make many virus particles that are released from the cell and proceed to infect neighbouring cells (lower row). These cells, however, have been primed for impending viral attack as they have received interferon from neighbouring infected cells. Interferon signals neighbouring cells to begin the production of anti-viral proteins that inhibit viral replication in these cells.

students perform a type of 'pin the tail on the donkey'. I have produced a pack of 25 A4-sized images that are laminated in plastic and are taken from previous lectures. The laminated prints are quick to attach at eye-height to the lecture walls before the lecture starts. Around the 20-minute mark of a lecture when interest is starting to wane, I let students know that we will perform an activity that will help them check their understanding of microbiology images we have previously seen and what they represent. I ask all students to leave the lecture hall and line up again at the entry door where I give them a labelled sticky note and they need to find one of images that corresponds to their label and attach their sticky note. I do a quick tour of the images and return to the lectern to recap on where misconceptions may have occurred (Grando, 2019). Buxton (2003) provides another example of how images can be used for a classroom activity. In the activity, photographs of a family dining along with descriptions of their individual characteristics form the basis for the exploration of risk factors, possible sources of disease and public health measures for prevention.

Creating a collection of multimedia resources

Before 2020, many academics began exploring the benefits of creating blended learning environments. For me this necessitated the creation of learning materials that engaged students and prepared them for face-to-face sessions where students could practice skills. Concepts encountered in pre-class blended learning activities were then reinforced and consolidated during valuable resource-intensive on-campus activities. In 2020, many of these on-campus activities were cancelled, hence the availability of well thought out, visually and mentally stimulating online exercises that were active learning exercises became even more essential for student education (Harris et al., 2020). Due to the elevated importance of these materials, it is important that we consider the attributes that make these resources a valued and effective teaching approach to online learning. In addition, as 2020 has shown, we live in a highly connected world so we can create a learning environment that supports students in several ways beyond the traditional learning management system.

Multimedia instruction as described by Mayer (2009) involves the use of 'presentations involving words and pictures that are intended to foster learning'. As microbiology educators, we are familiar with the importance of using images to supplement text or verbal explanations. Since these visual illustrations are such an important component in microbiology education, I wanted to evaluate some of the various strategies that I have used and align them with Mayer's 12 principles of multimedia learning as a way of explaining the success of these strategies. It is important to keep these principles in mind as our first aim must be to engage and educate but not be distracted by the lure to provide entertainment!

In Part 1 below I will look at examples of multimedia that are aimed at reducing cognitive overload by reducing extraneous processing.

Part 1: Strategies to reduce cognitive overload

Many of the multimedia resources I produce are accompanied by audio rather than text. This is because research shows that when text and images are presented in traditional lecture note format, the learner's attention is split between holding the text in their working memory whilst navigating to the corresponding image (Kalyuga et al., 1999). When text is used, it must be located as close as possible to the image. Kalyuga at al. also describes the redundancy effect. If the image already adequately describes the learning, then any accompanying text is redundant and only serves to place an additional cognitive load on the learner and interfere with their learning. Instructors may also be tempted to add multiple illustrations because students may find them interesting, even if irrelevant; however, research has shown that this may result in students performing less well on retention and transfer tests (Harp and Mayer, 1998).

Hand hygiene education

The teaching of hand hygiene has always been an important component of health science education and it has been found that techniques that go beyond simple lectures may be more effective at promoting changed behaviours (Martos-Cabrera et al., 2019). In fact, it is thought that this education should commence in early childhood as it will lead to life-long changes in habit (CDC, 2020a). There are ample resources available to help promote hand hygiene including those provided by the CDC (2020b) and divisions of government public health units, such as e-Bug (https://www.e-bug.eu/), who are solely devoted to promoting activities that support hand hygiene education. In 2020, Lin et al. (2020) also showed that Google searches for handwashing correlated with a lowered spread of SARS-CoV-2 within the jurisdictions they examined. Although the Centres for Disease Control reduce the steps of handwashing to five easy steps, it is important to note that more comprehensive descriptions and providing rationale can help individuals perform this task more safely and effectively. Amongst occupations who wash their hands often, over 20% of workers can suffer hand eczema (Thyssen et al., 2010). A study by Tauber et al. (2020) has shown that a specific training program improved hand hygiene behaviour, leading to improved skin outcomes in the individuals who took part in the study.

Before 2020, a hand hygiene education program was developed at our university and deployed across our biomedical science programs and specialist programs such as chiropractic, osteopathy, and Chinese medicine. This program involved a series of activities spread across 4 weeks of a practical program. In the first week, students would receive instruction in class on the important aspects of performing hand hygiene, including best practice for protecting skin health in the long term. Then students were asked to fingerprint one half of an agar plate marked 'before' then wash their hands and fingerprint the other half of the agar plate. The plates were incubated for a day and then left at room temperature for two additional days to enable colour development of colonies. The plates were then refrigerated until the students' next class. In the second session, their performance of hand hygiene was assessed by examination of the results of the handprint plates. The teaching staff would explain the significance of typical colony findings such a bacillus, micrococcus, *Staphylococcus epidermidis*, *Staphylococcus aureus* and mucoid colonies that were probably enteric bacilli. This was to reinforce the concepts of dirt, transient flora and resident flora. Students recorded the colony counts obtained in their pre- and post-handwashing halves of the fingerprinted agar plates. In the third session, students were asked again to reflect on the results from the previous session and decide on whether they needed to repeat the handwashing if colony counts were not significantly reduced or there was still evidence of 'dirt' or transient flora type colonies. Those students with the expected reduction of colonies and removal of target type organisms were then asked to repeat the experiment using alcohol hand hygiene. In their fourth session, they would again examine their handprint plates and record colony numbers. The students were asked to reflect upon the results and either make the association between sufficient handwashing and reduction of 'dirt'/transient colonies or in the case of students who had successfully reduced their hand colonies in the first exercise, evaluate the efficacy of the alcohol hand hygiene. Their teacher would then sign off on the students' successful completion of the hand hygiene training.

In 2020, we were unable to use the laboratory facilities so an alternative to performing the hand hygiene program was to have the student view an instructive video and then ask the students to fill in a table where they reflected on their handwashing practice before they had viewed the video and then after they had watched the video. The training video was created especially with this reflective activity in mind and it is important that you view the video to understand the importance of how it was recorded https://youtu.be/TgKYlBV3JxY.

Table 4.1 shows an example of how the students recorded their reflection:

TABLE 4.1 Student responses to video hand hygiene reflective exercise.

What I did before watching the video	What I will do following reflecting on the video
Applied soap directly to dry skin	Wet hands prior to applying soap
Spread the soap roughly between hands	Spread the soap gently between hands
Rinsed hands for a short period of time	Rinse hands for the same period of time that it took to lather hands with soap
Dried hands by rubbing the towel between hands	Dry hands by patting and squeezing hands with the paper towel

In designing the video content, I was mindful of the length of the video and taking care not to overload the video. In Mayer's book (Mayer, 2009, p.85), he describes the problem of extraneous overload. I could have included the use of hand products like moisturisers as an important post-washing step, but it was important to address the following barriers to learning effective handwashing namely:

1. Students perceive that handwashing is simple and apart from washing for 20 s, no other education is needed.
2. Students miss the importance of wetting their hands before adding hand wash products and the importance of proper rinsing to remove irritants.
3. Students do not realise that paper towel is abrasive to skin and with frequent handwashing, misuse could lead to dermatitis.

Thus, according to Mayer's first principle of reducing extraneous processing, the video avoided extraneous words, pictures and sounds. Another example of how effective visual demonstrations can be without words is that produced by Dr Rina Fu at https://www.youtube.com/watch?v=P81roNHPpb8.

Students have commented to us on how the hand hygiene education provided at our university in 2020 has been transformative and the following is an unsolicited example communicated to me:

I've always been big on washing my hands very often (if I cook dinner for an hour, I probably wash my hands about 10 times), and have also always been quite aware of how dry and wrinkly my hands are, alongside occasionally cases of dermatitis. Having started this assignment about a week and a half ago, I've noticed a marked change in my hands. Where I used to moisturize my hands 3–6 times daily, due to the dryness, I feel that I almost skip it entirely some days now. The major mistakes I was making, was putting soap onto dry skin, not rinsing for long enough, and not drying them completely. Thanks so much for teaching us how to wash our hands safely!

Reducing the incidence of respiratory infections

I have had a long-standing interest in promoting practices aimed at reducing respiratory infections. I ask my students at the start of semester, 'On average how many respiratory infections do people experience each year?' Estimates vary but it is probably 1–2 per year for adults and much more often for children. My students are often stunned to hear that I experience as few as 1 per 5 years. I have always taken a deliberate organised approach to teaching essential steps that must be adhered to in order to avoid contracting respiratory infections. By reducing the required actions into defined steps, I am able to design learning with another important principle of multimedia design, the Signalling Principle:

People learn better when cues that highlight the organization of the essential material are added

Mayer (2009, p. 108).

In 2020, my classroom approach to teaching preventative practices was also translated into a learning module. Students were instructed to watch the following video https://www.youtube.com/watch?v=nSp4Gc6HksU&t=7s and then report back on the four important steps to prevent respiratory infections. These four steps are presented as clearly defined cues in the video and each cue is supported by an image to help that cue stick in the mind of the viewer. One of the most important visual cues is the dual mode of transmission that occurs with a sneeze https://en.wikipedia.org/wiki/Sneeze and is reproduced in Fig. 4.2.

In designing this video, I decided that it should focus on promoting changes in behaviour that would result in a reduction in the transmission of all respiratory viruses, rather than focussing on Coronavirus. Thus, I was designing for sustainability in mind. In fact, if I had designed for prevention of coronavirus, then I would have to make a new video for each change in recommendations such as social distancing and when to wear a mask. It may be that going forward, mask wearing will be a stronger message as the public shifts its perception on its use as a preventative measure. As the current video

FIGURE 4.2 The expulsion of saliva droplets when sneezing. This illustrates the fine aerosol droplets causing airborne transmission of viral infections and the larger heavier droplets that will land on surrounding surfaces and cause contact transmission of viruses. *Photocredit James Gathany, CDC Public Health Image Library ID 11162.*

stands, it will be used again in forthcoming years, allowing me the time to focus on creating other new multimedia modules.

The design of this video illustrates an important aspect of good multimedia design as stated above by Mayer (2009), that cues are present to highlight the organisation of the essential material. The video is punctuated with the concept of focussing on four important points. Each image is used to reinforce the point and finally the video concludes with a recap of the four points. Students are then asked to write down these four points in response to the learning activity associated with viewing this video. As noted by a colleague (Megan Lloyd, personal communication, 2020), the principle of aiming for around four meaningful items in a lesson fits well with cognitive load theory (Cowan, 2010).

The use of animations

One of the most sustainable tools that I have found for use in my teaching is a set of animations that brilliantly and simply describe basic principles of microbiology (such as the action of antibiotics) and the immune response, produced by Gary Kaiser in 2009 for the American Society of Microbiology (ASM) Image library. Hartoyo and Abdul Gafur (2019) developed a system to enable the playing of Flash animations independently of a computer platform; however, Gary Kaiser has provided links to his animations as HTML 5 versions at his website (http://faculty.ccbcmd.edu/~gkaiser/). He has generously licenced all the visual material at his website as creative commons (free to use and adapt with attribution). In 2020, I fortuitously recorded voice-over movies of the animations so that my students could continue to learn from these excellent animations (an example recording can be found at: https://rmit-arc.instructuremedia.com/embed/86fa29d0-51e4-497d-92e7-cb83bb239cfe).

So, what is the evidence that animations are important learning tools? Berney and Bétrancourt (2016) performed a meta-analysis of published studies into the learning enhancement provided by animations and concluded a significant positive benefit to student learning, especially when animations are coupled with auditory descriptions and there is no accompanying explanatory text. There is also evidence that animations may detract from learning as they interfere with cognitive processing (Schnotz et al., 1999). So how do we ensure that the viewing of animations is not simply passive? Kombartzky et al. (2010) provide evidence that strategies such as scaffolded activities that surround the animation will ensure active engagement.

In order to increase cognitive processing of the immunology images, I have embedded static snapshots from these animations into a classroom activity where students use the walls of the lecture theatre to assemble the sequence of frames from the animation into a storyboard. You may like to listen to a podcast that was produced at RMIT University that highlights my experiences with using the classroom walls to run activities to convert the passive viewing of animations during lectures into an active inquiry of what the visuals represent (Grando, 2019). As students enter the classroom, I form groups, give each group a pack of laminated animation frames and laminated text boxes that need to be matched to the animation frames. Students brainstorm the matching and I rotate around the walls in case there are questions. I have created 25 packets of cards for around 150 students (students work in groups of around six). In 2020, however, all my classes needed to be converted to online. As an interim measure, I look screenshots and quizzed the students on what the screenshots represented. Even though it is possible to create break out rooms for students to work in groups online, it is difficult without teaching assistants to manage to supervise large numbers of students working in separate online rooms. Today these online rooms come with whiteboards so it should be possible to develop virtual cards sets in HTML 5 (H5P) that can be arranged sequentially on the virtual whiteboard.

In the past I have created interactive animations using Flash and interactive E-learning cards for other teaching academics. These were standalone objects and I noticed that if these were not integrated into a lesson, they became redundant. Those students who would stumble onto these objects would write notes of appreciation to me; however, these objects have largely disappeared from learning management systems and thus the enormous effort to create them has been lost. This can also be the case for video. For instance, I have created a video on pipetting technique for use in other classes (https://www.youtube.com/watch?v=weBuqelKwP4), but with the changing workforce, it has also been neglected for continuing use. This highlights two points: firstly, the need for repositories. These repositories should be held by societies and support given for their maintenance and upkeep. Institutional repositories suffer from the whims of management at universities that will change access to learning object repositories. Also, institutions tend to lock up assets behind firewalls thus limiting the educational use of some of these valuable artworks and learning tools. The second point is that the objects need to be indexed so that they are searchable and their function clearly identifiable and linked to specific learning outcomes. This will ensure that these valuable tools are also available for academics teaching in resource-poor countries. These materials should also be licenced by creative commons that then makes it clear that they can be adapted for local use, i.e., the addition of subtitles in other languages.

PowerPoint design

The use of PowerPoint to accompany lectures is ubiquitous and is often synonymous with information transmission (Jones, 2003). However, it can also be a powerful tool, and when used wisely, it can be a valuable resource. In fact, in 2020 many of my PowerPoint slides from previous years' lectures were repurposed into mini learning units and used to create short video segments. One of my first PowerPoint lessons that was converted into a video learning package is called the ABC of Microscope Setup. It has been created as a series of graphics using the letters A through to F to serve as a mnemonic to help students remember a sequence of steps that will ensure correct microscope set up. Regular use of mnemonics has been found to change brain architecture and improve memory (Dresler et al., 2017). It was first produced around the year 2010 as an introduction to using the microscope for my introductory microbiology class (https://youtu.be/JFBO4e_cMCs). The PowerPoint is delivered at the start of the practical class and then the concepts are reinforced by the individual group demonstrators. We have large classes of up to 120 students and we only have 60 microscopes. Working in pairs, one student has a dedicated 30 min on a microscope to learn their 'ABCs', whilst the other student in the pair performs an online learning package that teaches how to recognise different microbes – the 'Who am I quiz'. After 30 min, the students swap activities. Students are assessed on their microscope set up proficiency in the following week's lab class and this is done by the demonstrator altering all the microscope settings and the student then systematically checks and adjusts the settings to ensure they can proficiently find the microbe in a provided smear.

Although only originally intended to be used at the start of my first practical class before getting students to practice their learning on actual microscopes, its effectiveness as a learning tool soon became evident as instructors in other schools requested to use it to prepare students before similar activities in their classes. We have also added a video version of the ABC of Microscope Setup to many of our course materials of higher-level classes in microbiology as part of microbiology skills refresher units. We find this particularly useful for students entering higher level courses from other universities who have not provided enough training for students in microscope setup.

What are the characteristics of a PowerPoint that make it a sustainable resource? As Jones (2003) points out, slides should have minimal text. The subject material should be kept brief as most explanation can occur in the video recording of your delivery of the PowerPoint. There are many other pedagogical issues for consideration, and I refer readers to review Jones (2003) for further information and useful links. Encourage students to take notes. A quiz should follow PowerPoint delivery to help students understand the importance of note taking as students who simply view the slides will find they do not have enough depth of understanding to successfully answer quiz questions. Avoid mentioning events that will date the video, such as providing an annual incidence of disease as this will require annual updating of the PowerPoint. Although PowerPoints can be easily updated, recording the presentation is time consuming and it is more sustainable to use the same recording in subsequent years. If you wish to cover incidence of disease, then this can be made into a standalone learning exercise. For instance, in an online course called Outbreak (Calonge and Grando, 2012), I ask students to create a graph of the annual incidence of a particular disease by using publicly available notifiable diseases data (e.g., data from the Australian Government Department of Health National Notifiable Diseases Surveillance System: http://www9.health.gov.au/cda/Source/CDA-index.cfm). They then write a newspaper article that explains the annual and seasonal trends of the disease. The students graph comprises half of the assessable marks available for this assignment.

Clarke and Mayer (2016, p. 174) list important points to consider in lesson creation using multimedia:

1. Avoid including interesting points (including stories and details) not essential to the learning objectives
2. Use a wide range of types of graphs. Avoid including 'illustrations, photos and video clips that **may** be interesting' but not central to the learning
3. Avoid sounds and music unrelated to the learning
4. Consider simple line drawings to replace complex illustrations as these help the learner **build** their understanding
5. Use the minimal number of words and graphics required to help with the understanding of core concepts

Using these important points in lesson design will result in students' ability to maintain focus, progress quickly though the lesson and retain more information. This will enable more time for the students to practice their application of the learning through the assessment tasks that should accompany each lesson.

Copyright limitations on non-original material

For many educators in clinical microbiology, it will never be possible to amass the wide range of images required to provide comprehensive quality computer assisted learning. Also describing new technologies that students will experience in the workplace requires the use of images that are unlikely to be in the educator's collection. The first option to find images is to use the Google search engine. Type in a keyword and when the results are returned, click on Images (Fig. 4.3). Fig. 4.3 was quickly created using the Microsoft Snipping Tool, one of the quickest and most efficient ways to create images from the internet that are low resolution but of enough quality for generating computer-assisted learning materials. Please note by using this tool you are creating a derivative work, so you must check that you are not infringing copyright usage of the original image when you are taking a snapshot.

Once you have gone to the images page, you then need to do an advanced search (Fig. 4.4 - select Advanced search from the dropdown box that appears by selecting Settings) and then navigate down to the Usage rights option and select Creative Commons licences. This will then narrow your search to images more likely to be able to be used in your teaching.

As an example of appropriate use of the excellent images here would be to create a PowerPoint slide so that it is ready for use and complies with the Creative Commons usage for the image (Fig. 4.5).

FIGURE 4.3 Snapshot of the Google search engine used to find images for teaching. Note using 'All' (underlined in blue) will give websites but choosing the next option 'Images' will only give illustrations. Fig. 4.4 will demonstrate how to restrict the search results to those images with creative commons licences.

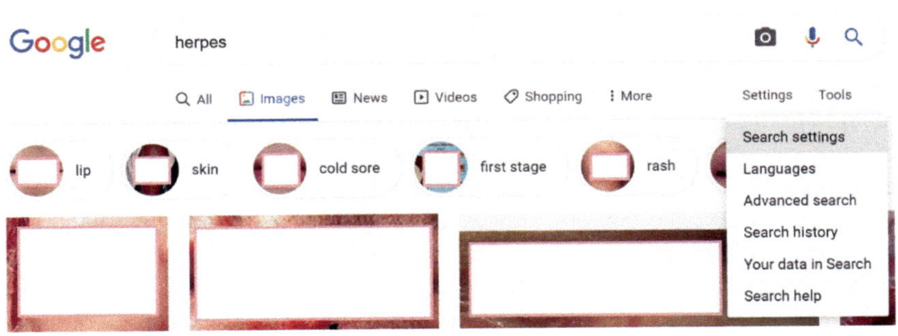

FIGURE 4.4 The search results following from the search shown in Fig. 4.3. The next important step is to use 'Settings' to open choices and from that dropdown box shown, select 'Advanced search' to restrict images to those licenced in Creative Commons. Images indicative rather than actual.

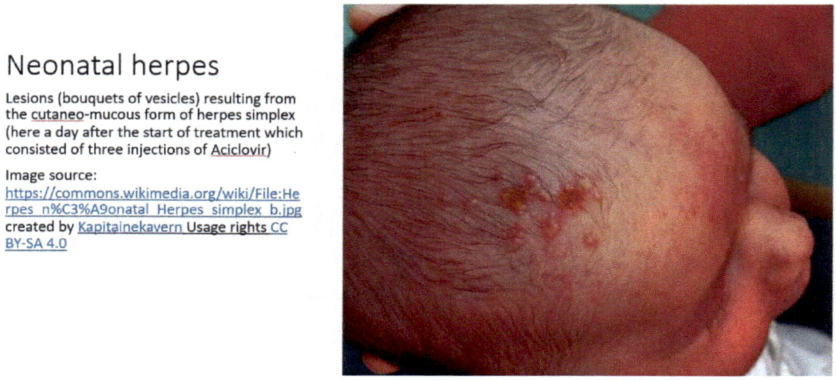

FIGURE 4.5 PowerPoint slide for repository of useful images. Note the important elements of this particular Creative Commons licence that informs those that may wish to reuse this image that they must give attribution (BY) and that they must add the same licence to any works that are created from this original image (SA - share alike).

Footnote: Usage rights can change with time, so it is important to periodically check that the image is still licenced in the same way as when you originally started using the file.

Part 2: How do we manage our multimedia resources?

The image library

An exciting initiative in the early 2000s was the American Society for Microbiology's Microbe Library. In its original inception, it was easy to navigate, and educators could quickly identify useful images. One of the evocative images from this collection that I have used in multiple lessons is that depicting the mucoid colonies of *Streptococcus pneumoniae* (Buxton, 2005) and available at https://asm.org/Image-Gallery/Blood-Agar-Plates-and-Hemolysis-Streptocpccus (Slide 12, at the line of writing the linking URL contained a spelling error for Streptococcus!). These bacteria produce this mucoid material because they can produce a very large capsule and this is seen in the Gram stain picture of the organism (Delisle and Tomalty, 2009), that beautifully illustrates how these capsules can be anti-phagocytic (available at https://asm.org/Image-Gallery/Encapsulated-Streptococcus-pneumoniae). Unfortunately, at the time of writing these images were all moved to a new portal called https://asm.org/Browse-By-Content-Type/Image-Gallery and they are only free to use for classroom materials and not for publication. In order to now find these images, you need to know what you are looking for as previews are not available, for instance, the Buxton (2005) image can only be found if you search for streptococcus as well as the Delisle and Tomalty (2009) image, and then by downloading PowerPoint files and viewing each of these presentation packages! It is becoming exceedingly difficult to find the original picture by Buxton and thus many educators will now miss using this wonderful resource. Another issue is that both these images are labelled with ASM www.MicrobeLibrary.com which has now become https://asm.org/a/ASMScience making it difficult for new educators to find these valuable images. The Creative Commons licence chosen for these images means that derivative works cannot be made from these images, so they are destined to remain labelled with a non-functional URL!

There are so many different images that can be used to illustrate microbiology that no single library is likely to be fully comprehensive for one educator's needs. In the section above, I covered how images can be found on Google. It is common for you to find an image on Google but later fail to find the same image. Thus, it is important to have a sustainable system for organising valuable images and every microbiology educator needs to record details that accompany each image so that correct attribution and labelling of the copyright is used. Over the last 20 years of teaching, I have found that I have repurposed images numerous times and quick retrieval is needed to keep abreast of the numerous new learning tools that need to be created year after year.

I have found that images are best saved by organising into folders that align with their source. In Windows Explorer it is possible to view the contents of folders by selecting in the view menu item by selecting view large icons. This enables a quick perusal of the folders for the types of images held there. Since every source has different usage rights, there needs to be a system for quickly accessing these details. In my case for each folder, I have a word document that lists each image in that folder and details the image attributes and usage rights. I have also found it useful to have a master word document that combines all the word documents into one so that I can easily search across my collection. Even more useful is to have thumbnails of the images on the word document. To create a thumbnail, open image editing software and resize to 100×100 pixels. As an alternative to keeping documents describing image attributes, PowerPoint slides can be created and then the attributes including copyright details can be recorded in the notes section.

In 2020, the use of images became even more important in order to illustrate techniques that could only be experienced virtually. To ensure students complete blended learning activities, quizzes can be created using still images of technique videos. The advantage of doing this is that if you need to create formative assessment to assess student learning outcomes, it is harder for students to search the internet for answers whilst doing their tests.

GERMM

In 2009, I received a Program and Courseware Enhancement Grant from RMIT University's College of Science Engineering and Technology to develop an electronic visual glossary to illustrate microbiological techniques (GERMM – Glossary and Electronic Resources of Microbiology Methods). This was to replace the need for students to buy a techniques manual that was largely descriptive and devoid of illustration and colour. It was also a resource for the newly established digital wet laboratories developed for biology teaching at RMIT University (Grando, 2010). The aim of this student aid was to enable students to quickly and easily access interpretative guides for the many technical methods used in microbiology. This resource lies behind the firewall of an institutional learning management system, so some screenshots of GERM are presented in Figs. 4.6 and 4.7.

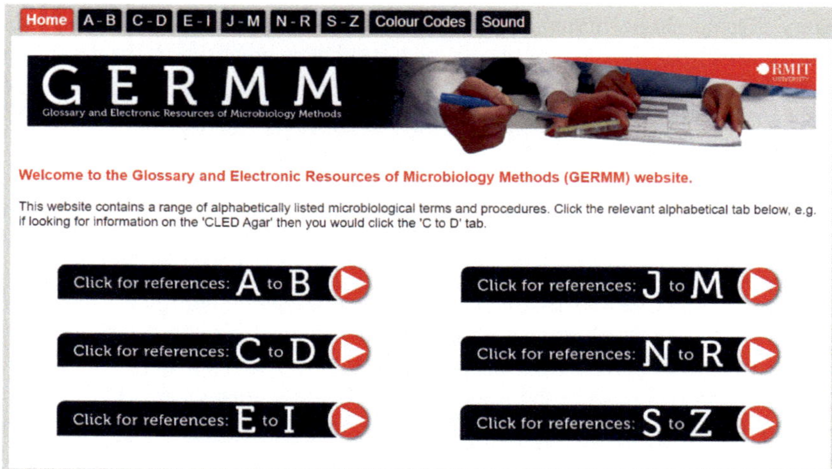

FIGURE 4.6 Screenshot of the opening page of GERMM. This repository is used to help students to understand and interpret tests used in the identification of microorganisms as shown in detail in Fig. 4.7.

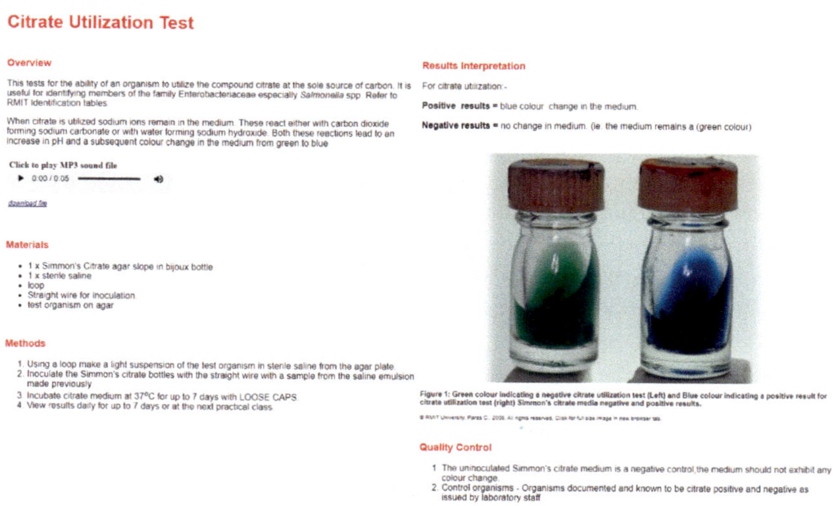

FIGURE 4.7 A sample page from GERMM. These pages also include audio files to help students understand how to pronounce microbiological terms.

It is probable that there are many outstanding resources such as GERMM hidden from broader use as they lie behind institutional fire walls. In 2020, the American Society for Microbiology published a list of publicly accessible resources to help online learning of practical techniques in response to the loss of face-to-face practical classes during the COVID pandemic (Horak, 2020).

Segmenting the learning

According to Mayer (2009), 'People learn better from a multimedia lesson when it is presented in user-paced segments rather than a continuous unit'. In the past it has become evident that many students will cram for exams rather than engage with learning materials throughout the semester and this is supported in the research literature (McIntyre and Munson, 2008). The authors go on to say that it is not just the fault of running a final exam; it is also the way the material is presented, 'The all-too-common use of PowerPoint slide lectures ..., does not engage students to take notes in their own language and handwriting, which shunts their processing of the material, leaving all effective learning to the cramming period at the end of the term' (McIntyre and Munson, 2008). Interactive video vignettes have been used by Cardinale et al. (2020) to engage students in self-paced learning by giving them the opportunity to see through short segments of video how they can problem solve experimental results. They have some example vignettes that can be viewed and are free to use at https://www.rit.edu/cos/interactive/MINT.

In 2020, I decided to try a different approach to organising the learning for the semester. The final exam was removed, and the learning organised into three modules, each of 3 weeks duration. In the week following the finish of each module, a 40-minute timed-online test was held worth 20%. Learning of new material was confined to the first 2.5 weeks of the 3-week module with review of material held in the latter half of the second week. Much of the previous lecture material was repurposed into four case study assignments each worth 10%. Segments of the repurposed lectures were recorded to inform the student responses to the case study questions (Fig. 4.8). Providing links to the recorded lecture segments helped keep student workload manageable; however, some questions were also designed to exercise students' lateral thinking. The students were given a 2-week period to submit their responses to these case study sets, and a tutorial was held 2 weeks after each case study set submission (to allow for 1-week extensions to assignment due dates requested on an individual basis). Students were nominated in the tutorial to give their explanation of the answer to one of these tutorial questions with the requirement that they were explaining their answer to a patient. This exercise was to stimulate the student into thinking how they would help someone else understand. This activity is analogous to peer-to-peer teaching with oral presentation which is found to be more effective for learning than simply preparing an answer (Fiorella and Mayer, 2013). Students were not graded on

Case Studies Number 2

These case studies will focus on Respiratory infections and also the prevention of spread. The answers may form the basis of some questions on the next test as we will have a meeting with your tutors in advance of your next test 😊

These case studies are worth 10% of your total assessment so expect 10 hours of work.

1. I made the following video of tips to prevent coronavirus at the start of the Coronavirus at the start of March 2020.

https://www.youtube.com/watch?v=nSp4Gc6HksU&t=7s

(viewing time 7 minutes)

What were the 4 strategies to prevent coronavirus in that video?

FIGURE 4.8 A snapshot taken of the first question in a case studies set provided to students. Each case study sheet had 15 case studies (only the first shown in this sample). For those wishing to view the video it can be accessed at https://www.youtube.com/watch?v=nSp4Gc6HksU&t=7s.

their oral response; however, if they did not participate in the tutorial, the maximum mark achievable on their marked responses to the case study assignment was 50% (Fig. 4.8).

There needs to be accountability by academics for how much work is required of students to complete these types of in-semester assignments. A common reason for students cramming for exams is lack of time (Loomis, 2014). Thus, it was important to segment the learning so that students can manage the competing demands on their time during the teaching period.

Part 3: How can we ensure that our multimedia learning is transformative?

Cranton (2006) uses the definition of Mezirow (2000, 2003) of transformative learning 'as a process by which previously uncritically assimilated assumptions, beliefs, values and perspectives are questioned and thereby become more open, permeable and better validated'. In this respect, the design of multimedia learning should engage cognitive processes and help students develop rationale for practices that have previously been performed by rote. Handwashing is an excellent example of such a process that is often taught without consideration for the importance of teaching practices that will lead to transformative learning. Students are taught the importance of handwashing, but they perform handwashing without thinking that the steps involved may be important to preventing damage to their skin. This was evidenced by the example shown above in Table 4.1 where a student added liquid soap to their hands before wetting their hands and can also be seen in examples in the media where school children are also incorrectly taught to add liquid soap to dry hands. This is often because the entire focus is on the application of soap and its distribution (as an example https://www.youtube.com/watch?v=WVqZP6F5IWo) copyright rather than the entire process that includes the essential thorough rinsing and careful drying of hands.

As with the example of handwashing, the use of case studies is also an excellent example of how their use engages students on a cognitive level that goes beyond the rationale. As Cranton states (2006, p. 2), the process of transformative learning is also enabled when the activity incorporates the stimulation of the students' imagination and affect. Many experiences in medicine are outside of the students' experiences, so it is important to provide these stories. As example for the importance of hand hygiene, the Australian National Hand Hygiene Initiative (NHHI - https://www.safetyandquality.gov.au/our-work/infection-prevention-and-control/national-hand-hygiene-initiative) incorporates into its online learning modules a very personal story on the effect of hospital acquired infections – Glen's story, viewable at https://www.youtube.com/watch?v=RIsBB6TmZvA&feature=emb_logo.

Case studies help students imagine what the clinical condition is like and leave the student affected by this understanding. The use of visual representations is an important component of any case study presentation. When a clinician presents a case study for an audience, they already have a picture in their mind of how the condition manifested and the impact on the patient. We often assume that our students can see what we see. We need to remember in our presentation of case studies that there needs to be visual representations that evoke an emotional response from our students. This can be a difficult choice as the representation of the effect of infectious disease on the human body can be graphic. It is important thus to ensure that a warning accompanies and precedes the showing of graphic images and that this warning explains why the image is being shown. As an example, I show images of immunocompromised patients, post-mortem who have died after contracting herpes simplex virus from health professionals who were unaware that their innocuous finger blister was a form of herpes infection (herpetic whitlow). An excellent source of clinical photographs is the 11 volume '*Atlas of infectious Disease*' edited by Gerald L. Mandell; however, this is no longer available to purchase. For those without access to this resource, an internet search using 'herpes immunocompromised' will lead to suitable images. A useful online source of clinical images is the MSD Manual Professional version and the following link is an example of the types of important impactful images that can be used https://www.msdmanuals.com/en-sg/professional/pediatrics/infections-in-neonates/neonatal-herpes-simplex-virus-hsv-infection.

A valuable resource for case study teaching for science courses is the collection housed at the University of Buffalo Library National Center for Case Study Teaching in Science (https://sciencecases.lib.buffalo.edu/). The usage rights are very reasonable, being that reproduction of the case studies can only be in a password protected learning system. After paying a small annual subscription, case studies and their accompanying resources can be downloaded. The resources can be modified for an educator's own use. Searching for the term 'microbiology' at the start of 2021 returned 99 case studies. These case studies also have freely available access to recommended supporting learning videos which can save an instructor much time trying to find suitable video links for use in their own teaching. I encourage readers to consider submitting their case studies to this well curated portal.

The creation of case studies that are transformative learning exercises needs to follow the sequence as outlined by Clarke and Mayer (2016) of 'explanation-example-question-feedback'. The multimedia presentation is provided to the student and presents **explanation** and **example**. A **question** or questions is then given to the student, and in the example above of how case studies are used in my course, students come prepared to class to explain their answer to their peers. During class, the tutor provides constructive **feedback** on whether the student's explanation of the answer enables fellow students (or in real life, the patient) to understand the situation, and make memory associations so that their learning experience is transformative. Often the tutor will add further examples to solidify the students' picture of the problem at hand. The following are comments from various students received as part of anonymised class feedback at the end of the course:

> *I loved the case studies they were very useful in consolidating information without the undue stress of the exam, it also provoked my curiosity a lot, made me think outside the square.*

> *The mini videos for the tutorial are great way of learning the course content.*

> *The worksheets were a great way of learning, with videos that correlated to the questions directly, and having a tutorial where we go over the answers and are asked to share as well. I found I retained a lot more information this way.*

> *I've really enjoyed the layout of the course in regard to assignments. Having case studies and quizzes at intervals and thus not having an exam has allowed us to better focus on the material that we are learning, thus having a better understanding, then forgetting then cramming at the end of semester but retaining nothing.*

> *The case studies were done really well, especially when you added the short videos for us to watch.*

Skills for the future: effective communication using multimedia

There are many guides to the recommendations for how multimedia resources are constructed to maximise the aesthetics and impact of the object. The creation of these multimedia objects is beyond the scope of this chapter and the tools available to produce these images are constantly changing. Teaching academics can tap into the new tools available by watching how research students use these tools to produce quality scientific illustrations and then integrate exercises that replicate the production of quality images into classroom activities. Although the software tools change with time, what does not change is the need to critically appraise multimedia and continue to strive for improvements in the way we create and use images that aim to convey a 1000 words. As we move forward, we should look at new ways of immersing our students in the world of microbiology.

Research groups are beginning to publish easy-to-use software that can take data and create interesting 3D graphs. For example, Hartmann et al. (2021) have produced 3D visualisations of biofilm communities. Biofilms are microbial communities that grow on surfaces and are often studied as they pose problems when these collections of microorganisms foul both industrial and medical devices. They can be difficult to remove, and prevention is better than cure. The cells in these communities form a hierarchy and their functions vary in space and time. By monitoring how these hierarchies and functions vary overtime, 3D representations can show a dynamic picture of how these communities change when exposed to treatments. An example of graph showing how these communities change with time can be seen at https://www.youtube.com/watch?v=wlX0cWTM4v0.

The use of 3D imagery is also becoming more common with the use of tools such as SketchFab (https://sketchfab.com/).

Microbiological 3D pictures and browser add-ons for viewing enable students to rotate the objects and view from various angles (example T1 bacteriophage at https://sketchfab.com/3d-models/t7-bacteriophage-01-c7cb5c078a4e423f8cbbf8a7db32db65).

I look forward to the future when we can immerse students into virtual reality representations of bacterial communities and their interactions, especially their interactions with the host (both positive and negative!). Researchers at the University of Toronto have begun the exploration of virtual reality as a microbiology teaching tool (https://www.utoronto.ca/news/u-t-researchers-explore-virtual-reality-microbiology-teaching-aid).

Summary of useful points to note when designing multimedia

The following points form my working checklist and are described in detail in Clarke and Mayer (2016), *E-learning and the Science of Instruction Proven Guidelines for Consumers and Designers of Multimedia Learning,* Fourth edition.

Choice of visuals

Choose carefully, do not use graphics for purely decorative purposes. Look at the graphic and make notes on the elements of the graphic that you would wish the student to notice. Your notes will ensure that these elements are included in the accompanying audio.

Sparing use of text

Use words sparingly, located close to the graphic. Check to see if any wording is redundant because the graphic already contains text. Is the text (or even the graphic for that matter) essential to the learning outcome or has it been added merely for interest?

Audio is better than text

The examples provided in 2020 of needing to move our teaching online have forced us to create learning materials that are engaging. Many educators have replaced lengthy text descriptions with bite sized audio descriptions of graphics and lecture snippets.

Use a personable style

Students appreciate warmth and caring both in their interpersonal communications with their academics but also in the way the audio descriptions are delivered in the multimedia learning.

Test students' application of the knowledge

Has your multimedia exercise promoted thinking skills in your students? This is part of the sequence of 'explanation-example-question- feedback'.

Conclusion

> *This course has been one of my favourites to date and I have loved taking it! This is the only subject this year where I have not felt severely disadvantaged by not being able to go to campus due to COVID. Danilla has a magical ability of making complicated content simplified, comprehensible and I love how she includes real life stories and events to make it easy to remember.*

As we have seen from the example of developing GERMM, there are probably many excellent resources that have been developed but are not accessible to the broader educational community as they are currently behind institutional firewalls. However, as 2020 has shown us, many of these materials can be repurposed and given new life. I agree with the sentiments of Bhattacharya et al. (2009):

> *Excellent teachers are successful because of the ways in which they mediate content and place the content within the context of the subject matter. It is not the specific content or images the successful teacher presents, but rather the way they are presented and framed within the scope of the topic area.*

Armed with the knowledge of the principles that are important for multimedia design, we can look for or create new avenues where this content can be shared. I call on teachers in microbiology to band together to create new multimedia tools for microbiology education that can be widely shared, especially so that access to and equitable use of these transformative tools is enabled across the world. Careful creation of these tools will ensure their long-term sustainability as teaching aids and will enhance the sustainable provision of microbiology education into the future.

Acknowledgements

I would like to acknowledge the brilliant teachers in clinical microbiology that have inspired me throughout my career. The flame that is my love of storytelling was ignited by my first great teacher in Medical Microbiology, Mr James Chapman. I wish to also acknowledge that great work of Gary Kaiser who has generously shared his brilliant animations for generations of teachers.

References

AAAS, 2011. Vision and Change in Undergraduate Biology Education: A Call to Action. American Association for the Advancement of Science, Washington, DC.

Berney, S., Bétrancourt, M., 2016. Does animation enhance learning? A meta-analysis. Comput. Educ. 101, 150–167. https://doi.org/10.1016/j.compedu.2016.06.005.

Bhattacharya, S., Dunning, J., Kaur, A., Daniels, D., 2009. Re-purposeable learning objects based on teaching and learning styles. In: Pagani, M. (Ed.), Encyclopedia of Multimedia and Technology, second ed. IGI Global Publishing, pp. 1224–1231.

Bjorngard-Basayne, E., Reyes, M., Kaeppel, K., 2018. Memes and GIFs as Powerful Classroom Tools. https://www.facultyfocus.com/articles/teaching-with-technology-articles/memes-and-gifs-as-powerful-classroom-tools/. (Accessed 17 December 2020).

Buxton, R., 2003. Lunch on the Run: Using Photographs to Identify Possible Sources of Disease. Curriculum Archive. American Society for Microbiology, Washington, DC. https://www.asmscience.org/content/education/curriculum/curriculum.40. (Accessed 10 January 2021).

Buxton, R., 2005. Blood Agar Plates and Hemolysis: Streptococcus and Other Catalase-Negative Gram-Positive Cocci. Image Gallery. American Society for Microbiology, Washington, DC. https://www.asmscience.org/content/education/imagegallery/image.2881. (Accessed 10 January 2021).

Calonge, D.S., Grando, D., 2012. Reality-based learning: outbreak, an engaging introductory course in public health and epidemiology. Health Educ. J. 72 (5), 584–600. https://doi.org/10.1177/0017896912452072.

Cardinale, J.A., Newman, D.L., Wright, L.K., 2020. An online interactive video vignette that helps students learn key concepts of fermentation and respiration. J. Microbiol. Biol. Educ. 21 (2), 1–8. https://doi.org/10.1128/jmbe.v21i2.1895.

CDC, 2020a. Handwashing a Family Activity. Centers for Disease Control and Prevention. https://www.cdc.gov/handwashing/handwashing-family.html#:~:text=Handwashing%20can%20become%20a%20lifelong,the%20bathroom%20or%20before%20eating. (Accessed 26 December 2020).

CDC, 2020b. Life Is Better with Clean Hands. Centers for Disease Control and Prevention. https://www.cdc.gov/handwashing/campaign.html. (Accessed 20 December 2020).

Clark, R.C., Mayer, R.E., 2016. E-learning and the Science of Instruction: Proven Guidelines for Consumers and Designers of Multimedia Learning, fourth ed. John Wiley & Sons, San Francisco, CA.

Cowan, N., 2010. The magical mystery four: how is working memory capacity limited, and why? Curr. Dir. Psychol. Sci. 19 (1), 51–57. https://doi.org/10.1177/0963721409359277.

Cranton, P., 2006. Understanding and Promoting Transformative Learning: A Guide for Educators of Adults, second ed. Jossey-Bass, San Francisco, CA.

Delisle, D., Tomalty, L., 2009. Encapsulated streptococcus Pneumoniae. https://www.asmscience.org/content/education/imagegallery/image.2654. (Accessed 10 January 2021).

Dresler, M., Shirer, W.R., Konrad, B.N., Müller, N., Wagner, I.C., Fernández, G., et al., 2017. Mnemonic training reshapes brain networks to support superior memory. Neuron 93 (5), 1227–1235. https://doi.org/10.1016/j.neuron.2017.02.003.

Fiorella, L., Mayer, R.E., 2013. The relative benefits of learning by teaching and teaching expectancy. Contemp. Educ. Psychol. 38, 281–288.

Grando, D., 2010. Digital wet laboratories: blended learning to improve student learning. Microbiol. Austr. 31 (1), 18–20. Available from: https://www.publish.csiro.au/MA/issue/6826/.

Grando, D., 2019. RMIT Science Engineering and Health Open Classrooms Podcast: Episode 2: Danilla Grando. Soundcloud. Available from: https://soundcloud.com/rmit_seh/episode-2-danilla-grando.

Harp, S.F., Mayer, R.E., 1998. How seductive details do their damage: a theory of cognitive interest in science learning. J. Educ. Psychol. 90, 414–434.

Harris, B., McCarthy, P., Wright, A., Schultz, H., Boersma, K., Shepard, et al., 2020. From panic to pedagogy: using online active learning to promote inclusive instruction in ecology and evolutionary biology courses. Acad. Pract. Ecol. Evol. 10 (22), 12581–12612. https://doi.org/10.1002/ece3.6915.

Hartmann, R., Jeckel, H., Jelli, E., Singh, P.K., Vaidya, S., Bayer, M., et al., 2021. Quantitative image analysis of microbial communities with BiofilmQ. Nat. Microbiol. https://doi.org/10.1038/s41564-020-00817-4 (Epub ahead of print).

Hartoyo, Abdul Gafur, D., 2019. Development and testing of biology learning multimedia effectiveness. J. Phys. Conf. 1233 (1), 12004. Available from. https://iopscience.iop.org/article/10.1088/1742-6596/1233/1/012004.

Horak, R., 2020. Virtual resources to teach microbiology techniques and experiments. American Society for Microbiology. https://asm.org/Articles/2020/December/Virtual-Resources-to-Teach-Microbiology-Techniques?MessageRunDetailID=4000628825&PostID=23343184&utm_medium=email&utm_source=rasa_io. (Accessed 2 January 2021).

Jones, A.M., 2003. The use and abuse of PowerPoint in teaching and learning in the life sciences: a personal overview. Biosci. Educ. 2 (1), 1–13. https://doi.org/10.3108/beej.2003.02000004.

Kalyuga, S., Chandler, P., Sweller, J., 1999. Managing split-attention and redundancy in multimedia instruction. Appl. Cognit. Psychol. 13 (4), 351–371.

Kombartzky, U., Ploetzner, R., Schlag, S., Metz, B., 2010. Developing and evaluating a strategy for learning from animations. Learn. InStruct. 20 (5), 424–433. https://doi.org/10.1016/j.learninstruc.2009.05.002.

Lin, Y.H., Liu, C.H., Chiu, Y.C., 2020. Google searches for the keywords of "wash hands" predict the speed of national spread of COVID-19 outbreak among 21 countries. Brain Behav. Immun. 87, 30–32. https://doi.org/10.1016/j.bbi.2020.04.020.

Loomis, D., 2014. Ramming for exams? Join the crowd. Hawk Eye. https://thehawkeyeonlinenews.wordpress.com/2014/06/04/cramming-for-exams-join-the-crowd/.

Martos-Cabrera, M.B., Mota-Romero, E., Martos-García, R., Gómez-Urquiza, J.L., Suleiman-Martos, N., Albendín-García, L., Cañadas-De la Fuente, G.A., 2019. Hand hygiene teaching strategies among nursing staff: a systematic review. Int. J. Environ. Res. Publ. Health 16 (17), 3039. https://doi.org/10.3390/ijerph16173039.

Mayer, R.E., 2009. Multimedia Learning, second ed. Cambridge UP, New York, NY.

McIntyre, S.H., Munson, J.M., 2008. Exploring cramming: student behaviors, beliefs, and learning retention in the principles of marketing course. J. Mark. 30 (3), 226–243.

Mezirow, J., 2000. Learning to think like an adult. In: Mezirow, J., Associates (Eds.), Learning as Transformation: Critical Perspectives as a Theory in Progress. Josey-Bass, San Francisco, CA.

Mezirow, J., 2003. Transformative learning as a discourse. J. Transformative Educ. 1 (1), 58–63.

Michael, J., 2006. Where's the evidence that active learning works? Adv. Physiol. Educ. 30, 159–167. https://doi.org/10.1152/advan.00053.2006.

Michael, J.A., Modell, H.I., 2003. Active Learning in Secondary and College Science Classrooms: A Working Model of Helping the Learning to Learn. Erlbaum, Mahwah, NJ.

Schnotz, W., Böckheler, J., Grzondziel, H., 1999. Individual and co-operative learning with interactive animated pictures. Eur. J. Psychol. Educ. 14, 245–265. https://doi.org/10.1007/BF03172968.

Stark, L., 2018. Facial recognition, emotion and race in animated social media. Clin. Hemorheol. and Microcirc. 23 (9). https://doi.org/10.5210/fm.v23i9.9406.

Tauber, M., Lourari, S., Bérard, E., Questel, E., Redoules, D., Giordano-Labadie, F., et al., 2020. Positive change in hand care habits using therapeutic patient education in chronic hand eczema. Contact Dermatitis 82, 10–17. https://doi.org/10.1111/cod.13390.

Thyssen, J.P., Johansen, J.D., Linneberg, A., Menné, T., 2010. The epidemiology of hand eczema in the general population - prevalence and main findings. Contact Dermatitis 62, 75–87.

Wiley, D.A., 2000. Connecting learning objects to instructional design theory: a definition, a metaphor, and a taxonomy. In: Wiley, D.A. (Ed.), The Instructional Use of Learning Object. Online version, New York, NY. http://www.reusability.org/read/. (Accessed 13 January 2021).

Further reading

American Society for Microbiology. Image Gallery. Available from: https://asm.org/Browse-By-Content-Type/Image-Gallery. (Accessed 13 December 2020.

Beakes, G., 2003. A picture is worth a thousand words. A personal view of using images in the teaching of the biological sciences. Biosci. Educ. e-J. 1 (1), 1–15. https://doi.org/10.3108/beej.2003.01010003.

Grando, D., Calonge, D.S., 2013. Digital wet laboratories: transforming biological science with engaging blended learning and online support. In: Gosper, M., Ifenthaler, D. (Eds.), Curriculum Models for the 21st Century. Springer, New York, NY.

Chapter 5

Improving public understanding of microorganisms by integrating microbiology concepts into science teaching throughout the education system

Megan L. Lloyd[1] and Jemma A. Berry[2]
[1]*The University of Western Australia, Crawley, WA, Australia;* [2]*Edith Cowan University, Joondalup, WA, Australia*

Chapter outline

Introduction

Germs, bugs, pathogens, plague, parasites and disease: the words commonly used to describe microorganisms have many negative connotations. The fear of 'bad' microorganisms is used to teach children important fundamental hygiene practices, e.g., washing hands after toileting and before eating or when visibly dirty. The fear of germs is also exploited by advertising agencies who market cleaning products for use around the home, where the germ-killing (and family

Importance of Microbiology Teaching and Microbial Resource Management for Sustainable Futures
https://doi.org/10.1016/B978-0-12-818272-7.00003-1

protecting) properties are praised. This situation potentially creates 'microphobia' – a fear of small things, particularly germs, and can result in excessive cleaning, a situation that has been associated with the hygiene hypothesis and the increase of some allergic diseases (Fotopoulou et al., 2018). In hospitals, the fear of pathogenic microorganisms is valid and constant, and equipment and surface cleaning are routinely carried out using sterilisation techniques such as autoclaving (reusable equipment), chemical disinfection with bleach and other chemical agents on surfaces (floors and benches) and sanitisation of hands with alcohol-based sanitiser or thorough hand washing before and after personal care (e.g., the World Health Organisation (WHO) – endorsed five moments of hand hygiene (Chou et al., 2012)). Hospital-acquired infections such as *Clostridioides* (formerly *Clostridium*) *difficile*, *Staphylococcus aureus*, and vancomycin-resistant enterococci are feared by clinicians and administrators, particularly as antimicrobial resistance increases in many of these pathogens, and the cost in terms of patient health and economic burden is substantial (Graves, 2004). However, in most circumstances, the microorganisms that populate our body cavities are protective rather than potentially dangerous, and risks posed by environmental microorganisms are countered by basic hygiene and routine vaccination (e.g., *Clostridium tetani* endospores are ubiquitous in dirt, and vaccination with the tetanus toxoid is the only completely effective protection from the neurotoxin produced by this microorganism). Understanding that most microorganisms are not pathogenic and that microorganisms are ubiquitous, meaning they are *in* us, *on* us and *around* us, is an important lesson.

The microbiome, the collection of microbes that populate various bodies or environmental niches, and the chemical activities these microbial populations undertake are increasingly becoming part of mainstream medical terminology, particularly where health supplements and products such as probiotics are concerned (Caron, 2020). As noted by Shan et al. (2019), 'The once tenaciously held belief that the only good microbe is a dead one has mostly been dispelled and replaced by the growing appreciation that many microbes are beneficial'. An example of this is the growing appreciation for the role of probiotics to lessen the risk of diarrhoea that can occur with taking antibiotics, suggesting a paradigm shift towards the protection of our gut flora and recognition that antibiotics do not just kill the 'bad' bugs (Neunez et al., 2020). Although such treatment may not be advised by all practitioners, probiotics are used as standard care for preterm infants in some neonatal intensive care units and show efficacy in reducing late onset sepsis and necrotising enterocolitis without a significant risk of adverse outcomes (Athalye-Jape and Patole, 2019). Thus, the medical personnel that distribute advice on probiotics (predominantly pharmacists and others aligned with the pharmaceutical sale of probiotic products as well as some general practitioners and other allied health practitioners such as naturopaths) are persuasive sources of information, albeit on a one-to-one basis.

An important source for information on health for the general public is the mainstream media, which is undeniably useful for distributing information about important health issues. In Australia, this has been demonstrated repeatedly during the SARS-CoV-2 pandemic, where important health messages have been distributed quickly and effectively via televised press conferences and targeted government advertising. Nevertheless, the mainstream media can be unreliable and sometimes advertises commercial products in a format disguised as medical commentary or may include professional commentary outside of individual expertise or provided by pseudo-professionals with self-appointed titles. Headlines or complete articles are written to accentuate the most provocative result from a scientific paper rather than presenting this in the context of a larger experiment (Shan et al., 2019). Without adequate science education, it is difficult for the public to judge the validity of information sources – particularly where an individual media supplier can be responsibly reporting some stories with excellent commentary from genuine experts in some circumstances but may use inflammatory headlines or unqualified 'experts' in others. In medical care, the common use of 'Doctor Google' has exacerbated these problems, and sources of 'research' can be anti-scientific at worst and misleading or misunderstood at best. This is particularly evident when evaluating information produced by anti-vaccination organisations, that may be seen as scientifically valid by a proportion of the population but is often quoting unreliable sources and is full of inaccuracies or lies (Hotez, 2020). New medical treatments are carefully evaluated and it may take some time for successful experimental treatments to transition from the laboratory to a clinical trial or from a clinical trial to mainstream medical treatment – a factor sometimes not divulged by the mainstream media or institutional press releases. This can lead to distrust and disappointment for those hoping for access to a new treatment or a potential cure. Sometimes promising clinical treatments are misappropriated by those who believe medical advice is outdated or medical professionals are too risk-averse regarding new therapies. There are potential dangers that stem from the appropriation of medical procedures from clinical research to mainstream use. One such example is faecal microbiota transplants (used to treat intractable *Clostridioides difficile* infections) which have been employed as a do-it-yourself treatment (Ekekezie et al., 2020). Another example is the ad hoc sharing of breast milk between mothers via the internet, which misapplies the careful practices of hospital-based breast milk banking (Hartmann et al., 2007) and with potentially devastating consequences (Keim et al., 2013). Having an improved understanding of microbial concepts would clearly be beneficial in these circumstances.

Until recently, the contribution of microorganisms to environmental and ecosystem health has largely been overlooked. Lessons about ecosystems featuring interactions between the environment and 'macro' organisms such as animals and plants are delivered from preschool age through formal education or popular culture in the form of films and other print and digital media. Rachel Carson, in her extraordinary book, Silent Spring which was first published in a series of New Yorker articles and then as a novel in 1962, noted that 'Life not only formed the soil, but other living things of incredible abundance and diversity now exist within it ... By their presence and by their activities, the myriad organisms of the soil make it capable of supporting the earth's green mantle' (Carson, 1962). This work is now nearly 60 years old and, in the past decade, where molecular biology and the development of rapid and cheap sequencing technologies have changed the microbiology landscape completely, this has also substantially extended scientific knowledge of soil organism populations and the profiles of microorganisms that inhabit different ecological niches in contact with various plant life. The contribution of microorganisms to ecosystems is mostly ignored by organisations that oversee projects with the potential to impair the macro-environment. It is important that the influence of human activity on microbial populations is assessed, for example, the evaluation of soil microorganisms should be included in mine bioremediation activities and may be useful as a measure of bioremediation success (Thavamani et al., 2017). It is increasingly apparent to microbiologists that the micro-environment both underpins and reflects ecological health and needs to be acknowledged, evaluated and maintained. These concepts also should be understood by decision makers at all levels of government, but improved microbiological education will be required to achieve this.

Microbiology teaching at all ages is the logical conduit between scientists and the public. In the past few years, there have been some excellent books written for primary school-age children that have plots based on microbial population dynamics and the essential interactions that take place between different organisms, including within coral reefs ('Zobie and the Zooks') and between soil organisms ('Nema and the Xenos' (Wild et al., 2019, 2018)). Other books such as the award-winning 'Do not lick this book' and the e-book 'Paul has measles' (distributed via Virology blog as a free pdf at https://www.virology.ws/2018/11/20/paul-has-measles/, available in 16 languages) provide important information on vaccination and disease as well as fostering interest in the microbial world around us (Ben-Barak and Frost, 2017; López et al., 2019). There are also many organisations that promote science to children such as local museums like the Micropea Museum in Amsterdam or the 'Meet your Microbiome' exhibition at the American Museum of Natural History, professional organisations through science events such as the Australian National Science Week (https://www.scienceweek.net.au/) or university promotions (e.g., presenting information about the utility of microorganisms in producing food described in Verran et al., 2019) and Institutes such as Centre of the cell (https://www.centreofthecell.org/) which is the first science education centre in the world to be located within working biomedical research laboratories, and local Science Technology Engineering and Mathematics (STEM) ambassador programs linking professional scientists with the general public, generally focussing on school students.

STEM literacy

Increasing STEM-associated knowledge has been a popular ideology that aims to produce technology-ready citizens with skills that will readily translate to employment in new industries providing thoughtful solutions to problems that have not yet been defined (Zollman, 2012). This requires more than just being able to regurgitate facts; it requires the linkage of knowledge from different areas and collaborative activities, potentially across different disciplines (for example, the 'Technology' moniker is arguably applicable to all areas of STEM). A barrier to increasing STEM literacy is the dogma that these are 'difficult subjects' and are only suited to some students, resulting in some competent students failing to progress from secondary STEM subjects to tertiary study (De Meester et al., 2020). Improved STEM education has been a goal of governments in many countries, and the requirement for improved education of teachers and continual improvement of teaching skills, particularly via peer-to-peer learning, has been noted (Milner-Bolotin, 2018). Some helpful steps for establishing STEM partnerships between universities and industry have recently been outlined by Knippenberg et al. (2020) and include ideas for utilising existing resources to connect with community partners and evaluation of outreach activities.

Microbiological literacy

Microbiology affects every human on earth every day of their lives, whether they realise it or not. This gives instructors a plethora of ways to make the study of microbiological concepts relevant and interesting. Self-interest and intrinsic knowledge of the role played by microorganisms in human and animal disease and food and beverage development should be leveraged to increase public awareness (Verran et al., 2019). The importance of improved microbial literacy in the general population has been noted, and a call to arms has been raised: 'Microbiology literacy needs to become part of the world citizen job description' (Timmis et al., 2019). This is particularly important where changes to the global climate have repercussions that may seem unconnected to microbial populations at first glance, for example, changes to mosquito geographical distribution associated with changing climates leading to an increasing disease range (Andersen and Davis, 2017). This concern has become more pronounced during the current global SARS-CoV-2 pandemic, where a virus originating in bats has influenced human activity across the world (see Sahin et al., 2020 for a comprehensive discussion of the interconnectedness of environmental, economic, social and health systems and the myriad effects of government interventions to these areas). Poor scientific literacy within populations has been associated with inadequate responses to COVID-19 and been directly responsible for increased human morbidity and mortality (Miller, 2020). However, the problem of oversimplification of complex subjects and subsequent inaccuracies being touted as fact can be exploited by those at the fringes of science to increase confusion and turn individuals against mainstream science and scientists, with increasingly wild conspiracy theories being developed (Thomas and Zhang, 2020).

Effective teaching is an important step in improving the microbial literacy of the general population. Unfortunately, education about microorganisms is often focused on 'germs' or so-called 'bad bugs', giving microorganisms a negative association rather than presenting the dichotomies of health and disease and environmental health and risk that actually exist. An appreciation of the importance of microorganisms to our health, food supply and ecosystems is a less clear narrative, and a broadening of the teaching about germs is also important for public health initiatives such as vaccine promotion and antibiotic stewardship. For this message to become dominant or at least for some acknowledgement of the importance of microorganisms to humanity and ecosystems to become more widely understood, it is important to ensure that microbial concepts are taught throughout formal education from primary through to secondary schooling and beyond.

Microbiology teaching in primary school

Primary school students often have a negative opinion of microorganisms as the majority of lessons in this area aim to keep them safe (washing hands and food spoilage/safety, for example), and they are less likely to have been taught that bacteria and other microorganisms are also required for food production and health more generally. One of the best microorganisms for primary school students to work with is the yeast used for making bread, *Saccharomyces cerevisiae*, and wonderful teaching opportunities can be safely realised with minimal equipment (Connor et al., 2020). Misconceptions are common in primary school students (for example, not understanding the ubiquity of microorganisms), but this may also be due to misconceptions held by their teachers as few primary school teachers will have had an opportunity to study microbiology during their tertiary studies because most teachers are generalists at this level (Marcos-Merino et al., 2020). Including some practical activities in primary school science will improve understanding and may improve the ability to communicate with others about microbial functions (Ruiz-Gallardo and Paños, 2018). This is important as the effort required to relearn misconceptions in secondary studies and beyond is substantial.

Microbiology teaching in secondary school

Microbiology education at secondary school is often provided solely in the context of human health (particularly pathogens/commensal organisms). However, general microbial concepts and an understanding of the complexity of microbial populations (for example, the microbiome) are increasingly being taught at a secondary school level, and many of these concepts can be applied to other circumstances (e.g., soil organism communities). The understanding that students obtain from this often-brief interaction with the microbial world can be the only opportunity, within their formal education, to become familiar with microbiological concepts. Therefore, teaching methods that incorporate practices that improve the embedding of concepts and improve concept recall are important.

Microbiology is increasingly a feature of many upper/senior school science curricula, and practical activities such as student-led investigations are becoming more frequent. Safety is of great concern for these activities in secondary school settings as schools are unlikely to have equipment or facilities suitable for potential human pathogens (Biosafety level 2), and comprehensive risk analysis should be performed for all laboratory activities. A lack of appropriate equipment was seen as a major limitation in a study assessing microbiology in UK schools, but only around 12% listed a lack of confidence in microbiological techniques as a reason for not incorporating microbiology into school classes (Redfern et al., 2013). Agencies such as the Australian Science Teachers Association (https://asta.edu.au/), the National Science Teaching Association in the US (https://www.nsta.org/) and the Association for Science Education (https://www.ase.org.uk/) in the UK are currently available to provide teachers and technicians with information about appropriate experiments for requirements the secondary school system that fulfils curriculum requirements for different countries. Some agencies are able to respond to queries and safety rulings about existing or new experimental plans (for example, Science ASSIST-Australian School Science Information Support for Teachers and Technicians (https://assist.asta.edu.au/) and CLEAPSS (http://www.cleapss.org.uk/)). In some countries, there are strict guidelines and rules aimed at reducing risk for students – for example, not allowing subculturing, not allowing enriched media to be used, ensuring that environmental samples are not taken at sites that are likely to harbour pathogens (e.g., bathrooms). The training of Science Technicians (those who set up and pack down/sterilise after microbiology experiments have taken place) in secondary schools is often provided on an ad hoc basis. In a recent survey of science technicians and teachers throughout Australia, 55% of 610 teachers/technical assistants (19% and 81% of respondents, respectively) from around Australia reported that they had not been offered any microbiology training at their secondary school, and 76% of total respondents indicated that they would like more training in this area (Personal communication, Teresa Gigengack, Science ASSIST, Australia). This is a concerning situation as the qualifications possessed by science technicians can range from no formal qualifications to having completed a PhD, which may or may not include training in microbiology. Training is often provided by their predecessor at the school and may not be adequate to allow realistic risk assessments of class activities.

One solution to this problem is to partner with tertiary institutions to increase the range of activities available (a successful model for this activity is described in Caine et al., 2015). Closer links are of increasing interest to both sectors as tertiary institutions compete for local student populations, and secondary schools look to incorporate more risk-heavy activities into their classes. One way to improve the public visibility of tertiary institutions is to provide on-site workshops for upper-level secondary students. In a workshop for upper secondary school students carried out at an Australian tertiary institution, one of the typical participant comments was: 'Well I experience what University life is like and in the labs and I now know what I want to do and what I'm working for (sic)', a clear benefit to both parties (participant at workshop for Year 10 and 11 students, 2017, Edith Cowan University).

A wonderful example of collaboration between secondary schools and tertiary institutions is discussed by De Meester et al. (2020) and was implemented in Belgium between 2014 and 2018. The STEM at school program developed integrated STEM (iSTEM) educational activities designed to allow the students to identify where science and technology were interacting with their lives and encouraging the recognition of linkages between the different STEM disciplines. Teachers from different schools and teaching different grades were set up in teacher design teams incorporating all STEM disciplines and integrating tertiary educators to develop learning units. Importantly, time was allowed to develop the program before implementation, and feedback from teachers was used to revise and improve the units for delivery the following year. Learning units included assignments that challenged students to design an autonomously driven car (a scale model) that was able to negotiate through a defined series of green traffic lights. This Grade 9 assignment utilised and integrated complex concepts such as modelling, scale, programming, time and velocity. Another such assignment asked students to develop an 'eco-friendly application of algae cultivation' (De Meester et al., 2020). Assessment strategies were evaluated, and student learning outcomes were assessed for student attitudes towards STEM as well as for more traditional assessment of student learning. Student engagement was measured and was found to be higher than for more traditional mathematics and science classes. Interestingly, students who participated in iSTEM classes had a reduced 'self-efficacy' regarding mathematics and science within the integrated subject, postulated to be due to the perception of these integrated concepts as being more complicated (compared to straight mathematics and physics classes where one subject at a time is studied in detail).

Although this example is not strictly microbiology based, it provides a template for integration of secondary and tertiary education that could easily incorporate tertiary premises, e.g., PC2 laboratory conditions in modern tertiary 'super' labs providing safe conditions allowing processes such as culturing from the environment with evaluation of organisms through staining and subculturing.

Another example of a valuable interaction between secondary schools and tertiary institutions is described by Coy et al. (2019) where a tertiary institution (the University of the West Indies) partnered with secondary schools in

Jamaica to provide opportunities – and equipment – for students to enhance their learning in STEM and to 'support critical analyses and the learning of specific laboratory techniques' to allow students to complete the practical requirements for the completion of senior science subjects. Again, this is not a microbiological example, but shows how such integration of secondary and tertiary education can improve outcomes for participants and can contribute to continuing education for teachers.

Many tertiary institutions in Australia currently have short 'courses' for upper-level secondary school students that focus on the microbiology laboratory. These are designed both to introduce students to microbiology concepts (mostly disease related, but sometimes emphasising the ubiquity of microorganisms) and to excite them about the specific institution (often recruitment activities are integrated). However, this rather limited activity could be enhanced by careful integration of curriculum-appropriate and potentially interest-driven (student focussed, student designed) activities that take advantage of the increased safety provided by laboratories at tertiary institutions and utilise the expertise of lecturers. In the Belgian example, the integration of these different teaching levels clearly had significant governmental financial support, and this would be required for such integration to work, particularly considering increased workloads in all teaching sectors. Secondary school science teachers arguably can have the greater impact on population literacy overall by virtue of the number of students they teach. For real change to occur, it is necessary that secondary school science teachers have themselves been exposed to quality microbiology teaching at a tertiary level (Al Sultan et al., 2018).

Tertiary microbiology teaching

Microbiology is featured in multiple undergraduate science degrees, including biological science, medical/nursing and other allied health fields, as well as teaching degrees and agriculture/animal sciences. Typically, tertiary science subjects are delivered using a modified didactic methodology – formal lectures with lecture slides provided and recorded lectures are increasingly common. This teaching mode is generally used for logistical reasons, such as large undergraduate student numbers, rather than for good pedagogical motivation to ensure optimal student learning. Active learning, where students actively participate with the subject matter rather than passively being provided with facts, is being integrated into teaching practice, but strategies need to be put in place to ensure effectiveness (Tharayil et al., 2018). Active learning practises have been shown to improve student outcomes by reducing failure rates, presumably by improved engagement and increased interest in the subject (Freeman et al., 2014). Active learning has always been a part of microbiology teaching due to the inclusion of practical classes, which will be mentioned in detail later. For the lecture component of microbiology subjects, some universities utilise flipped classrooms where resources (readings, multimedia presentations, quizzes) are provided through Learning Management Systems (LMS) before classes and, in some cases, assessments are flexible, allowing students to have some interaction with the curriculum and to choose assessment formats and times (Wanner and Palmer, 2015; Cavert, 2020). Blended learning (a mix of online and face-to-face learning) allows students to manage the pace of their learning outside class time and means that face-to-face time can be used for higher-order problem solving activities and collaborative group interactions. This type of class is often popular with students but requires significant input both by the institution in terms of time required to effectively convert lessons to this format as well as by the students who are unable to participate fully if they have not completed the set work (Khan and Ibrahim, 2017). In a systematic review of papers on flipped classrooms (71 articles), advantages in student learning performance, satisfaction and flexible learning were noted; however, so were disadvantages such as teacher workload and time and problems with student preparation for classes (Akçayır and Akçayır, 2018). There are also potential problems with university workload calculations as the time required to 'flip' a single class and prepare additional resources for a flipped class is more than is required to prepare a didactic lecture-based class – a factor that may not be included in workload calculators used to allocate academic subjects and the increased hours of preparation significantly increase instructor stress (Wanner and Palmer, 2015). If instructor face-to-face hours are reduced, this can lead to faculty demands for individuals to teach more subjects, and the prospect of overwork leading to burnout of educators becomes a real concern. It is important for tertiary institutions to recognise that time and self-evaluation are required to effectively transition from face-to-face teaching to an effective online delivery that plays to the individual strengths of instructors and provides a good teaching experience for the student and maximises student results (Maggio et al., 2018).

In microbiology units, practical classes provide valuable hands-on opportunities to enhance and expand on formal lecture content through experiential learning and allow students to experience the ubiquity of microorganisms realistically. Many microbiologists fondly remember their tertiary practical classes and recall the thrill of microbial identification and the lifelong love affair that this engendered. In undergraduate classes, there may be opportunities for student-led investigations, particularly in the final year of undergraduate degrees, and these provide a wonderful opportunity for students to explore whether they are suited to academic research and can also allow talented students to be identified to potential supervisors. Senior student projects are probably most at risk of institutional cost-cutting as they require the availability of relatively diverse laboratory resources and equipment as well as confident supervision to allow student creativity to be encouraged and enhanced. Subject groupings that should be included in microbiology majors and may be defined by professional societies demonstrate the importance of practical training and include both basic laboratory skills (Merkel and ASM, 2012) and the integration of undergraduate capstone projects that require integration of theory and practical skills (Burke et al., 2016). Practical laboratory assessment also contributes to the real-world experiences for students and provides good evidence of competency when they are seeking employment after completion of their degree (Wang et al., 2018). Importantly, practical training also allows standard precautions to be modelled in a real-life setting and underscores safe working in the laboratory (Fleming and Hunt, 2006).

Student safety in the laboratory

Student safety in the microbiology laboratory is extremely important; however, this topic is often marginalised as 'boring' by the students and is a problem where the same laboratory safety advice is repeated constantly at the start of each science subject every semester with little adjustment made for individual subjects. Safety training in the microbiology laboratory is imperative at all levels of education. Students must understand that growth on an agar plate is not equivalent to growth in the environment and that microorganisms must be respected for their ability to infect humans via mucous membranes or hand-to-mouth transmission. One way to make students take an interest in this topic is to test students on safety and to allocate these marks towards their final grade. Table 5.1 shows the results of a quiz administered after an online test that was allocated to a forensics class who had watched the standard safety audio-visual of a PowerPoint presentation provided during the first practical class in all science subjects at a tertiary education institution ($n = 29$). The quiz was worth 5% of the final subject mark, and students who did not achieve a mark of 80% or more after three attempts were asked to watch the safety video again.

The results indicate that more than 80% of students thought that the test was a good way to ensure that students were aware of safety issues and indicated that the safety test was a good idea (Q1 and 7, respectively). Q3 shows the utility of a summative test to ensure that the students participated, a problem that others have highlighted in circumstances where results did not contribute to a final mark (Stevens et al., 2017). Interestingly, Q4 indicated that the students did not feel more confident in the laboratory because of the test, which is encouraging as the safety video does not itself provide laboratory competence. This should be achieved through the practical classes that follow the safety assessment. Q5 and Q6 suggest that the importance of laboratory safety was well communicated by the slide presentation and that the concept of laboratory safety was effectively associated with the workplace.

Increasingly, there is pressure on subject coordinators to reduce practical learning experiences within tertiary institutions and to replace these with a

TABLE 5.1 Student response (%) to a laboratory safety test carried out in class. Responses were scored using a Likert scale from strongly agree to strongly disagree. Number of participants = 29.

Question (strongly agree to strongly disagree)	Broad agreement	Neutral	Broad disagreement
1: The laboratory safety test is a good way to make people aware of laboratory safety issues.[a]	86	11	4[b]
2: Because of the laboratory safety test I read the safety handouts.	72	14	14
3: Knowing there would be a test involved made me pay attention to the safety induction video.	69	14	17
4: Because of the laboratory safety test, I feel more confident in the laboratory.	48	34	17[b]
5: The laboratory safety test made me more aware of the importance of laboratory safety.	76	14	10
6: The laboratory safety test highlighted the value of good laboratory safety practices in the workplace.	79	14	7
7: I think having a laboratory safety test is a good idea.	86	7	7

[a]*28 responses received.*
[b]*Rounding effects ≠ 100.*

combination of online activities, group discussion/result interpretation and demonstration-only laboratory experiences. This is usually due to perceived costs of practical classes; however, the costs of replacing outdated technology hardware and annual software licences, as well as increased tutorial payments due to higher duties, may not be more cost effective than standard laboratory classes. Additionally, the reduction in practical skills learnt by students may reduce the perceived value of the subject and potentially reduce student numbers in associated degrees. For some students, such as undergraduate nursing students, the inclusion of limited microbiology practical activities has been shown to improve awareness of bacterial traits and how to prevent infections (Yano et al., 2019). However, it is regrettable that the inclusion of practical sessions in nursing training is limited due to 'busy class schedules and a focus on theoretical teaching' and confronting to recognise that these professionals are at the frontline of infection control.

The benefits of practical classes are often profound, not the least for skill acquisition (Noel et al., 2020), and these benefits are commonly mentioned positively in student comments at the end of the semester via standardised unit evaluation. What may save practical classes is the imperative need to develop useable skills for employment such as aseptic technique, which is applicable in careers ranging from food and beverage preparation and serving to skincare and pharmaceutical manufacture and standard microbiology skills such as plating for single colonies and activities confirming that populations are clonal rather than mixed and are essential for research. Isolation of specific bacterial genera is easy in theory; however, good technique is required to ensure that what you have (in terms of microbiological culture) is what you want, and reliable practical skills are still required in the research laboratory as well as a variety of government and private laboratories. The integration of some online activities into the practical class schedule provides benefits for students, particularly the ability to repeat activities and to integrate them into their schedules as required (Brockman et al., 2020). This is an excellent way of introducing practical exercises to students to maximise the use of limited laboratory time but will never provide the same experience as physically completing the lab exercise on the bench. Additionally, to be able to provide evidence of the mastery of basic laboratory skills to future employers as well as the ability to work in teams, it is important that practical classes remain part of the microbiology landscape.

The recent global pandemic caused by SARS-CoV-2 has caused a general re-evaluation of face-to-face learning, particularly for lectures, and more resources appropriate to remote learning are currently being developed (discussed in the context of medical degrees in Cavert, 2020). University lectures in the post-COVID-19 world may look quite different. Already some universities have stated that their commitment to online education has increased, and some institutions are suffering shortfalls in funding due to a loss of international students and have reduced their academic workforces accordingly (Witze, 2020). The ability for students to effectively participate in online activities may also depend on external features such as internet speed and cost (Dutta and Smita, 2020). Recent evidence (collected pre-COVID-19) suggests that online learning disproportionately benefits top students and disadvantages poor students (Cacault et al., 2021). There will undoubtedly be more research on online delivery of education as the SARS-CoV-2 pandemic continues.

Mechanisms of learning science

Early scientific learning relies heavily on rote learning, with each year's new information building on the foundation of information already committed to memory. Yet, ironically, secondary school students have embraced experience-led investigations, even as they are discarded by tertiary institutions due mostly to associated costs (reagents, tutors, other infrastructure). The benefits of experiential learning are well established, and the improved learning that such experiences provide has been noted (Redfern et al., 2013).

How do we learn?

Our understanding of how we acquire knowledge is improving. As de Bruin and colleagues suggest, 'in monitoring their learning, learners have to deal with the problem that the mental effort they invested in object level processing is a highly salient but not necessarily a predictive cue of their actual learning' (de Bruin et al., 2020). An important limitation with traditional didactic teaching (an expert 'with' knowledge imparts it to students in a relatively static setting) is that the cognitive load is very large – it has been suggested that young adults have between 3 and 5 active memory items or 'slots' available (Cowan, 2010) where previous estimates were up to 9 (Miller, 1956). Techniques to reduce cognitive load include presenting new information in the context of existing knowledge allowing the 'chunking' of information, considering slide design to reduce distraction and emphasising that information needs to be reviewed several times by the students to ensure that knowledge is retained in long-term memory (Gobet, 2005). This requires repetition, which can be provided by quizzes constructed by the instructors or by freely accessible smart device apps, which will be discussed in more detail later.

Unfortunately, effective knowledge acquisition is rarely at the forefront of curriculum development. Reduced face-to-face hours with restricted experiential and co-operative learning experiences through reductions in practical class time as well as reduced allocated lecture times (in hours per week) contribute to the increased cognitive loads being experienced by modern undergraduate medical/science students as instructors struggle to cover the same content. This was already a concern before COVID-19 restrictions reduced science practical class sizes and eliminated most group work, and face-to-face learning was moved online. However, this problem may now be exacerbated by universities attempting to reduce the loss of income stemming from the loss of international students. A commitment to majority online education will likely be the new paradigm for much university teaching in the post-COVID-19 era (for an example, see Turner, 2020) although some Australian universities are recognising the changes to university life that this will entail and are making efforts to encourage students to return to campus (Zhou, 2020). It is imperative that educators consider how best to help students navigate online education and blended learning more broadly, and this includes considering how to reduce cognitive loads and improve information retention without requiring increased lecturer facilitation and face-to-face time. The online component is likely to include mobile learning (m-learning) using personal devices (computers, tablets and smartphones) and ideally will help embed basic concepts to allow classes to move together towards more complex tasks within the curriculum framework.

Forming a sound knowledge basis to allow higher-order activities

Most education theory acknowledges that the initial acquisition of knowledge is critical for higher-order thinking, such as problem-solving; however, this situation is exacerbated in science subjects where personal experience is less available to help with the learning process than other subjects. The fundamental differences in requirements for learning science subjects compared to subjects like the humanities are often glossed over in texts discussing pedagogy, but they are evident to teachers of these subjects who rarely come directly from a teaching and learning background. Often, educators are building on prior learning and utilising past experiences to scaffold new knowledge acquisition. Bloom's taxonomy has fallen out of favour with many educators, but effectively describes the skills required to master scientific concepts and to move from learning facts to more creative and innovative activities that investigate the natural world. Lower order skills such as knowledge acquisition and recall (tested by simple questioning and requiring the regurgitation of facts) underpin higher-order skills that require student comprehension (for example, linking scientific facts with practical experiences), interpretation of results (using established scientific concepts to explain positive and negative results or failure of experiments) and the formulation of new hypotheses (discussed in Adams, 2015). It is imperative that students have a sound understanding of basic concepts to enable them to progress to the more complex tasks requiring the application of knowledge and the linking of different concepts together to produce hypotheses. Key to this is the ready acquisition of vocabulary, which must be mastered to allow unambiguous communication and higher-order activities such as analysis and evaluation.

There are plenty of inefficient ways to tackle learning, and an excellent monograph by Dunlosky et al. (2013) evaluated common techniques used by students such as (i) summarisation of text, (ii) highlighting text, (iii) use of mnemonics (although useful in some circumstances), (iv) using images to aid text learning (again, quite useful where images can be readily associated with information to be learnt, but not necessarily as the major methodology) and (v) re-reading notes, which were all identified as being of low utility to the majority of students due to the short duration of memory. The same monograph suggested that practice testing and distributed practice – a study schedule that is distributed across a time frame, meaning that knowledge is acquired more slowly but is retained for longer – were the most effective learning techniques (Dunlosky et al., 2013). Both of these practices involve retrieval-based learning. The retrieval of facts has been shown to provide far better memory retention than simply re-reading texts, although perceptions held by students about their ability to recall information were higher for reading than for active retrieval — which was unjustified but effectively demonstrated defects in the way students instinctively evaluate the effectiveness of different study techniques (Karpicke, 2012).

Although many tertiary educators are reluctant to participate in helping students with learning which is seen as 'spoon-feeding', it is far more efficient to help students to master the terminology and basic background knowledge so that the class can attempt more complex tasks without the inherent risk of leaving students behind intellectually as they move through an undergraduate degree. There are a multitude of subject-specific quiz applications available on personal devices that could be tasked independently by students to help them with vocabulary and knowledge acquisition, but it has been noted that students will tend not to use them unless they are suggested by the instructor (Wilkinson et al., 2020).

One way of helping students to master the learning of basic concepts is via the development of simple online activities accessed via the LMS used by the school/university. Such systems often have their own embedded methods for developing online quizzes and can have HTML activities integrated into their interface, allowing students to complete these tasks online from personal devices. All activities are not equal, though, and it is important to recognise that even the best-planned curriculum-integrated activities may not be accessed by the students who would benefit from them the most. It is, therefore, useful to have an idea of which online activities are used and enjoyed by students and which activities are perceived as being helpful to learning.

Which online activities engage students?

Medical microbiology is a subject that requires a subject-specific science vocabulary as well as mastering genus and species names as well as other important characteristics of different types of pathogens. Vocabulary is usually mastered by students using standard rote learning techniques. However, increasingly, modern students engage enthusiastically with online activities, particularly where these are provided in one-stop style within the LMS. Students in a second year medical microbiology class where Medical Microbiology was an elective (part of a Bachelor of Medical Sciences degree with no associated microbiology major, assessed by practical and theoretical examinations and a literature review) were enrolled in a study that asked them to indicate which of the online learning activities accessible via the LMS they had interacted with over the semester. The students were asked about their interaction with the following activities (Box 5.1):

BOX 5.1 Online activities assessed in a medical microbiology class via the LMS (Blackboard)

1. PDF copies of PowerPoint presentations (slides).
2. Recordings of lectures (recorded as lectures were delivered live, not prepared as online-only activities).
3. Pre-lecture PowerPoint slides integrating hyperlinked activities – YouTube clips, TED Talks, other institutional-associated online activities.
4. Drag and drop word quizzes made using H5P (Singleton and Charlton, 2019) and integrated into the LMS, Blackboard as HTLM.
5. Multiple choice quizzes made using H5P and integrated into the LMS.
6. Kahoot homework quizzes made on https://kahoot.com/ (Jones et al., 2019).
7. Checklists – generated using Microsoft Word, listing all topics covered during lectures. These allowed students to fill in information on the topics.

The survey was in the form of three checkbox questions to eliminate the risk of identifying student answers. The questions were:

i. Which of the following resources did you **use**?
ii. Which of the following resources did you **enjoy** using?
iii. Which of the following resources did you feel helped with your learning (a surrogate marker for **effective** learning).

The one-page (single side) questionnaire was distributed at the beginning of the practical class in week 12 of a 13-week semester and students were able to return the survey (marked or unmarked) to a box at the side of the class at any time during the class. Students could tick more than one checkbox and a 'none' option was available, which was not selected by any participant for any question. One student chose to return the unmarked survey to the box anonymously and was not included in the overall numbers. All other surveys were completed, $n = 55$. Results are presented in Fig. 5.1.

The most commonly used online activity was downloading PowerPoint lecture slide PDF files which were reported to have been downloaded by all students who participated ($n = 55$). This was not seen as a particularly enjoyable activity (noted as being enjoyable by 28/55, 51% of students), but the slide handouts were seen as helpful for learning by 45/55 students. The least utilised activity, the pre-lecture slides, were also the least popular activity for those that used them, and were listed as useful to learning by 7/18 students, 39% of students that used the resource. The most popular activities for the students that used them (the percentage of students who used a resource that reported that they enjoyed using it) were the quick electronic activities – Drag and drop quizzes (mostly used to teach vocabulary, used by 46/55 students and

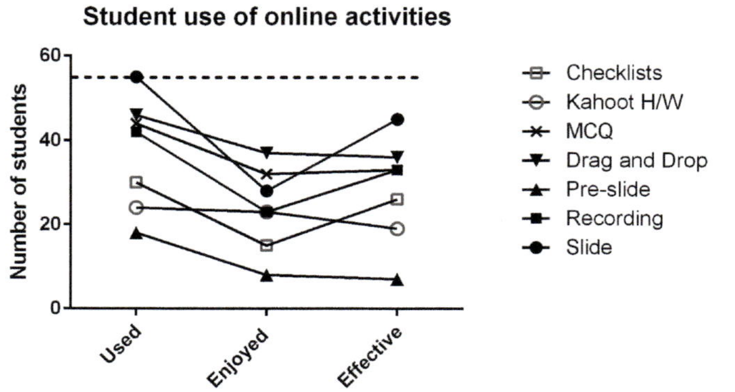

FIGURE 5.1 Number of students who i. used, ii. enjoyed using and iii. found on online activity helped with their learning (labelled 'Effective', n = 55). (----- = 55 students).

enjoyed by 37 of these students – 80% of the students that used the resource), MCQ (mostly used to test understanding, used by 44/55 students and enjoyed by 32 of these students, 73% of the students that used the resource) and Kahoot homework quizzes (used to test understanding) that were accessed through an app on personal devices were used by 24/55 students, 23/24 students (90%) of the students who used the Kahoot homework quizzes reported enjoying them.

The difference between the percentage of students that enjoyed using a resource and those that thought that it was helpful for learning is shown in Fig. 5.2. Activities fell into three broad categories – those that were regarded as being more effective learning tools but were less enjoyable forms of learning (1. lecture slides, 2. lecture recordings and 7. checklists – all more traditional forms of learning), those that had little difference between enjoyment and effective learning (3. pre-lecture slides, 4. drag and drop quizzes and 5. multiple choice quizzes – all of these were accessible electronically) and finally, 6. Kahoot homework quizzes were broadly seen as being more enjoyable than effective for learning content.

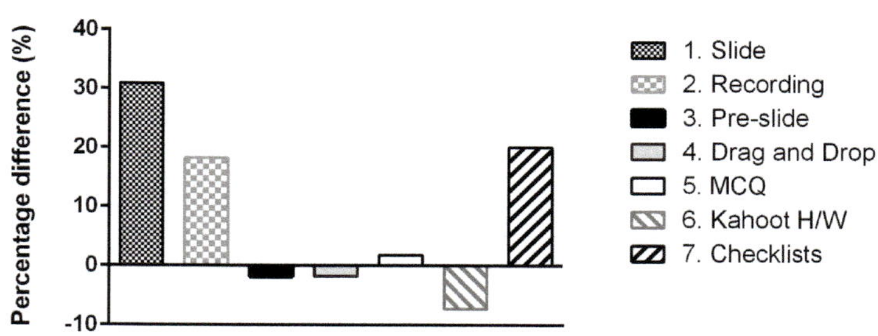

FIGURE 5.2 Student perception of activities that were effective at improving their learning. This was calculated by the difference between the percentage of students that used a teaching activity and enjoyed using it and those who reported that it was an effective learning tool (value could be positive or negative).

In this survey, the **lecture recordings** were standard PowerPoint presentations with audio recordings embedded using Techsmith Camtasia and delivered via the LMS. The format of these lectures did not include visual images of the lecturer. Lecture recordings can have different formats – for example, they can simply be a recording of a lecture recordings or standalone podcasts and used as supplementary resources for learning or as part of a flipped learning format. In all of these circumstances, the engaging quality of the lecturer is important (Hadjianastasis and Nightingale, 2016). Although it has been reported that the provision of lecture recordings does not significantly reduce student attendance (Cacault et al., 2021), this was not the experience at this university with fewer than 30% of students attending lectures – even where the lecturer was also the tutor and had regular access to the students.

The **online quizzes** provided in this study were all utilising retrieval learning – information needed to be recalled to answer each question (there was a time limit for the Kahoot quizzes). There is evidence that continuous testing is itself a good learning technique and that incorrectly answering questions improves learning where feedback on the correct answers is provided (Roediger and Pyc, 2012). This suggests that students can improve their learning by attempting tests where they do not expect to know the answer (Kornell et al., 2009). Testing in this way can also highlight areas of deficiency and allow study to focus on these areas. To improve learning, the interleaving of topics should be considered to add 'desirable difficulty'; it has been shown to reduce test performance in practice tests but to improve the final test result compared to providing questions in single topic tests (Taylor and Rohrer, 2010). Arguably, online tests can help to contend with procrastination (the voluntary choice of one task over another) by stopping task delay and reducing task aversiveness where tasks are seen as enjoyable and useful for learning (Steel, 2007).

The **pre-lecture PowerPoint slides** were designed to reduce the cognitive load by introducing concepts in an informal way before the more formal lecture with a similar effect to a flipped classroom – where information is provided to students as either reading, pre-recorded lectures or other audio-visual formats. These slides could be downloaded and had hyperlinked activities embedded – e.g., TED talks (www.ted.com/) or online activities such as PHeT interactive simulations (https://phet.colorado.edu/) or simulations of gene transcription and translation (https://learn.genetics.utah.edu/). This is particularly useful where the class has come from diverse backgrounds (for example, students may have studied different subjects in senior secondary school or at first year university) and some topics have been extensively covered in other subjects by some students and are very difficult to approach at a higher level (e.g., DNA replication in second year microbiology classes can be very challenging for teaching students who may not have learnt about this previously). In this class (second year microbiology students who had all completed a foundation microbiology subject), this was not a successful strategy on the surface but may have reduced cognitive load in a less obvious way than targeted knowledge provided by the short electronic activities, which is much more difficult to assess. Others have found that activities that have been designed to be used before the lecture have been utilised by many students after the lecture or multiple times throughout the semester, demonstrating that newly introduced resources are not always used in the way (or at the time) that is intended. Use of the additional resources did not increase the grade of one cohort compared to the previous cohort (no additional online resources), even though anecdotal evidence suggested that students were more pro-active in class and asked more questions. This was thought to be due to an improved understanding of the topic (Kinsella et al., 2017). In another study that introduced additional resource activities (podcasts, MCQs, and pathogen-focussed clinical cases), student marks increased in relation to time spent online in short note questions, but there was no correlation with MCQ achievement (Stevens et al., 2019).

The **checklists** provided to the students were word documents that listed the main topics covered by each lecture. They allowed students to print and write on them or to type in answers on their computers. Anecdotally, some students used these to guide their study. Checklists are used in aviation and in healthcare to reduce mistakes and in education have been shown to improve the marks of students who used them for practical anatomy classes, where they were perceived as being more useful as hard copy sheets rather than as lists on a whiteboard (Hofer et al., 2011). Checklists were not seen as being enjoyable but were identified as being good for learning.

Integrating fun into lectures

Using the homework capacity of the popular online quiz activity, Kahoot is a less-common means of integrating this activity into the curriculum. Kahoot can also be used during lectures to engage students (for example, where lectures run over multiple hours, a quiz can provide some active learning interaction between each hour or can be used to wrap up a lecture series), and most students generally embraced them. When participating in Kahoot quizzes, students use personal computers, tablets or smartphones and answer multiple-choice questions in real time. Clicker questions could be used during lectures in a similar way, although this infrastructure needs to be supplied by the institution (Solomon et al., 2018). The questions generally have a multiple-choice format (2–4 answers), and individuals get real-time feedback on their personal device about whether they answered each question correctly or not. The quiz has a musical backing that sometimes increases in intensity as the time allowed to answer each question elapses (the lecturer can change the time given) and the colourful interface and the ability to integrate images adds to the visual interest. Points are allocated depending on the speed that the question is answered by an individual and whether the answer is correct. There is a leader board, and as students use a nickname to log on for each quiz, individual results are not necessarily identifiable, which allows some anonymity. Kahoot quizzes are excellent when used for revision of a previous week's content or to determine whether frequently misunderstood concepts have been adequately covered, and there is a quiet break between questions that allows the lecturer to explain why the correct answer is correct.

Students are generally very positive about Kahoot quizzes in class. To test this, a cohort of students ($n = 66$) was questioned about the use of Kahoot during their lectures. The questions are provided in Box 5.2, and answers were given on a Likert scale from Strongly Agree to Strongly Disagree:

Results showing the Likert scale responses are presented in Fig. 5.3A. The proportion of students responding either strongly agree or agree to each question are shown in Fig. 5.3B with the only question receiving less than 80% being Q9 suggesting that student attendance was not improved by the integration of a Kahoot quiz into the lecture (again, this tertiary institution has typically low lecture attendance). The most variable feature of Kahoot was the competitive aspect, which was enjoyed by many students, but disliked by others (5/66 answered strongly disagree).

There is evidence that students happily accept the integration of Kahoot games because they are easy to use and are good learning tools that they think will improve their academic performance (Adeleye et al., 2020).

The use of gameplay or game informed learning (using quiz games to learn subject content, which is a term that is distinct from game-based learning where the game is the subject of learning, e.g., Plague Inc.) has been shown to increase students' basic knowledge and to improve their quick recall of

BOX 5.2 Questions asked about in-class Kahoot quizzes

Q1: The Kahoot in-class quiz system was a good way of testing what I learned in the lecture.
Q2: The Kahoot in-class quiz system made the lectures more fun and engaging.
Q3: The Kahoot in-class quiz system encouraged me to pay more attention in class so that I would be able to answer all the questions.
Q4: I enjoyed using the Kahoot in-class quiz system.
Q5: I think the Kahoot in-class quiz system helped me in my understanding of the topics.
Q6: I liked the competitive nature of the Kahoot in-class quiz system.
Q7: The Kahoot quizzes gave me confidence in my knowledge.
Q8: The Kahoot quiz system helped me to identify areas where my understanding was lacking, so I could focus my studies on these areas, specifically.
Q9: I attended more lectures because of the Kahoot in-class quiz system.

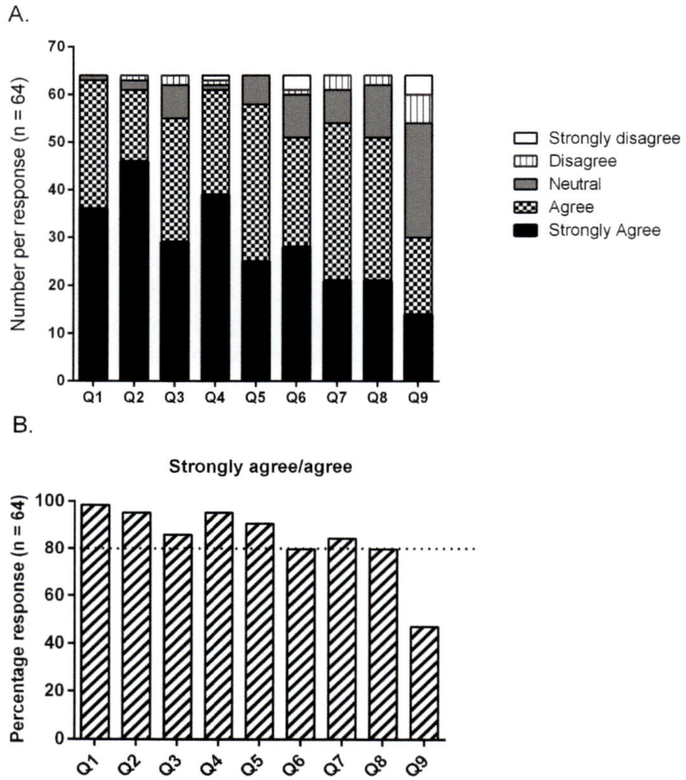

FIGURE 5.3 Responses to Kahoot quiz questions (n = 64). (A) Likert scale responses. (B) Percentage of strongly agree + agree for all questions (- - - = 80%).

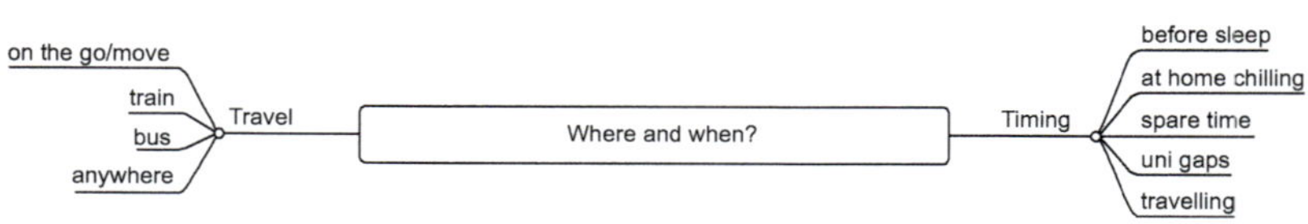

FIGURE 5.4 From (Wilkinson et al., 2020). Where students reported using m-learning.

concepts (Vinney et al., 2016). There is evidence that using game-based approaches can improve knowledge retention and that this type of learning helps with revision, potentially determining which subject areas need further study. Activities that can be accessed on mobile devices are particularly useful in this context and have been used during commuting and spare time (Fig. 5.4). The ability to do short bursts of repeated quizzes will help knowledge retention (Wilkinson et al., 2020).

Other activities

Online games are increasingly popular pastimes for modern students, and the integration of gamification into coursework will provide increased interest for a proportion of the student population. One game that has been used for some time and has evolved to be available on personal devices is Plague Inc., where participants can either act as the disease agent or as authorities trying to control the disease (described in Robinson et al., 2018). This will undoubtedly be interesting to compare to actual pandemic mitigation processes used by different countries with differing success in controlling infection.

Conclusion

The culture of microbiology teaching needs to be updated to ensure that as many global citizens as possible have access to knowledge that will allow them to make intelligent life choices, e.g., opting into vaccination programs, trust in medical advice, the ability to ascertain reliable sources of information and an ability to think of the microenvironment as well as more visible environmental features. Understanding the relatedness of microorganisms to every aspect of our lives and our healthy environment as well as the sustainability of all life on Earth will not be possible without improvements in understanding microbiological growth, the rapidity of evolution in response to environmental pressures and recognising the effects of changing microbial population profiles. To do this, effective education at all levels is crucial, and we must be prepared to use best teaching practices to ensure optimal retention of information. Advocates for microbiology must ensure that experts teach these subjects to the teachers of the future to ensure that common misconceptions are reduced, and expert advice is valued. Every decision made about land use, human population health and environmental care should be overseen by those with microbiological knowledge to ensure the health of microbial populations can contribute to our sustainable future. One microbe is currently dictating our lives and better education may have improved the global situation by ensuring that individuals within leadership groups had a better understanding of infection and more respect for the pathogen. Microbiology is an important discipline, and it is time that everyone knows it!

References

Adams, N.E., 2015. Bloom's taxonomy of cognitive learning objectives. J. Med. Libr. Assoc. 103, 152–153.

Adeleye, I., Muesser, N., Philip Siaw, K., 2020. Student perception of usefulness and ease using Kahoot, a free web-based tool in a tertiary education setting. Acta Sci. Technol. 43.

Akçayir, G., Akçayir, M., 2018. The flipped classroom: a review of its advantages and challenges. Comput. Educ. 126, 334–345.

Al Sultan, A., Henson Jr., H., Fadde, P.J., 2018. Pre-Service elementary teachers' scientific literacy and self-efficacy in teaching science. IAFOR J. Educ. 6, 25–41.

Andersen, L.K., Davis, M.D.P., 2017. Climate change and the epidemiology of selected tick-borne and mosquito-borne diseases: update from the international society of dermatology climate change task force. Int. J. Dermatol. 56, 252–259.

Athalye-Jape, G., Patole, S., 2019. Probiotics for preterm infants – time to end all controversies. Microbial. Biotechnol. 12, 249–253.

Ben-Barak, I., Frost, J., 2017. Do Not Lick This Book. Allen and Unwin Children's. ISBN 9781760293055.

Brockman, R.M., Taylor, J.M., Segars, L.W., Selke, V., Taylor, T.A.H., 2020. Student perceptions of online and in-person microbiology laboratory experiences in undergraduate medical education. Med. Educ. Online 25, 1710324.

Burke, C., Cain, H., Coleman, N., Grando, D., Hughes, M., Johanesen, P., Lategan, J., Lloyd, M., Markham, J., Mohideen, M., Waller, K., Wang, J., 2016. Threshold learning outcomes for a microbiology major. Microbiol. Austr. 37, 93–97.

Cacault, M.P., Hildebrand, C., Laurent-Lucchetti, J., Pellizzari, M., 2021. Distance learning in higher education: evidence from a randomized experiment. J. Eur. Econ. Assoc. 19 (4), 2322–2372.

Caine, M., Zuchuat, S., Weber, A., Ducret, V., Linder, P., Perron, K., 2015. BiOutils: an interface to connect university laboratories with microbiology classes in schools. FEMS Microbiol. Lett. 362.

Caron, C., April 16, 2020. Why do parents keep on hearing about the microbiome? N. Y. Times.

Carson, R., 1962. Silent Spring. Penguin Books Ltd, Harmondsworth, Middlesex, England.

Cavert, M.S., 2020. Why flip the medical resident classroom? A pilot study in Singapore. In: Walker, Z., Tan, D., Koh, N.K. (Eds.), Flipped Classrooms with Diverse Learners: International Perspectives. Springer Singapore, Singapore, pp. 1–13.

Chou, D.T.S., Achan, P., Ramachandran, M., 2012. The world health organization '5 moments of hand hygiene'. J. Bone Joint Surg. 94-B, 441–445.

Connor, C.R., Watkins, M., Walter, B., Harper, J.D.I., 2020. Food for thought: bringing primary school microbiology to life. Teach. Sci. J. Aust. Sci. Teach. Assoc. 66, 20–28.

Cowan, N., 2010. The magical mystery four: how is working memory capacity limited, and why? Curr. Dir. Psychol. Sci. 19, 51–57.

Coy, A., James-Williamson, S., Bramwell-Lalor, S., Sadler-Mcknight, N., Mclean, N., Penugonda, V.R., Rainford, M., 2019. A review of a tertiary level institution's initiative for enhancing education in STEM at the secondary school level. J. Educ. & Develop. Caribb. 18, 132–182.

De Bruin, A.B.H., Roelle, J., Carpenter, S.K., Baars, M., Efg, M.R.E., 2020. Synthesizing cognitive load and self-regulation theory: a theoretical framework and research agenda. Educ. Psychol. Rev. 32, 903–915.

De Meester, J., Boeve-De Pauw, J., Buyse, M.-P., Ceuppens, S., De Cock, M., De Loof, H., Goovaerts, L., Hellinckx, L., Knipprath, H., Struyf, A., Thibaut, L., Van De Velde, D., Van Petegem, P., Dehaene, W., 2020. Bridging the Gap between secondary and higher STEM education – the Case of STEM@school. Eur. Rev. 28, S135–S157.

Dunlosky, J., Rawson, K.A., Marsh, E.J., Nathan, M.J., Willingham, D.T., 2013. Improving students' learning with effective learning techniques: promising directions from cognitive and educational psychology. Psychol. Sci. Publ. Interest 14, 4–58.

Dutta, S., Smita, M.K., 2020. The impact of COVID-19 pandemic on tertiary education in Bangladesh: students' perspectives. Open J. Soc. Sci. 08, 53–68.

Ekekezie, C., Perler, B.K., Wexler, A., Duff, C., Lillis, C.J., Kelly, C.R., 2020. Understanding the scope of do-it-yourself fecal microbiota transplant. Off. J. Am. Colleg. Gastroenterol. 115, 603–607.

Fleming, D.O., Hunt, D.L., 2006. Chapter 29. Biosafety in the teaching laboratory. In: Biological Safety - Principles and Practices, fourth ed. American Society for Microbiology (ASM).

Fotopoulou, M., Iordanidou, M., Vasileiou, E., Trypsianis, G., Chatzimichael, A., Paraskakis, E., 2018. A short period of breastfeeding in infancy, excessive house cleaning, absence of older sibling, and passive smoking are related to more severe atopic dermatitis in children. Eur. J. Dermatol. 28, 56–63.

Freeman, S., Eddy, S.L., Mcdonough, M., Smith, M.K., Okoroafor, N., Jordt, H., Wenderoth, M.P., 2014. Active learning increases student performance in science, engineering, and mathematics. Proc. Natl. Acad. Sci. USA 111, 8410.

Gobet, F., 2005. Chunking models of expertise: implications for education. Appl. Cogn. Psychol. Appl. Cogn. Psychol 19, 183–204.

Graves, N., 2004. Economics and preventing hospital-acquired infection. Emerg. Infect. Dis. 10, 561–566.

Hadjianastasis, M., Nightingale, K.P., 2016. Podcasting in the STEM disciplines: the implications of supplementary lecture recording and 'lecture flipping'. FEMS Microbiol. Lett. 363.

Hartmann, B.T., Pang, W.W., Keil, A.D., Hartmann, P.E., Simmer, K., 2007. Best practice guidelines for the operation of a donor human milk bank in an Australian NICU. Early Hum. Dev. 83, 667–673.

Hofer, R.E., Brant Nikolaus, O., Pawlina, W., 2011. Using checklists in a gross anatomy laboratory improves learning outcomes and dissection quality. Anat. Sci. Educ. 4, 249–255.

Hotez, P.J., 2020. Anti-science extremism in America: escalating and globalizing. Microorg. & Infect. 22, 505–507.

Jones, S., Katyal, P., Xie, X., Nicolas, M., Leung, E., Noland, D., Montclare, J., 2019. A 'KAHOOT!' Approach: the effectiveness of game-based learning for an advanced placement biology class. Simulat. Gaming 50, 104687811988204.

Karpicke, J.D., 2012. Retrieval-based learning: active retrieval promotes meaningful learning. Curr. Dir. Psychol. Sci. 21, 157–163.

Keim, S.A., Hogan, J.S., Mcnamara, K.A., Gudimetla, V., Dillon, C.E., Kwiek, J.J., Geraghty, S.R., 2013. Microbial contamination of human milk purchased via the internet. Pediatrics 132, e1227.

Khan, M., Ibrahim, M., 2017. Flipped classroom in technology courses - impact on personal efficacy and perception based on learning style preferences. In: 2017 IEEE Integrated STEM Education Conference (ISEC), 11-11 March 2017, pp. 135–142.

Kinsella, G.K., Mahon, C., Lillis, S., 2017. Using pre-lecture activities to enhance learner engagement in a large group setting. Act. Learn. High. Educ. 18, 231–242.

Knippenberg, M.T., Leak, A., Disseler, S., Segarra, V.A., 2020. Establishing partnerships for science outreach inside and outside the undergraduate classroom. J. Microbiol. Biol. Educ. 21, 21.22.60.

Kornell, N., Hays, M.J., Bjork, R.A., 2009. Unsuccessful retrieval attempts enhance subsequent learning. J. Exp. Psychol. Learn. Mem. Cognit. 35, 989–998.

López, S., Zárate, S., Yocupicio, M., Lobatón, E., 2019. Paul Has Measles. Independent.

Maggio, L.A., Daley, B.J., Pratt, D.D., Torre, D.M., 2018. Honoring thyself in the transition to online teaching. Acad. Med. 93, 1129–1134.

Marcos-Merino, J.M., Gallego, R.E., De Alda, J.A.O., 2020. Training pre-service teachers to approach microorganisms through an active practice based on homemade culture: role of students' emotions. In: Puig, B., Blanco Anaya, P., Quílez, M.J.G., Grace, M. (Eds.), Biology Education Research. Contemporary Topics and Directions. Servicio de Publicaciones. Universidad de Zaragoza.

Merkel, S., ASM Task Force on Curriculum Guidelines for Undergraduate Microbiology, 2012. The development of curricular guidelines for introductory microbiology that focus on understanding. J. Microbiol. Biol. Educ. 13, 32–38.

Miller, B.L., 2020. Science denial and COVID conspiracy theories: potential neurological mechanisms and possible responses. JAMA 324, 2255–2256.

Miller, G.A., 1956. The magical number seven, plus or minus two: some limits on our capacity for processing information. Psychol. Rev. 63, 81–97.

Milner-Bolotin, M., 2018. Evidence-based research in STEM teacher education: from theory to practice. Front. Educ. 3.

Neunez, M., Goldman, M., Ghezzi, P., 2020. Online information on probiotics: does it match scientific evidence? Front. Med. 6.

Noel, T.C., Rubin, J.E., Guerrero, Y.A., Davis, M.C., Dietz, H., Libertucci, J., Sukdeo, N., 2020. Keeping the microbiology lab alive: essential microbiology lab skill development in the wake of COVID-19. Can. J. Microbiol. 66, 603–604.

Redfern, J., Burdass, D., Verran, J., 2013. Practical microbiology in schools: a survey of UK teachers. Trends Microbiol. 21, 557–559.

Robinson, L.A., Turner, I.J., Sweet, M.J., 2018. The use of gamification in the teaching of disease epidemics and pandemics. FEMS Microbiol. Lett. 365.

Roediger, H.L., Pyc, M.A., 2012. Inexpensive techniques to improve education: applying cognitive psychology to enhance educational practice. J. Appl. Res. Memory & Cogn. 1, 242–248.

Ruiz-Gallardo, J.-R., Paños, E., 2018. Primary school students' conceptions about microorganisms. Influence of theoretical and practical methodologies on learning. Res. Sci. Technol. Educ. 36, 165–184.

Sahin, O., Salim, H., Suprun, E., Richards, R., Macaskill, S., Heilgeist, S., Rutherford, S., Stewart, R.A., Beal, C.D., 2020. Developing a preliminary causal loop diagram for understanding the wicked complexity of the COVID-19 pandemic. Systems 8, 20.

Shan, Y., Segre, J.A., Chang, E.B., 2019. Responsible stewardship for communicating microbiome research to the press and public. Nat. Med. 25, 872–874.

Singleton, R., Charlton, A., 2019. Creating H5P content for active learning. Pacif. J. Technol. Enhanc. Learn. 2, 13–14.

Solomon, E.D., Repice, M.D., Mutambuki, J.M., Leonard, D.A., Cohen, C.A., Luo, J., Frey, R.F., 2018. A mixed-methods investigation of clicker implementation styles in STEM. CBE-Life Sci. Educ. 17, ar30.

Steel, P., 2007. The nature of procrastination: a meta-analytic and theoretical review of quintessential self-regulatory failure. Psychol. Bull. 133, 65–94.

Stevens, N.T., Holmes, K., Grainger, R.J., Connolly, R., Prior, A.-R., Fitzpatrick, F., O'neill, E., Boland, F., Pawlikowska, T., Humphreys, H., 2019. Can e-learning improve the performance of undergraduate medical students in clinical microbiology examinations? BMC Med. Educ. 19, 408.

Stevens, N.T., Mcdermott, H., Boland, F., Pawlikowska, T., Humphreys, H., 2017. A comparative study: do "clickers" increase student engagement in multidisciplinary clinical microbiology teaching? BMC Med. Educ. 17, 70.

Taylor, K., Rohrer, D., 2010. The effects of interleaved practice. Appl. Cognit. Psychol. 24, 837–848.

Tharayil, S., Borrego, M., Prince, M., Nguyen, K.A., Shekhar, P., Finelli, C.J., Waters, C., 2018. Strategies to mitigate student resistance to active learning. Inter. J. STEM Educ. 5, 7.

Thavamani, P., Samkumar, R.A., Satheesh, V., Subashchandrabose, S.R., Ramadass, K., Naidu, R., Venkateswarlu, K., Megharaj, M., 2017. Microorganisms from mined sites: harnessing their potential for reclamation of derelict mine sites. Environ. Pollut. 230, 495–505.

Thomas, E., Zhang, A., 2020. In: ID2020, Bill Gates and the Mark of the Beast: How Covid-19 Catalyses Existing Online Conspiracy Movements. Australian Strategic Policy Institute, JSTOR.

Timmis, K., Cavicchioli, R., Garcia, J.L., Nogales, B., Chavarría, M., Stein, L., Mcgenity, T.J., Webster, N., Singh, B.K., Handelsman, J., De Lorenzo, V., Pruzzo, C., Timmis, J., Martín, J.L.R., Verstraete, W., Jetten, M., Danchin, A., Huang, W., Gilbert, J., Lal, R., Santos, H., Lee, S.Y., Sessitsch, A., Bonfante, P., Gram, L., Lin, R.T.P., Ron, E., Karahan, Z.C., Van Der Meer, J.R., Artunkal, S., Jahn, D., Harper, L., 2019. The urgent need for microbiology literacy in society. Environ. Microbiol. 21, 1513–1528.

Turner, R., 2020. Curtin University Plans to Ditch In-Person Lectures and Exams, Even after Coronavirus Pandemic Ends. Australian Broadcasting Corporation Online.

Verran, J., Redfern, J., Moravej, H., Adebola, Y., 2019. Refreshing the public appetite for 'good bacteria': menus made by microorganisms. J. Biol. Educ. 53, 34–46.

Vinney, L.A., Howles, L., Leverson, G., Connor, N.P., 2016. Augmenting College students' study of speech-language pathology using computer-based mini quiz games. Am. J. Speech Lang. Pathol. 25, 416–425.

Wang, J.T.H., Huston, W.M., Johanesen, P., Lloyd, M., Waller, K.L., 2018. A laboratory competency examination in microbiology. FEMS Microbiol. Lett. 365.

Wanner, T., Palmer, E., 2015. Personalising learning: exploring student and teacher perceptions about flexible learning and assessment in a flipped university course. Comput. Educ. 88, 354–369.

Wild, A., Reed, A., Barr, B., Crocetti, G., Stock, S.P., 2019. Nema and the Xenos: A Story of Soil Cycles. CSIRO PUBLISHING.

Wild, A., Reed, A., Crocetti, G., Barr, B., 2018. Zobi and the Zoox: A Story of Coral Bleaching. CSIRO PUBLISHING.

Wilkinson, K., Dafoulas, G., Garelick, H., Huyck, C., 2020. Are quiz-games an effective revision tool in Anatomical Sciences for Higher Education and what do students think of them? Br. J. Educ. Technol. 51, 761–777.

Witze, A., 2020. Universities will never be the same after the coronavirus crisis. Nature 582, 162–164.

Yano, R., Okubo, T., Shimoda, T., Matsuo, J., Yamaguchi, H., 2019. A simple and short microbiology practical improves undergraduate nursing students' awareness of bacterial traits and ability to avoid spreading infections. BMC Med. Educ. 19, 53.

Zhou, N., 2020. Australian Universities Plan to Ramp up In-Person Learning in Early 2021. Guardian Australia. https://www.theguardian.com/australia-news/2021/jan/20/australian-universities-plan-to-ramp-up-in-person-learning-in-early-2021. (Accessed 20 January 2021).

Zollman, A., 2012. Learning for STEM literacy: STEM literacy for learning. Sch. Sci. Math. 112, 12–19.

Chapter 6

Data management in culture collections

Lorenz Christian Reimer and Andrey Yurkov
Leibniz Institute DSMZ-German Collection of Microorganisms and Cell Cultures, Braunschweig, Germany

Chapter outline

Objectives of data management in culture collections

Culture collections routinely preserve microbial diversity and collect the associated data. Collection holdings undergo a complex process of authentication using a diverse set of tests and tools that analyse physiological and biochemical properties as well as determine molecular biological characteristics. Thus, strain characteristics are made up of results of tests performed in collections and the information provided by depositors or published in the literature. In addition, to the primary literature, data collection in culture collections is the second most comprehensive source for research data on microbial strains. Culture collection data is often systematically collected and stored in a more or less standardised format. Information about microbial

Importance of Microbiology Teaching and Microbial Resource Management for Sustainable Futures
https://doi.org/10.1016/B978-0-12-818272-7.00004-3

strains in culture collection then flows in collection catalogues. The evolution of catalogues went a long way from hand-written lab books to sophisticated electronic databases. But the aim of a catalogue remained fundamentally the same – to provide a systematically collected information about collection holdings. With the advance of tests used to identify and characterise microbial diversity, the amount of information accumulated in collection catalogues grew substantially. In order to get through that huge amount of often heterogeneous information, data management soon became an important task for collections. Data available in collection catalogues and online databases often show only a tip of an iceberg – the majority of data are not visible to external users. Thus, it is important to distinguish between data collected in culture collections and the limited amount of data displayed in catalogues.

The modern catalogue evolved from printed books into an electronic source of data. Often available on a public website, catalogue moved away from an inventory of holdings towards the interface that globally shares data about strains. Keeping these data up to date requires a constant effort. Various sources and types of data make it even more challenging. Collections routinely collect, store and share highly heterogeneous information such as culture collection identifiers, taxonomic assignments, cultivation conditions, isolation sources, biosafety and legal restrictions, strain history and links to the literature. Altogether this makes an incredible amount of information stored and systematically processed in a daily collection business. But internal files make the lion's share of data obtained and stored in collections. The great majority of data from internal analyses is rarely shared. That constantly growing array of information comes from collections' background processes associated with the authentication and quality control and can comprise various kinds of tests, like metabolic profiling (e.g., API, BIOLOG), molecular fingerprints (e.g., ribotyping, AFLP), MALDI-TOF spectra, fatty acid profiles, antibiotic susceptibility testing, microscopic and macroscopic pictures and, of course, nucleotide sequencing data. Culture collections have increased their skills and their quality standards following Best Practice Guidelines (OECD) and thanks to ISO 9001, ISO 17025 and NF S 96–900 certifications. This also changed analyses that are routinely performed in culture collections making them more uniform and, thus, better suitable for an automated recording. The major objectives and services require culture collections to obtain and provide data for the following needs:

- *Basic strain data in the catalogue. Data to verify if this is the correct strain: strain number, numbers in other collections, origin of strains (including country), history* (e.g., *depositor, other collections*)
- *Information and resources for a web-shop: availability of strains, form of dispatch, risk assessment and pathogenicity, import and export restrictions, CBD/NP restrictions and MTA*
- *Cultivation conditions: cultivation temperature, medium, oxygen relation, light, etc.*

- *Quality management: viability tests and a large diversity of tests used to authenticate strains*
- *Experimental results from the literature, both own data mining and information from external sources, depositor and other culture collections*
- *Taxonomic information and the associated literature*
- *Application of strains: data from the literature and/or own results*

History of data collection (from paper over Excel to complex databases)

The complexity of collection catalogues grew along with our knowledge of biology of microorganisms. With the establishment of very first techniques to obtain pure cultures, the information associated with microbial cultures was collected and analysed. When properly recorded, the primary data provided with old strains was often limited to the isolation source and locality. Subsequent studies using collection isolates increased that information to include growth tests, physiology, biochemical properties, cell ultrastructure and DNA properties. Cultures that represent nomenclature types of microorganisms were studied in more details.

The diversity of accumulated data was growing in response to ever-developing taxonomy and systematics of microorganisms. Shortly after the scientific community adopted a new laboratory technique for species identification, that technique was applied to study reference cultures in culture collections and later to other cultures.

The first available tools to distinguish and classify microorganisms were the light microscopy invented by Leeuwenhoek in the 1670s and cultivation itself. From that time, the following characteristics of cultures are still in use: morphological appearance (colour, colony shape), cell shape (also size, flagellation), motility and production of spores (including sexual and asexual reproduction in *Fungi*) and filaments. Gram staining (developed in 1884) is another useful test of the structural characteristics of bacterial cell walls. The thick layers of peptidoglycan cell walls of 'Gram-positive' bacteria stain purple, whilst the thin 'Gram-negative' cell wall appears pink. Other kinds of staining were targeting bacterial endospores and capsules. Scientists began cultivating microorganisms on agar media in the 1880s (Hitchens and Leikind, 1939). Starting from the end of the 19th century (Beijerinck, 1889), microorganisms were characterised according to their metabolic properties, which substantially broaden our understanding of microbial metabolism. These tests included growth in selective media (aerobic and anaerobic; differential media), at different temperatures, in the presence of certain compounds, and at various concentrations of salt or sugar, as well as some simple enzymatic reactions (e.g., catalase reaction), utilisation of exogenous compounds and fermentation. Some of these tests are still in use, though in a modified form (e.g., Wickerham and Burton, 1948; Kurtzman et al., 2011).

Characterisation of cell components, including amino acids, cellular lipids (and their component fatty acids), pigments (bacteriochlorophylls, carotenoids), peptides, polysaccharides and respiratory quinones was gaining its importance in the 1960–70s (Goodfellow, 2000). The widespread application of DNA characteristics, such as DNA base composition (guanine–cytosine content), DNA–DNA hybridisation kinetic, partial sequencing of ribosomal genes (16S rRNA, 18S rRNA), genome size and whole-genome phylogenies, brought identification and systematics of bacteria and fungi to its present state (e.g., Nouioui et al., 2018; Lücking et al., 2020). Additionally, a few rapid identification tools such as automated ribotyping and MALDI-TOF have been utilised for the identification of microorganisms, also in culture collections.

The oldest working service culture collections are recorded as being the Mycothèque de l'Université Catholique de Louvain (MUCL) and the Centraalbureau voor Schimmelcultures (CBS), established in 1894 and 1906, respectively. But even younger collections have strain records spanning over several decades. In times when strain descriptions were merely limited to drawings and hand-written observations, books were the source of the information. Later on, typewriter and note cards were used and became a useful and flexible way to manage strain data in culture collections. With the evolution in computer technology and the ongoing increase of data, first digital storage systems were employed. Today data management in culture collections ranges from simple spreadsheets (for smaller collections) to highly complex databases with hundreds of data fields. The evolution of data in culture collections experienced several break points in technology, but only rarely old data were dismissed. This leads to an enormous aggregation of data, some of which is often 'out of fashion', both in terms of its nature and obsolete data format, but still highly valuable for future generations of researchers. On the one hand, this makes internal files of culture collection a kind of treasure chest for microbial research – collections accumulated many experimental measurements and records over several decades. On the other hand, this long tail of often heterogeneous data from the past poses a certain challenge in the ever-increasing and evolving process of data collection.

In order to get digital catalogues, whether simple or complex, collections would need to start with the digitalisation (see Wu et al., 2017). The very first step is to convert in a digital format the list of collection holdings from either books or cards. Strain numbers become unique alphanumeric identifiers that will link all associated metadata to the strain, internal or external. Catalogues can benefit from external sources of information, nomenclatural (names of organisms) and genetic (sequences in public databases). Digitalisation of collection records and data mobilisation are extremely time-consuming and challenging processes, but they are essential for any future analyses and data exchange (Fig. 6.1).

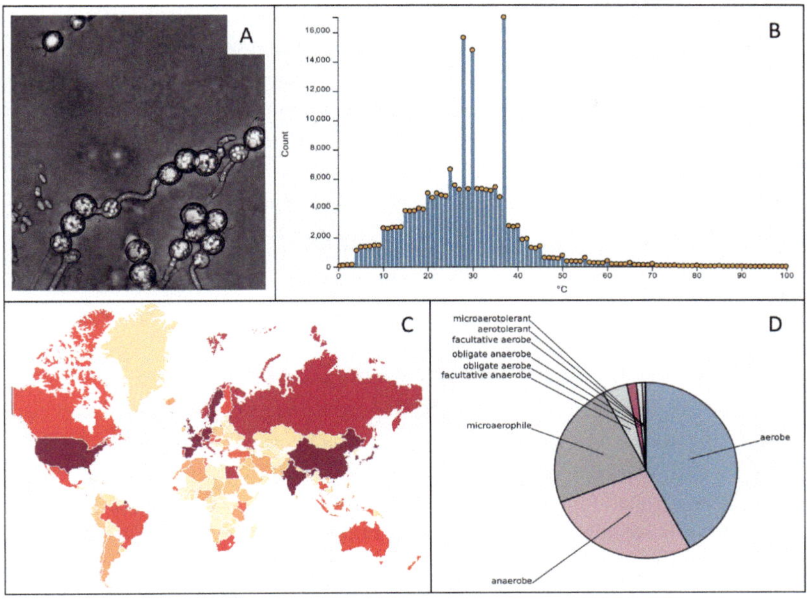

FIGURE 6.1 Overview showing different examples of data types in culture collections. (A) microscopic picture of a fungus with hyphae and teliospores (DSM 27755), (B) growth temperatures of 39,759 bacteria, (C) country data counts for the origin of 45,076 bacterial strains (darker colour represents higher numbers), (D) oxygen relation for 17,251 bacterial strains. (B), (C) and (D) show data from the Bac*Dive* database.

Types of data

Over the time, a wide array of methods and tools have been employed for species descriptions. Identification approaches were modified to meet modern criteria for a more accurate species delimitation. Consequently, standard procedures routinely used for strain authentication in collections adapted modern tools. Although sequencing of single genes (or gene regions) and entire genomes strongly dominates the modern era of research, older tools are still in use and only rarely abandoned. Microphotographs using either light or electron microscopy introduce the studied organism in every description. Growth on solid or in liquid media and its visual appearance also provide useful characteristic traits to identify the target organism. Although the information is often limited in the case of microorganisms, researchers are used to looking at cultures with their own eyes. Being probably an innate part of human nature, microscopic observations were essential to document the development, e.g., sexual propagation, under laboratory conditions. Physiological tests, e.g., for enzyme activity, have a long standing in species descriptions and are typically performed in a standardised miniature approach like the Analytical Profile Index tests. These tests were first introduced in the laboratories in the 70s of the last century and have only little evolved since then. Because of the low price (no expensive machine needed) and the easy handling, they are still widely spread. Still, more modern and complex approaches are available, e.g., phenotyping using high-throughput systems like the BIOLOG phenotype MicroArray system, to test hundreds of carbon or nitrogen sources at once.

This array of information makes a diverse potpourri of data types:

1. Graphic data as microscopic images as files.

   They cannot be analysed as is and require metadata to include, for example, cultivation conditions (e.g., media, temperature, duration), the microscopy technique and magnification.
2. Binary or mostly binary data of physiological tests

   Several relevant morphological properties were recorded as binary data, for example, as yes/no, present/absent or produced/not produced. This could be information about the sexual reproduction, spore production, synthesis of capsules, and motility to name a few. Utilisation of a certain substrate is usually collected in a binary type of data (growth or no growth). However, in addition to positive and negative tests, there are a few commonly used characters reflecting the growth qualitatively (weak) or timely (delayed). Phenotypic MicroArray tests could be either binary (growth/no growth) or numeric (optical density, enzymatic reaction) data.
3. Numeric

   Growth temperature and pH are good examples of simple numeric data. Although it sounds trivial, mobilisation of that data may require some attention. It is not unusual when pH values were recorded as a range (e.g., pH 8–10) or an ecological category (e.g., alkaliphile). Growth temperatures naturally differ between minimum, maximum and optimal. Thus, it is important to distinguish them and, when needed, separate them into different fields. For numerical data like concentrations of substances (e.g., salt) and time measurements (e.g., growth kinetic), the units may vary and, therefore, have to be taken into account. Several widely used DNA characteristics are other frequently used numeric data, e.g., DNA–DNA hybridisation results and diverse sequence similarity values, including ANI and GGDC (Meier-Kolthoff et al., 2013). When it comes to larger numeric values beyond four digits like geographic coordinates, this data becomes very prone to errors that only can be avoided by logic checks using additional data or checksums (see Harmonisation of data section).
4. Textual data

   Text fields are avoided as much as possible, as the degree of freedom is too high for standardised comparison and exchange. Language barriers, misspelled words and synonyms largely complicate the use of this data. Still, some data is traditionally collected in the text format as for example isolation source, isolation location (toponyms), strain history (persons, institutions) and any other observations that do not fit into the standard fields. Whilst some data like strain history might be transformed into a more structured format, other data like isolation source is more challenging as the text can be anything between a single word and a long sentence describing different environmental parameters. Moreover, the isolation source description is a historic document that cannot be customised. This problem might be overcome by assigning predefined keywords to enrich these textual data (see Harmonisation of data section).

5. Time and date

   A few years ago, date information was only collected as an additional note and was mostly roughly estimated. Today, that information is of high importance because both deposition and distribution of strains depend on a national CBD/NP legislation and the date when it came into force.

6. Categorical data

   Wherever possible, data is collected as categorical data. Precondition is a widely accepted consensus of the applied categories. This is typically the case for data like oxygen relation, cell shape, Gram staining, type of hemolysis, nutrition type, biosafety level, risk assessment (pathogenicity and host range) and country data. Most common ecological classifications of microorganisms (autotroph, chemotroph, phototroph, psychrophile, symbiont, parasite, xerophile etc.) are categorical data. Still, there is a chance to deal with data that does not fit into the categories and therefore has to be dismissed (see Harmonisation of data section).

7. Specific tests including metadata

   Multiple tests are typically conducted that consist of a series of numerical data, that are dependent on each other and that need additional metadata to record conditions and parameters that influence the result. As for example fatty acid profiles, that are collected together with cultivation parameters, as well as instrument and software information. Nucleotide sequences have largely facilitated identification of microorganisms. A typical identification step includes a comparison of newly generated nucleotide sequences with sequences available in external or internal databases. An identification result would typically include three parameters, sequence identifier (e.g., accession number), sequence similarity (e.g., % identity) and overlap (e.g., coverage). Additionally, the version of the database (release) and a particular sequence (like in GenBank) determine reproducibility of the result. For sequencing, 16S rRNA was a long-standing gold standard for the identification and phylogenetic classification of bacteria. Likewise, ribosomal ITS is the first common barcode for *Fungi* (Schoch et al., 2012). Though sequences of these two DNA fragments prevail over other sequences in public databases, the number of available whole genome sequences (particularly of prokaryotes) and sequences of other fungal DNA markers is steadily growing (e.g., Lücking et al., 2020).

8. Literature sources

As in the whole science, literature referencing in culture collections is of utmost importance. Literature data sources in culture collection are not limited to the taxonomic literature, descriptions of new taxa and combinations, but also include publications reporting properties of species and strains (growth, metabolites) and new methodologies. Offering a selection of the literature to a strain in the catalogue is a much-appreciated service to the users.

The assessment of the literature is challenging for several reasons. Before advanced literature search engines PubMed, Scopus and Google Scholar facilitated our access to the literature, records have been maintained manually by collection curators. It was not unusual that depositors provided their published and unpublished publications to strains. Nowadays search engines provide immediate access to the literature but data mining is still inefficient mostly because of inconsistent use of strain identifiers. Rules for authors determine formatting of citations, biological names, chemicals and nucleotide sequences but not collection identifiers. Therefore, errors in collection acronyms and strain numbers are common, like twisted numbers and missing spaces and thereby the link to the original bioresource gets easily lost. Strain numbers are not necessarily included in keywords or tables. Many online journals do not convert strain identifiers to links to collection websites and do not include tags to readily distinguish strain numbers. As a result, automated data mining from literature sources is poor. Besides being inconvenient for users, this also creates a non-transparent system where utilisation, or any other use, of strains cannot be traced.

9. Persons and institutions

Collections display history of their holdings and acknowledge persons who contributed to the isolation and investigation of strains. Most common records include, but are not limited to, collector, isolator and depositor. In most cases, this information is static and does not account for changing of name and affiliation of these persons. Since some collections introduced electronic accession forms submitted from user accounts, this data can be automatically collected and updated. But updating historical records would require a manual intervention. Ideally, an identifier system could be used to match collectors, isolators, depositors and other researchers who work with a strain. As an option, Open Researcher and Contributor ID (ORCID) can be used to match the majority of active researchers and contributors. That alphanumeric identifier received broad recognition from journals and is used as a tool to increase the recognition of people who contributed to publications by providing datasets, unpublished drafts and other sources, like Wikipedia entries (ORCID Open Letter - Publishers, n.d.). Taking into consideration that isolation of fastidious is extremely time consuming and costly (Overmann et al., 2017), it would be valuable to acknowledge people who isolated (including sampling and purification) and characterised strains. Authors of taxonomic novelties are associated with names of scientists who proposed nomenclatural names, e.g., *Streptomyces* (Waksman and Henrici, 1943). Collection records extend this information to people who assisted the research by isolating and providing cultures, including non-type or non-reference.

Harmonisation of data

Culture collections have increased their skills and their quality standards. After a successful implementation of ISO 9001, ISO 17025 and NF S 96–900 certifications, some European culture collections, such as BCCM, CBS, CECT, CIRM and DSMZ, have acquired the status of Biological Resource Centre (Smith, 2003). To comply with the high standards, data in culture collections was transferred into safeguarded databases. As a result, there was quite some unification of data fields in collection catalogues. Thanks to exchanging strain data between culture collections on a regular basis, data structure and data fields between collections are largely consistent but might be adapted and extended to the different needs in the institutions.

As the guarantee of strain preservation and identity, culture collections play the central role in solving issues regarding the identification of microorganisms and relation of physical and virtual entities of microbiota. Whilst the preservation and availability of the material are anchored in the *International Code of Nomenclature of Prokaryotes* (ICNP, Rules 18a, 30) and the *International Code of Nomenclature for algae, fungi, and plants* (ICNafp, Art. 8.4, Rec. 8B.1), the identity of the deposited culture to its published description is solely the responsibility of a culture collection. By controlling the identity of the incoming material, collections contribute to reproducibility of all later published results. Collection identifiers and history of strains allow scientists to track the material. In other words, a properly authenticated collection strain is the answer to the question 'Do I work with the same strain as the one in the experiment that I try to replicate?' Although no global strain identifier yet is available, culture collections are a most valuable source to prove strain identity. Culture collections ensure accessibility of the reference material and the associated data by improving their preservation techniques, adapting new quality standards and by exchanging strains as well as by matching strain identifiers in different culture collections. Transfers between collections are tracked within the strain history, a very simple but important system to trace back changes that may occur in strains over time or by accident loss. Still, the matching of culture collections numbers is limited and at the moment an overarching system offering the resolution across all service culture collection numbers is missing.

A major issue in harmonisation of culture collection data is the frequent changes in taxonomic classification and the resulting changes in species names. In the effort to keep species names up to date, collection staff needs to screen the literature in regular intervals, which is becoming more labour intensive with the rising numbers of new discoveries over the last decade. A wide range of literature sources that publish new taxa (like *Fungi*) makes manual and semi-automated mining even more time consuming. The task of tracking nomenclatural changes can be achieved by using external dedicated nomenclature services such as LPSN (Parte et al., 2020), free for academic users or NamesForLife (namesforlife.com, commercial) for bacteria and an electronic nomenclatural repository (Fungal Names, MycoBank, Index Fungorum; all free) for *Fungi*. Renaming is not only a matter of offering the up-to-date nomenclature. Name changes impact legal and safety regulations that apply to strains. Thereby, an ordinary taxonomic rearrangement (renaming) can strongly impact strain usage, such as acquisition, export and import. Another side of the same problem is the information about risk assessment. This is probably one of the most heterogeneous and troublesome fields to harmonise. Although changes in national and international regulations regarding the risk assessment are not as frequent as taxonomic reclassifications, the immediate impact of a modification or even an error can be high because culture collections and end users must comply with the regulations. The problem concerns, but is not limited to, pathogens (animal, human, plant), quarantine species, industrial strains and Generally Recognized As Safe (GRAS) agents. Recent findings like the production of toxins have to be adapted and communicated to both users and responsible authorities.

Other data fields also pose challenges, particularly those related to the metabolism of the strains. Whilst the classification and nomenclature of enzymes are commonly made according to the IUBMB in EC classes and well organised like in the database BRENDA (brenda-enzymes.org), a correct naming of metabolites is more problematic. Although several databases for metabolites exist (e.g., ChEBI, PubChem), the variety of names used by researchers is high and in contrast to enzymes; the identification of the metabolites using an identifier (e.g., a ChEBI-ID) is uncommon. Additionally, a metabolite can occur in different forms (e.g., status of protonation, isomerism) that need to be taken into account. Another challenge is data that is dependent on certain cultivation conditions or instrumental parameters, like, for example, fatty acid profiles. Measurements and metadata need to be linked to ensure a correct interpretation of data. However, when dealing with data from non-active sources, metadata gets sometimes lost and thereby devalues the data.

A trivial problem at first sight is geographic data. The vast majority of data can be assigned to a country as a sovereign state. But assigning some areas to a sovereign state is not always easy: old country data (e.g., USSR, Republic of Yugoslavia); international waters (and countries EEZ); regions with an uncertain political status (e.g., Taiwan); disputed territories (e.g., Crimea, Gibraltar, Falkland Islands) and unincorporated (USA), external (Australia), overseas (UK, France, the Netherlands) territories. But also, recent geographic data fields like coordinates can rather be problematic. First of all, many different formats of recording geographic coordinates exist and several formats are equally wide distributed: degree, minutes, seconds (DMS), degree and decimal minutes (DMM), decimal degree (DD). Fortunately, several online tools convert between formats easily. Even more problematic is the high susceptibility to errors of the numeric format, where errors and twisted numbers occur frequently and that can hardly be traced back without additional information, like the exact location. Large numeric formats are missing a checksum that allows to detect errors easily. Despite technical errors that occur, geographic coordinates also provide another issue: missing information on accuracy. Independent from the applied format, coordinates always represent an exact point on earth. Researchers round up coordinates, as they do not want to expose the exact location or because text fields are limited in length. Thereby rounded coordinates may point to a totally different location. Ideally, coordinates data should be accompanied by estimates of accuracy, for instance, in meters or kilometres.

Among different types of data (see above), textual data is extremely challenging, when not based on a conserved format. A trivial example is the exact location and toponyms that can be spelled differently and contain uncommon symbols. Another example for highly valuable data stored in text form is the isolation source. The isolation source can be described in many ways – from meagre 'air', 'soil', 'plant leaf' to very detailed records made of long sentences describing anything the researcher has noticed during the sampling. It is important to note that in addition to typos, misspelled words and orthographic variants, isolation sources may contain words from different languages (e.g., species names). Although hard to process automatically, the isolation source information is important to illustrate the ecology of a strain. To overcome this problem, certain databases categorise isolation sources into large categories or classes like 'soil', 'marine', 'host associated' (e.g., Groenewald et al., 2017). This approach is limited as it only allows a coarse classification and requires a binning into a certain category which is not always possible. Another common problem is that categories are not of the same size, some are more detailed than others. A more flexible way is the annotation with several keywords which makes this type of data systematically searchable.

Approaches for data harmonisation across several collections

Data harmonisation across several collections can be achieved in several ways. The first one is to use pre-defined templates to export and import data for a few desired fields in a specific format. That approach is similar to an online submission tool, like the GenBank BankIt, that is importing data in specific fields from either a table or annotation file. Information extracted from culture collection catalogues and databases can be later uploaded, linked and displayed in a large database. The Global Catalogue of Microorganisms (GCM, WDCM) and now discontinued StrainInfo.net obtained data from member collections (Wu et al., 2013). The GCM collected strain catalogue information from collections by online submission (Wu et al., 2013). After the information (like strain number(s), origin, taxonomic status, medium and temperature) is collected, a unique strain identifier was used to link data from other external sources, such as nucleotide sequences (e.g., source modifiers 'strain' and 'culture_collection') and literature using text mining. This system comes with several limitations. A first particular limitation of any crowdsourcing system is that the information in fields may vary in terms of units and values format that, in turn, would complicate big data analyses. A specific challenge related to the collection of data from culture collections is to correctly link essentially the same culture media and strains shared between datasets. The same medium may have several different labels. Likewise, records of strain history are not always uniform. A second limitation is the dependency on regular contributions from participants. Data submission and homogenisation can be a tough task. But keeping existing data sets up-to-date requires continuous inputs from partner collections and ultimately depends on their interest and capacities. Several currently discontinued database projects faced this problem in the past, e.g., StrainInfo.net (Verslyppe et al., 2010) and digital-protologue (Rosselló-Móra et al., 2017).

The second approach is to use the same software or tool to maintain collection catalogues in different collections. But collections use diverse, often exclusive, software packages and tools. Probably the only successful commercial software developed to meet needs of a culture collection is the BioloMICS (Robert et al., 2011). The software offers the possibility to publish selected tables and records online, creating, thus, a dynamic website and catalogue. The possibility to exchange and analyse data within the single software environment is especially interesting in the context of harmonisation data. Matching strains, fields and data of strains and specimens from different collections is obviously easier to do within the same environment. Any commonly used field can be analysed across several collections. Although managing collection data in a single software solves the problem of matching records, its full functionality will be limited to collections using that specific software. Linking and matching data from collections, which use a different software, would need a manual intervention to harmonise data formats.

The third approach is to collect, curate and then link data from individual collections in a central database format, as it is the case for the database Bac*Dive* (see below). An important prerequisite for the task is the matching of identical strains (progenies of the same original culture) available from different collections. Pooling unmatched strains will exaggerate the total number of records and inflate data quality. Whilst matching type strains seems to be rather easy to implement (see GCM, Bac*Dive*), correctly linking isotypes (the same strain from different sources) and type strains of heterotypic synonyms requires a manual curation. The number of duplicate type specimens can be really large even in a single culture collection. Matching is prone to two types of errors: not-matching of identical strains will cause doubling of strains and, though much less frequent, mis-matching of different strains, which causes annotation of false data. As soon as strain identifiers are matched, other data fields can be pooled together following harmonisation, re-formatting and curation. Although data harmonisation centrally is a laborious task, it also gives the flexibility that is typically required to integrate different kinds of sources, as for example data from primary literature, and extend datasets with additional fields anticipating future needs.

The data treasures in culture collections and projects around it

The Microbial Information Network Europe (MINE) was probably the first large-scale activity to unify data across European culture collections. Collections of microorganisms agreed on recording their data in common formats to be integrated in a common system (Gams et al., 1988; Stalpers et al., 1990). Several national nodes have been established in member countries. Participating collections agreed to comply with minimum standards; criteria that were also suggested to serve as guidelines for newly established culture collections (Hawksworth and Schipper, 1989). These criteria included, amongst others, digital catalogues organised according to the MINE formats and minimum dataset (Gams et al., 1988). The MINE standards of data established a basis for subsequent exchange of information between culture collections. During the last two decades, several projects have approached different problems of data mobilisation and management in culture collections. Whilst the database itself is of no use anymore, the main merits of MINE lie in the development of an internationally agreed format — 135 fields for fungi and yeasts and 145 fields for bacteria which cover the most important aspects of microbial taxonomy, ecology, physiology, and biochemistry, including also data pertaining to the practical applications of microorganisms (Fritze, 2006).

The Global Catalogue of Microorganisms, GCM

The first version of the GCM database contained information of over 273,933 strains (both bacteria and fungi) from 52 collections in 25 countries and regions (Wu et al., 2013). The current version includes 460,491 strain records from 133 collections in 50 countries and regions. This comprehensive database provides a good overview of microbial ex situ diversity. The service is particularly valuable for collections with poor informatics infrastructures as their microbial resources become visible to users worldwide. As it has been mentioned before, this approach heavily relies on external sources, namely data mobilisation, curation and correct formatting of individual datasets are provided by member collections (Wu et al., 2018).

Microbial Resource Research Infrastructure, MIRRI

The Microbial Resource Research Infrastructure (https://www.mirri.org) is a pan-European association of more than 40 public culture collections and research institutes from 19 European countries (Casaregola et al., 2016) that offers different services like strain, literature or media databases. The strain database offers information on taxonomy, nomenclatural status of the name as well as literature for 8319 strains. The media database comprises 427 cultivation recipes. A thesaurus explains briefly 8011 terms and also gives information about the etymology of the term. The ambitious plan to build a united European microbial resource centre to have a central access point not only to the data about microbial resources, but also to microbial resources itself, did not yet yield the desired progress. To date, a common database service for European culture collection is still unavailable.

Bacterial Diversity metadatabase, Bac*Dive*

Bac*Dive* is the worldwide largest prokaryotic metadatabase, offering data on 82,892 strains comprising 14,208 type strains (Reimer et al., 2019). Founded in 2012 (Söhngen et al., 2014), its first mission was to mobilise, structure and publish data from the internal files of culture collections. Thereby, a broad range data covering taxonomy, morphology, cultivation, metabolism and molecular biology was made available for research. Whilst data from different culture collections (e.g., CIP, CABI, CCUG and DSMZ) made up the backbone of Bac*Dive*, the database evolved over the time and incorporated data from other sources, including the most comprehensive resource of all: primary literature. Today Bac*Dive* offers data within over 600 data fields, including comprehensive data from over 6900 species descriptions, as well as the worldwide largest set of physiological (API stripes) tests. By using global persistent identifiers, such as INSDC sequence accession numbers, EC numbers for enzymes, ChEBI-IDs for metabolites and NCBI taxonomy-IDs, Bac*Dive* offers a high degree of interoperability, which is the prerequisite to combine data obtained from other external sources and analyse them to generate new insights into research data. The data is not only available via the GUI, but also via a Web service for programmatic access (free after registration).

In the future, Bac*Dive* aims to incorporate more data from literature sources by using a supervised text mining approach. Furthermore, functional data from genome annotations with a high degree of confidence are going to be integrated in the database to enable genotype–phenotype comparisons. It is expected that by transforming Bac*Dive* data into machine-interpretable format and connecting it within the LinkedOpenData network the barrier of data integration over several sources will be significantly lowered.

Electronic nomenclature repositories for *Fungi*

The concept of some form of obligatory registration of newly proposed scientific names for fungi goes back to the 1950s (Ainsworth and Ciferri, 1955). The mandatory registration of names was a matter of ongoing debates since then but was only approved by the International Code of Nomenclature of algae, fungi, and plants (ICNafp) in 2011 (discussed in Hawksworth, 2011). In 2004, the CBS-KNAW Fungal Biodiversity Centre (presently Westerdijk Fungal Biodiversity Institute) in Utrecht established an online system to store key information on newly proposed names of fungi – MycoBank (Crous et al., 2004). One of the major problems experienced by mycologists at that time was that many newly published fungal names were not accessible to researchers or simply overlooked, because they were published in obscure sources (Robert et al., 2013). Since 2007, MycoBank has operated under the auspices of the International Mycological Association (IMA). The deposition of the key elements for the valid publication of new scientific names of fungi became mandatory in the ICNafp Melbourne Code (McNeill et al., 2012). The Nomenclature Committee for Fungi appointed three electronic nomenclature repositories, i.e., Fungal Names, Index Fungorum and MycoBank. Electronic repositories register a taxonomic novelty and issue a unique alphanumeric identifier, which must be cited in the publication describing that taxon. The three currently operating electronic repositories exchange the information. In the summer of 2013, a new typification registration system was added to MycoBank to register a type specimen for an existing taxon with a unique alphanumeric identifier.

Nomenclatural services for prokaryotes

A review of the currently valid names of bacteria was made and published under the title of Approved Lists of Bacterial Names (Skerman et al., 1980). These lists became effective on 1 January 1980 and determined priorities for names of new taxa under the ICNP. Numerous taxonomic experts worked towards the approved lists, reviewing about 40,000 bacterial species according to more stringent, modern taxonomic cognition and reduced that number to 4000 names (Fritze, 2006). Although the ICNP does not require a registration of names, two online services provided an overview over taxonomic changes, as a service to the research community.

Prokaryotic Nomenclature Up-to-Date (PNU) was established at the DSMZ in 1993 by Norbert Weiss, Manfred Kracht and Dorothea Gleim. At first it was developed as an internal service to keep species names in the DSMZ catalogue up to date. Over the time, this service was offered to researchers as an up-to-date alphabetical list of names as well as a monthly published list of changes in nomenclature. In 2019, Markus Göker took over and transferred this service on a whole new database system including a semi-automated workflow for processing new names.

The List of Prokaryotic names with Standing in Nomenclature (LPSN, n.d., also known as *bacterio.net*) was established in 1997 by Professor Jean Euzéby as the List of Bacterial names with Standing in Nomenclature (LBSN, Euzéby, 1997). The website developed by Euzéby soon became a key online resource for anyone interested in bacterial and archaeal nomenclature. In 2013, it changed ownership to Aidan C. Parte who continued Euzébys work. But with the rise in the frequency of new prokaryotic names, keeping nomenclatural data up to date became increasingly difficult.

In February 2020, the two nomenclatural services PNU and LPSN merged and published a new resource 'lpsn.dsmz.de'. The **new LPSN** service is based on the newly developed PNU database which was greatly extended. For the merged resource, a new website was developed that succeeds the old LPSN/bacterio.net webpage (Parte et al., 2020). This new website includes additional data like etymology and offers links to type strains in culture collections catalogues as well as to the database Bac*Dive* and the Type Strain Genome Server (tygs.dsmz.de), a tool for genome-based taxonomic analysis. The website offers a responsive design which is optimised for use on mobile devices and the data is also available via a Web service for programmatic access (free after registration).

BioloMICS and culture collections catalogues

Most researchers have no time to spend on the creation of databases and the necessary database creation software. Therefore, BioloMICS was developed as a system that allows any user to create their own custom databases without prior knowledge of databases or programming (https://www.bio-aware.com/). Since its launch in the Westerdijk Fungal Biodiversity Institute, the BioloMICS is used by several culture collections and natural history collections, Collections of the Institut Pasteur, Portuguese Yeast Culture Collection and Atlas of Living Australia, to name a few. The software offers the full spectrum of data fields and tools to cover collection needs.

MycoBank, an on-line mycological database and electronic and nomenclatural repository, is also using BioloMICS since 2010 (Robert et al., 2013). The advantage of the software is that the structure of the database could evolve according to the needs identified by the end-users and the curators of MycoBank (e.g., new tables, new fields). The new BioloMICS-based version of MycoBank has been regularly updated and improved since then. The new software interface was created in order to improve flexibility for queries, i.e., complex Boolean queries by combining AND, OR and NOT together with brackets. MycoBank is not only a repository of data associated with fungal names and vouchers, but also offers unique online pairwise sequence identification services. In addition to several above-mentioned mycological collections, identification services of MycoBank provide access to data from specific scientific databases, such as ISHAM Barcoding Database for Human and Animal Pathogenic Fungi, *Fusarium* MLST database, *Aspergillus* and *Penicillium* database, Indoor molds database, Dermatophyte database, The Yeasts – all running with the BioloMICS software. Users interested in identifying unknown sequences can compare them against all the wanted reference databases at the same time or separately and results are gathered centrally and proposed as a unique matching list (Robert et al., 2013). As a result of the single-platform unification, a sequence search query would return results from GenBank, but also UNITE and sequences from the aforementioned sources, sequences produced in collections and in collaborative scientific projects. When available, MycoBank provides onward links to other databases containing, for example, living cultures, DNA data, reference specimens and pleomorphic names linked to the same holomorph.

Database theyeasts.org. The taxonomic compendium *The Yeasts* was the source of the most up-to-date and complete taxonomic information on all published yeasts, from the first edition in 1952 to the current edition in 2011. The online database theyeasts.org is the successor of *The Yeasts* (Boekhout et al., 2018). The database will use public information on authenticated yeast strains from public culture collections in a species-based electronic information platform. Information from collection databases, clustered according to species boundaries, will include but not limited to physiological properties, most suitable identification tools, occurrence and susceptibility to the most commonly used antifungals. The information provided by the platform will be curated by the scientific community to include most up to date knowledge in the field. Libraries of molecular barcodes, nutritional growth data and MALDI-spectra may be useful for identification of unknown yeast isolates. Besides, protocols for the isolation and identification of yeast isolates will be provided.

CCINFO – a database of culture collections. The international database on culture resources worldwide was established by V. B. D. Skerman (University of Queensland, Australia) and his colleagues in the 1960s (Smith, 2003; Fritze, 2006; Wu et al., 2017) and surveyed culture collections of the International Association of Microbiological Societies (presently International Union of Microbiological Societies, IUMS), and later member collections of the World Federation for Culture Collections (a COMCOF of IUMS). The project resulted in the publication of the World Directory of Collections of Cultures of Microorganisms in 1972 (Martin and Skerman, 1972). The information given on each collection included the name of the collection, postal address, name of the director or head of department, and of the curator if appropriate, the types of culture held (e.g., bacteria, fungi, protozoa etc.) and the number of each type of culture held, together with the availability of the cultures. The list of collections was supplemented by a geographical index of collections and by a list of personnel in alphabetical order. Since the publication of the first World Directory, the database of culture collections has become a part of the World Data Centre for Microorganisms (WDCM) and established the basis for Culture Collections Information Worldwide, CCINFO database. At present, the database provides metadata information on 708 culture collections from 72 countries and regions, of which only 60 have published online catalogues and 96 have their catalogue computerised (Wu et al., 2017). CCINFO serves as a metadata recorder. It provides a unique identifier for each culture collection and lists the organisms in the collection's holdings. The service also provides an overview on preservation techniques used by culture collections. Acronyms of culture collections listed in the World Directory of Collections of Cultures of Microorganisms, presently CCINFO, can be used in abbreviated form to specify the type specimen under the ICNafp (Art. 40.7, Note 4).

In 2008, the web service **StrainInfo.net** was developed that addressed a central issue of microbial researchers: the matching of culture collection identifiers from different culture collections or in other words: 'is the strain in this publication with an ATCC number identical to the strain I got with the LMG number?'. By focussing on most important data like strain identifiers, sequence accession numbers and history data, StrainInfo.net soon became something like a central registry for microbial strains. In later developments, a crowd sourcing system for strain history annotation (Verslyppe et al., 2011) and an electronic passport system for microorganisms (Verslyppe et al., 2014) were developed and integrated. In the beginning StrainInfo.net actively collected data from culture collections, mostly by screen-scraping. Over the time, the workload of keeping the data up to date constantly increased. Therefore, the strategy was adapted and a submission format, the microbial common language (Verslyppe et al., 2010), was developed with the idea that culture collections submit their data on a regular basis. The time has shown that this hardly ever happened. From the year 2014 on StrainInfo.net was not updated anymore, but still heavily used by researchers. In the year 2018, StrainInfo.net went finally offline. This example dramatically shows the huge amount of work that is needed to develop and maintain central resources and the lack of long-term funding to sustain these resources.

Concluding remarks

In the present chapter, we reviewed the progress and challenges of data management in culture collections. Future activities should be directed towards conversion of data into machine-interpretable formats, which not only enable search engines to find data on a deeper level but also allow fast, flexible and dynamic synthesis of data over several sources. However, to achieve this goal, high-level standardisation of data that is currently not reached is essential. Matching strains in different collections will be an ongoing and laborious task that can only be overcome with a unified identifier system and a central registry. Also, journals can help to facilitate text mining by providing keywords, tags and links to collection strains, as, for instance, GenBank does with its source modifiers.

Acknowledgements

Authors thank Imen Nouioui and Brian J. Tindall for their assistance with the literature on history of microbiological research.

References

Ainsworth, G.C., Ciferri, R., 1955. Mycological taxonomic literature and publication. Taxon 4 (6), 3–6.

Beijerinck, M.W., 1889. L'auxanographie ou la méthode de l'hydrodiffusion dans la gélatine appliquée aux recherches microbiologiques. Archives Nérlandaises des Sciences Exactes et Naturelles 23, 367–372.

Boekhout, T., Daniel, H.M., Groenewald, M., Yurkov, A., Bai, F.Y., Robert, V., 2018. The Yeasts, a new online and updatable platform for yeast diversity. Med. Mycol. 56 (Suppl. 1_2). S153-S153.

Casaregola, S., Vasilenko, A., Romano, P., Robert, V., Ozerskaya, S., Kopf, A., Glöckner, F.O., Smith, D., 2016. An Information System for European culture collections: the way forward. Springer Plus 5 (1), 772.

Crous, P.W., Gams, W., Stalpers, J.A., Robert, V., Stegehuis, G., 2004. MycoBank: an online initiative to launch mycology into the 21st century. Stud. Mycol. 50, 19–22.

Euzéby, J.P., 1997. List of bacterial names with standing in nomenclature: a folder available on the internet. Int. J. Syst. Evol. Microbiol. 47 (2), 590–592.

Fritze, D., 2006. The European initiatives MINE, CABRI, EBRCN and ENBI. Microbiol. Austr. 27 (1), 28–30.

Gams, W., Hennebert, G.L., Stalpers, J.A., Janssens, D., Schipper, M.A.A., Smith, J., Yarrow, D., Hawksworth, D.L., 1988. Structuring strain data for storage and retrieval of information on fungi and yeasts in MINE, the Microbial Information Network Europe. Microbiology 134 (6), 1667–1689.

Goodfellow, M., 2000. Microbial systematics: background and uses. In: Priest, F.G., Goodfellow, M. (Eds.), Applied Microbial Systematics. Springer, Dordrecht, pp. 1–18.

Groenewald, M., Boundy-Mills, K., Čadež, N., Endoh, R., Jindamorakot, S., Pohl-Albertyn, C., Rosa, C.A., Turchetti, B., Yurkov, A., 2017. Census of yeasts isolated from natural ecosystem and conserved in worldwide collections. In: Buzzini, P., Lachance, M.A., Yurkov, A. (Eds.), Yeasts in Natural Ecosystems: Diversity. Springer, Cham, pp. 455–476.

Hawksworth, D.L., Schipper, M.A.A., 1989. Criteria for consideration in the accreditation of culture collections participating in MINE, the Microbial Information Network Europe. MIRCEN J. Appl. Microbiol. & Biotechnol. 5 (3), 277–281.

Hawksworth, D.L., 2011. A new dawn for the naming of fungi: impacts of decisions made in Melbourne in July 2011 on the future publication and regulation of fungal names. IMA Fung. 2 (2), 155–162.

Hitchens, A.P., Leikind, M.C., 1939. The introduction of agar-agar into bacteriology. J. Bacteriol. 37 (5), 485–493.

Kurtzman, C.P., Fell, J.W., Boekhout, T., Robert, V., 2011. Methods for isolation, phenotypic characterization and maintenance of yeasts. In: Kurtzman, C.P., Fell, J.W., Boekhout, T. (Eds.), The Yeasts: A Taxonomic Study, fifth ed. Elsevier, Amsterdam, pp. 87–110.

LPSN - List of Prokaryotic names with Standing in Nomenclature, n.d. Available at: https://lpsn.dsmz.de/. (Accessed 16, December 2020).

Lücking, R., Aime, M.C., Robbertse, B., Miller, A.N., Ariyawansa, H.A., et al., 2020. Unambiguous identification of fungi: where do we stand and how accurate and precise is fungal DNA barcoding? IMA Fung. 11 (1), 1–32, 14.

Martin, S.M., Skerman, V.B.D., 1972. World directory of collections of cultures of microorganisms. In: World Directory of Collections of Cultures of Microorganisms. John Wiley & Sons, New York.

McNeill, J., Barrie, F.R., Buck, W.R., Demoulin, V., Greuter, W., Hawksworth, D.L., Herendeen, P.S., Knapp, S., Marhold, K., Prado, J., Prud'homme van Reine, W.F., Smith, G.E., Wiersema, J.H., Turland, N.J. (Eds.), 2012. International Code of Nomenclature for Algae, Fungi, and Plants (Melbourne Code) Adopted by the Eighteenth International Botanical Congress Melbourne, Australia, July 2011. Regnum Vegetabile No. 154. Koeltz Scientific Books, Konigstein.

Meier-Kolthoff, J.P., Auch, A.F., Klenk, H.P., Göker, M., 2013. Genome sequence-based species delimitation with confidence intervals and improved distance functions. BMC Bioinf. 14 (1), 60.

Nouioui, I., Carro, L., García-López, M., Meier-Kolthoff, J.P., Woyke, T., Kyrpides, N.C., Pukall, R., Klenk, H.P., Goodfellow, M., Göker, M., 2018. Genome-based taxonomic classification of the phylum actinobacteria. Front. Microbiol. 9, 2007.

ORCID Open Letter - Publishers, n.d. Available at: https://orcid.org/content/requiring-orcid-publication-workflows-open-letter. (Accessed 16 December 2020).

Overmann, J., Abt, B., Sikorski, J., 2017. Present and future of culturing bacteria. Annu. Rev. Microbiol. 71, 711–730.

Parte, A.C., Carbasse, J.S., Meier-Kolthoff, J.P., Reimer, L.C., Göker, M., 2020. List of prokaryotic names with standing in nomenclature (LPSN) moves to the DSMZ. Int. J. Syst. Evol. Microbiol. 70 (11), 5607–5612.

Reimer, L.C., Vetcininova, A., Carbasse, J.S., Söhngen, C., Gleim, D., Ebeling, C., Overmann, J., 2019. Bac*Dive* in 2019: bacterial phenotypic data for high-throughput biodiversity analysis. Nucleic Acids Res. 47 (D1), D631–D636.

Robert, V., Vu, D., Amor, A.B.H., van de Wiele, N., Brouwer, C., et al., 2013. MycoBank gearing up for new horizons. IMA Fung. 4 (2), 371–379.

Robert, V., Szoke, S., Jabas, B., Vu, D., Chouchen, O., Blom, E., Cardinali, G., 2011. BioloMICS software: biological data management, identification, classification and statistics. Open Appl. Inf. J. 5 (1), 87–98.

Rosselló-Móra, R., Trujillo, M.E., Sutcliffe, I.C., 2017. Introducing a digital protologue: a timely move towards a database-driven systematics of archaea and bacteria. Antonie Leeuwenhoek 110 (4), 455–456.

Schoch, C.L., Seifert, K.A., Huhndorf, S., Robert, V., Spouge, J.L., Levesque, C.A., Chen, W., Fungal Barcoding Consortium, 2012. Nuclear ribosomal internal transcribed spacer (ITS) region as a universal DNA barcode marker for Fungi. Proc. Natl. Acad. Sci. USA 109 (16), 6241–6246.

Skerman, V.B.D., McGowan, V., Sneath, P.H.A., 1980. Approved lists of bacterial names. Int. J. Syst. Bacteriol. 30 (1), 225–230.

Smith, D., 2003. Culture collections over the world. Int. Microbiol. 6 (2), 95–100.

Söhngen, C., Bunk, B., Podstawka, A., Gleim, D., Overmann, J., 2014. Bac*Dive* - the bacterial diversity metadatabase. Nucleic Acids Res. 42 (D1), D592–D599.

Stalpers, J.A., Kracht, M., Janssens, D., De Ley, J., Van Der Toorn, J., Smith, J., Claus, D., Hippe, H., 1990. Structuring strain data for storage and retrieval of information on bacteria in MINE, the Microbial Information Network Europe. Syst. Appl. Microbiol. 13 (1), 92–103.

Verslyppe, B., Kottmann, R., De Smet, W., De Baets, B., De Vos, P., Dawyndt, P., 2010. Microbiological common language (MCL): a standard for electronic information exchange in the microbial commons. Res. Microbiol. 161 (6), 439–445.

Verslyppe, B., De Smet, W., De Baets, B., De Vos, P., Dawyndt, P., 2011. Make Histri: reconstructing the exchange history of bacterial and archaeal type strains. Syst. Appl. Microbiol. 34 (5), 328–336.

Verslyppe, B., De Smet, W., De Baets, B., De Vos, P., Dawyndt, P., 2014. StrainInfo introduces electronic passports for microorganisms. Syst. Appl. Microbiol. 37 (1), 42–50.

Waksman, S.A., Henrici, A.T., 1943. The nomenclature and classification of the actinomycetes. J. Bacteriol. 46 (4), 337–341.

Wickerham, L.J., Burton, K.A., 1948. Carbon assimilation tests for the classification of yeasts. J. Bacteriol. 56, 363–371.

Wu, L., Sun, Q., Sugawara, H., Yang, S., Zhou, Y., et al., 2013. Global catalogue of microorganisms (gcm): a comprehensive database and information retrieval, analysis, and visualization system for microbial resources. BMC Genom. 14 (1), 933.

Wu, L., Sun, Q., Desmeth, P., Sugawara, H., Xu, Z., et al., 2017. World data centre for microorganisms: an information infrastructure to explore and utilize preserved microbial strains worldwide. Nucleic Acids Res. 45 (D1), D611–D618.

Wu, L., McCluskey, K., Desmeth, P., et al., 2018. The global catalogue of microorganisms 10K type strain sequencing project: closing the genomic gaps for the validly published prokaryotic and fungi species. Giga Sci. 7 (5), giy026.

Chapter 7

Quality and competence management in microbial biobanks

Anabela Martins[1], Joaquim F. Martins da Silva[2], Paulo Sampaio[3] and Nelson Lima[1]

[1]*CEB – Biological Engineering Centre, Universidade do Minho, Braga, Portugal;* [2]*Colégio Luso-Internacional de Braga, Braga, Portugal;* [3]*Systems and Production Department, University of Minho, Braga, Portugal*

Chapter outline

Importance of Microbiology Teaching and Microbial Resource Management for Sustainable Futures
https://doi.org/10.1016/B978-0-12-818272-7.00009-2

Introduction

The invention of the microscope in the sixth century, and the report by Antoine van Leeuwenhoek about his first observations of microscopic single-celled organisms in October 1676, opened a window to a new and fascinating universe that has boosted the development of life sciences: *The World of Microorganisms*. If making microorganisms visible was a huge achievement, the long-term preservation of their unique characteristics was an even greater challenge which culture collections (CCs) first embraced 120 years ago (Santos and Lima, 2001).

CC, biological research centres on microbial domain and microbial biobanks are all designations for the systematised *ex-situ* repositories dedicated to the collection and preservation of microorganisms. CCs are a critical component of the scientific infrastructure as they allow us to learn more about microorganism forms, functions, origins and distribution and advance scientific discovery and innovation, whilst studying the impacts of humans on biodiversity, advancing biomedical research, developing improved crops, biocontrol agents and pharmaceuticals, enriching education, connecting communities to nature and science and preserving Earth's biological heritage.

The health of CCs, trustee of our legacy for future generations, relies on the underlying global network that assembles, maintains and provides access to these collections (NASEM, 2020). Networking implies connecting, harmonising, sharing and relying on competence and quality, which poses the question *how can we ensure it globally*?

This chapter presents the framework for the establishment, and permanent update, of the *International Quality Infrastructure*, the biobanking inclusion in such infrastructure and how biobanks can progress in it. It also addresses the added value of standards for CCs and reflects upon the current strategies that the collections are using to manage quality.

The quality infrastruture and role of standards

Being part of the international *Quality Infrastructure* (https://www.inetqi.net/) is one of the most positive and practical steps that a microbial biobank can take on the path forward global recognition of competence and quality. It implies compliance with international standards and is used as a means to harmonise practices, achieve consistent results and improve communication. Being part of the international *Quality Infrastructure* paves the way for cooperation and interoperability.

The *Quality Infrastructure* is a system created worldwide to conquer the confidence of companies/organisations, consumers and governments in each other's work. It covers essential aspects such as policies, high-level objectives, procedures, overseeing-institutions, independence and impartiality and the value-adding use of international standards.

Standards are a key mechanism for the diffusion of technological knowledge (Ernst et al., 2014) with macroeconomic impacts (Blind et al., 2011) by promoting productivity growth and economic advancement (Allen and Sriram, 2000; Tassey, 2000). They may spur innovation by establishing a baseline of accumulated efficient technological experience from which new technologies emerge and by increasing global competitiveness, which in turn boosts innovation.

Additional recognised purposes and advantages of standards are:

a) to enable or facilitate the access to markets in other countries, by reducing the number of tests required in national and international trading – because validity and global comparability of test results is ensured,
b) to transfer and render useful data and other information across geographically dispersed systems, organisations, applications or components (Gasser and Palfrey, 2013) making products compatible and able to interact with other products,
c) to safeguard consumer safety and simplify product development speeding up the time it takes for a product to get to market,
d) to ensure the quality and safety of products and services (CEN-CENLEC, 2013),
e) to help companies complying with relevant international legislation (CEN-CENLEC, 2013).

Standardisation is sometimes accused of decreasing flexibility and potentially blocking innovation; however, it seems that the potential disadvantages are outweighed by the benefits (Allen and Sriram, 2000; Raven and Blind, 2017).

There are two commonly used definitions of 'standard':

> The ISO/IEC Guide 2 defines a standard as a 'document established by consensus and approved by a recognised body, which provides, for common and repeated use, rules, guidelines or characteristics for activities or their results, aimed at the achievement of the optimum degree of order in a given context' (ISO/IEC, 2004). Standards are established by consensus; however, this does not imply unanimity. Consensus is a general agreement, characterised by seeking to take into account the views of all parties concerned and to reconcile any conflicting argument. It implies an absence of sustained opposition to substantial issues by any important part of the concerned interests (ISO/IEC, 2004).
>
> The Agreement on Technical Barriers to Trade (TBT) from the World Trade Organisation (WTO), in its turn, defines 'standard' by highlighting some aspects that are important for the Agreement, such as the fact that they are voluntary, and restricts its definition to products (WTO, 2012). Hence, from the *Quality Infrastructure* perspective, the ISO definition is probably more useful. Nevertheless, the fact that standards in themselves are considered voluntary, as defined in the WTO TBT Agreement, should always be kept in mind.

Several kinds of standards exist. They may be categorised according to the function they need to perform. According to ISO, standards may be categorised as follows: '*Specification*', which sets out detailed requirements often used for safety purposes; '*Methods*' another kind of standard also highly prescriptive that sets out a way of measuring, testing or specifying what is reliably repeatable in different circumstances and places; '*Codes of practice*', standards that recommend sound good practice and have a certain degree of flexibility in the application; '*Vocabulary*', that provides a set of terms and definitions aiming to harmonise language in different fields; '*Guides*', which give less prescriptive advice and reflect the current thinking and practice amongst experts in particular disciplines.

When developing a standard there are, at least, three issues that must be defined:

a) the field of standardisation, which is the domain or group of related subjects to standardise (e.g., engineering, transport, agriculture, quantities and units, biobanking),
b) the subject of standardisation or topic to be standardised (e.g., type of material, component, equipment, system, activity),
c) the specific aim(s) (e.g., protection of the environment, safety, compatibility, variety control) (ISO/IEC, 2004).

When the standardisation activity is open to relevant bodies from all countries, it is considered international standardisation.

How does the quality infrastructure work?

Accreditation bodies and conformity assessment bodies

In modern society, for reasons of safety, health, environmental protection, market fairness or fraud prevention, it is often required to assess and state objectively conformity of services and products, including microorganisms, with specified requirements. Such assessments and the resulting conformity statements are assigned to the *Conformity Assessment Bodies* (CABs). These organisations are responsible for testing products and services for conformity with established requirements and for issuing certificates and reports accordingly.

To ensure that all interested parties have confidence in both the CABs assessment and the assessment outputs (the results issued in CABs certificates and reports), an international accreditation infrastructure was established aiming to ensure CABs' competence, impartiality and capability of acting consistently.

For that purpose, the work performed by each CAB is assessed by an Accreditation Body (AB) which grants accreditation to the CAB. In turn, confidence in ABs' work around the world is achieved through a mutual recognition arrangement (MRA), also known as the *International Laboratory*

Accreditation Cooperation (ILAC) Arrangement. According to this agreement, all the ABs follow the same rules of operation and are peer-evaluated for them. ILAC MRA signatories[1] agree to accept the results of each other's accredited CABs so that the results of each CAB accredited by an ILAC MRA signatory are internationally recognised.

ILAC MRA underpins a global network of laboratories, inspection bodies, proficiency testing providers and reference material producers, that have been accredited to provide accurate and reliable results which are used extensively by regulators for the public benefit in the provision of many services that promote, for example, an unpolluted environment, safe food, clean water, energy, health and social care services. ILAC MRA also enhances international trade through the removal of technical barriers to trade, in the form of re-testing and re-inspection, every time a product enters a new market.

Accreditation is the top level in the Quality Infrastructure. It is ruled by two key international entities: the ILAC and the International Accreditation Forum (IAF). ILAC and IAF work together to coordinate efforts in managing the accreditation and conformity assessment worldwide.

Whilst IAF responsibilities fall in the programmes of the conformity assessment of management systems, products, services, personnel and other similar programmes, ILAC is the international authority on accreditation of laboratories, inspection bodies and other entities serving those through accreditation such as proficiency testing providers and reference material producers.

A diagram representing the international accreditation infrastructure is presented in Fig. 7.1.

The ILAC arrangement

ILAC counts as its members' ABs representing over 124 economies.[2] It manages the MRA, commonly known as the 'ILAC Arrangement', between its members. The agreement is based on the ABs peer-evaluation for compliance with established requirements (e.g., ISO17011 standard) and the mutual acceptance of the results issued by different ABs MRA signatories. In 2020, almost 82,000 laboratories, over 550 proficiency testing providers and almost 200 reference material producers were accredited by the ILAC MRA 102 signatories.[2]

For an AB to be accepted into the ILAC Arrangement it must have been successfully evaluated by peers from another AB for compliance with specific criteria: the ISO/IEC17011 standard and additional mandatory documents (not

1. The complete list of signatories and their scope of accreditation is available at https://ilac.org/ilac-mra-and-signatories/ (accessed 21.01.2021).
2. More information may be found at https://ilac.org/about-ilac/facts-and-figures/ (accessed 21.01.2021).

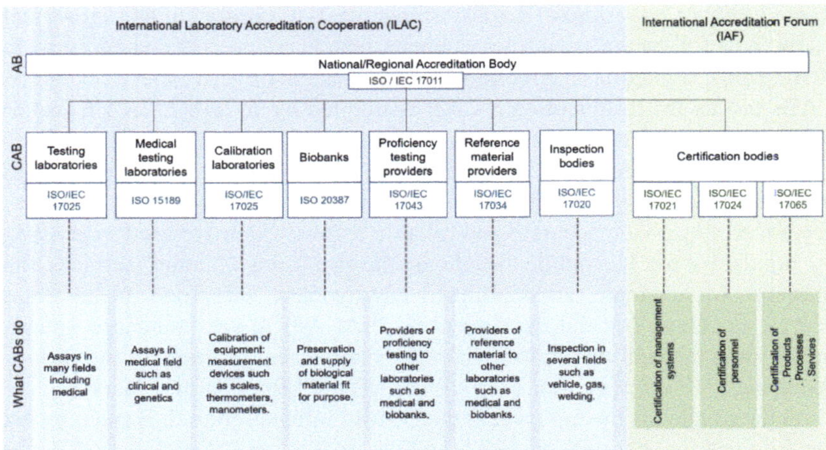

FIGURE 7.1 Diagram describing the international accreditation infrastructure: AB, CAB, the standards they need to conform with and the activities performed by CABs.

included in the ILAC Arrangement Structure [please see 3.3] defined by the IAF and ILAC).

The results of the ILAC MRA signatories routine peer-evaluations are reviewed by a special committee which decides on whether or not the body meets the requirements.

The accreditation programmes included in the Arrangement are under ILAC's responsibility. The procedures to be used by ILAC to consider and approve new international accreditation programmes for which ILAC Arrangement might expand, such as biobanking, are documented and communicated in the ILAC website (ILAC, 2019).

The ILAC arrangement structure

The Arrangement structure establishes the technical areas for which international agreements exist and the criteria for compliance under those agreements. ILAC controls the *Level 1* criteria, the *Level 2* activities and the *Level 3* standards of the Arrangement.

The infrastructure of the Arrangement is represented in Fig. 7.2.

Level 1 contains the criteria with which the AB must comply: the ISO/IEC17011 standard.

Level 2 refers to the conformity assessment activities of a CAB to which the AB grants accreditation according to the generic, normative documents listed in *Level 3*. These activities are Testing, Calibration, Inspection, Proficiency Testing Provision and Reference Material Production.

Level 3 contains the generic normative documents – recognised by the ILAC as meeting the accreditation requirements – used by the AB to assess

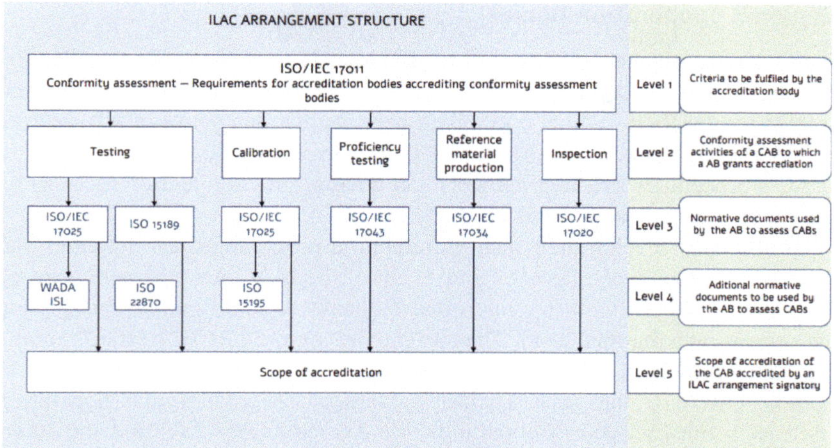

FIGURE 7.2 Structure of the ILAC arrangement. *Adapted from ILAC, 2019. Structure of the ILAC Mutual Recognition Arrangement and Procedure for Expansion of the Scope of the ILAC Arrangement, ILAC-R6:05/2019. ILAC, Silverwater, Australia. https://ilac.org/publications-and-resources/ilac-rules-series/. (accessed 21 January 21).*

CAB's competence for each activity in *Level 2*. These are the standards for CABs currently covered by the ILAC MRA (Fig. 7.2) that are indicated in the ILAC Arrangement Text (ILAC, 2019):

- For testing: ISO/IEC17025 and ISO15189,
- For calibration: ISO/IEC17025,
- For inspection: ISO/IEC17020,
- For proficiency testing providers: ISO/IEC17043,
- For reference material production: ISO17034.

MRA signatories may be recognised for one or more activities in *Level 2* and *Level 3* normative document.

In *Level 4* we have sector-specific normative documents which specify internationally recognised applications of the generic, normative document listed in *Level 3*. These application documents are used by the AB, in combination with the generic, normative documents listed in *Level 3*, to assess the CAB's competence in the relevant sector. To be used in combination with ISO/IEC17025, there is (1) the WADA ISL for anti-doping testing laboratories also accredited by the WADA and (2) the ISO15195 to be used by medical reference measurements laboratories.

For point-of-care testing, there is the ISO22870 to be used in combination with ISO15189.

Level 5 contains the scope of accreditation of the CAB accredited by an ILAC arrangement signatory. *Level 5* is maintained by each ILAC Arrangement signatory as well as *Level 4* to which, each ILAC Arrangement signatory, may add additional requirements to those established by the ILAC arrangement.

Regional cooperation bodies

ILAC builds on the activities of its Regional Cooperation Bodies (RCBs) members. The ILAC MRA builds on existing regional MRAs established worldwide by RCBs which coordinate peer evaluations between AB that are signatories to the regional MRA. RCB recognised by ILAC also adhere to ILAC's procedures and requirements and undergo routine peer evaluations by the members of another RCB or ILAC.

RCBs may develop their own accreditation programmes for activities and standards (respectively, *Levels 2* and *3*) out of the ILAC Arrangement opening a window of opportunity for microbial biobanks to build mutual recognition agreements at a regional level. The currently recognised RCBs of ILAC are the African Accreditation Cooperation (AFRAC), the Arab Accreditation Cooperation (ARAC), the Asia Pacific Laboratory Accreditation Cooperation (APLAC), the European Cooperation for Accreditation (EA) and the Inter-American Accreditation Cooperation (IAAC). The Southern African Development Community in Accreditation (SADCA) is currently in process of developing its MRAs and their associated evaluation procedures to further seeking recognition with ILAC.

The ILAC MRA mark

CABs of ILAC MRA signatories that have signed an agreement with ILAC for the use of the ILAC MRA Mark may use the Accredited CAB Combined ILAC MRA Mark (ILAC MRA Mark in combination with the accreditation symbol which the accredited CAB is entitled to use) (ILAC, 2015). This allows these CABs to provide an instantly recognisable link to the ILAC MRA on their reports and certificates containing the results of activities carried out under the scope of their accreditation as well as, for example, in online applications such as websites and newsletters.

Accreditation

General

Accreditation is a conformity assessment technique specifically related to the assessment of the conformity of a CAB by a third-party entity, the AB.

There is no competition between ABs to avoid the creation of a '*market for accreditation*, leading to the commercialisation of accreditation, which would jeopardise the added value and role of accreditation as a public authority activity and last level of control of the conformity assessment chain'.[3] ABs are

3. More information can be found at Regulation (EC) 765/2008 of the European Parliament and of the Council, of 9 July 2008, Article 7, p. 31. Available at https://eur-lex.europa.eu/legal-content/EN/TXT/PDF/?uri=CELEX:32008R0765&from=PT (accessed 21.01.2021).

though regulated at a technical level to ensure their competence is not compromised by cost optimisation resulting from market competition.

According to the Regulation (EC) 765/2008 of the European Parliament and of the Council, of 9 July 2008, CABs are required to request accreditation by the national AB (NAB) of the Member State in which they are established.[4] However, a CAB may request accreditation with a NAB in another Member State in the following situations:

- there is no NAB in its own Member State,
- the NAB does not offer the requested accreditation service,
- the NAB has not received a positive outcome in the peer-evaluation in relation to the conformity assessment activity for which accreditation is requested.

In such cases, the NAB of the Member State in which the requesting CAB is established may participate as an observer.

The impartiality rule[5] underlying the accreditation process requires that the AB is impartial in deciding on accreditation, meaning that the AB shall not have any conflict of interest with the CAB applying for accreditation, and that the AB shall not be subject to any other pressure of any nature, such as political, commercial or financial. Impartiality also means that the decision of accreditation shall be made by persons different from those having conducted the assessment.

CABs wishing to be accredited are required to comply with appropriate international standards and additional applicable documents for the consistent application of those standards (ILAC, 2019). Accreditation involves the use of auditing techniques by assessment teams which include expertise in both the organisational aspects – such as those related to the management system – and the specific technical activities of the CAB. At the end of the assessment, the evidence data of conformity, gathered by the assessment team, is summarised in a report in sufficient detail. The assessment report may also include the description of any major or minor non-conformities (for which corrections and corrective actions must be defined planned and undertaken by the CAB) and comments/concerns.

An accredited CAB will be provided with an accreditation certificate and an annexe to the certificate where the accredited activities whether they are assays, calibrations, exams, certifications or inspections are to be listed.

4. More information can be found at Regulation (EC) 765/2008 of the European Parliament and of the Council, of 9 July 2008, p. 36 and 27.
5. More information can be found at Regulation (EC) 765/2008 of the European Parliament and of the Council, of 9 July 2008, p. 31 and 37.

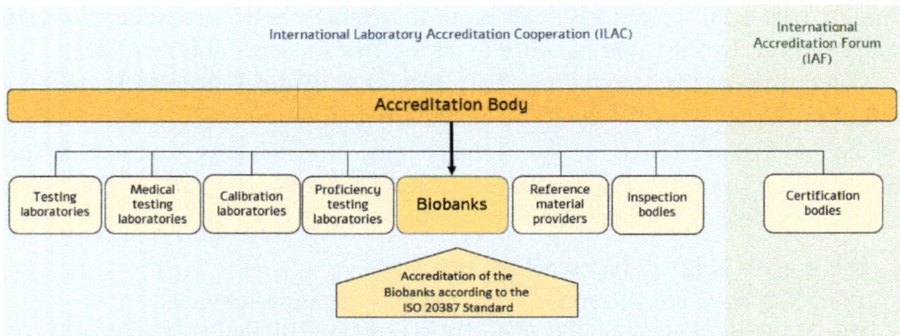

FIGURE 7.3 Diagram representing the entities involved in the accreditation of microbial biobanks under the framework of the international quality infrastructure.

The framework for the accreditation of biobanks is represented in the diagram in Fig. 7.3.

Standards for accreditation

Whilst the development of international standards is assigned to the ISO, the decision of adopting a certain standard for conformity assessment purposes lays on ILAC/IAF.

International standards, intended for both certification and accreditation, are developed in response to a request from industry or other interested parties such as consumer groups. In general, a sector or group communicates the need for a standard to its national standardisation body which in turn contacts ISO. If the proposal meets a market need within a specific area, the development of the standard is assigned to a group of experts from all over the world inside a technical committee. These experts negotiate all aspects of the standard, including its scope, terms and definitions and content. During the process, interested parties are invited to comment on the standard. Their comments are further analysed and taken into account.

The standards intended for accreditation are assigned to the ISO Committee on Conformity Assessment (CASCO). CASCO is responsible for the development of international standards intended for 'the practice of testing, inspection and certification and to the assessment of management systems, testing laboratories, inspection bodies, certification bodies, accreditation bodies and their operations and acceptance'.

However, in very particular cases, accreditation standards may also be developed by technical committees other than CASCO. In such cases, the standard is developed by the assigned technical committee in coordination with CASCO. Some examples include the ISO20387 (ISO, 2018b) and ISO15189 standards which were developed, respectively, by the technical committees (TC) 276-*Biotechnology*[6] and 212-*Clinical laboratory testing and* in vitro *diagnostic test systems*.

The standards intended for conformity assessment have:

- a common structure (general headings/main clause sequence), that should be kept unchanged whilst adapting the sub-headings (ISO, 2020a),
- six common elements ('impartiality', 'confidentiality', 'competence', 'complaints', 'appeals' and 'management system') which include obligatory wording (ISO, 2020b).

Certification

General

Certification is granted by a CAB (certification body) to an entity that proves to comply with the conformity assessment criteria (a specific standard and/or a relevant document). The conformity assessment – which is carried through an audit – is performed to obtain objective evidence of compliance with the auditing criteria and may cover, for a laboratory, more processes than those directly related to assays and tests, considering the total business including strategy and planning.

ISO19011 standard provides guidance on planning and conducting management system [please see 3.7.2] audits (internal or external) including the principles of auditing, how to manage an audit programme and how to evaluate the competence of auditors (ISO, 2018a).

The framework for the certification of biobanks under the ISO9001 (the ISO standard most sought by microbial biobanks) is represented in the diagram in Fig. 7.4.

Standards for certification

There are three main groups of standards liable to certification:

(a) management system standards (MMS),
(b) standards providing certification of people,
(c) standards providing certification of products, services and processes.

MMS are the most sought for certification. They cover many different technical areas[7] namely Quality, Industry, Safety and Security, General Management, Health and Medical, Environmental and Energy, Information Technology and Services.

---

6. More details can be found at http://www.iso.org/iso/home/standards_development/list_of_iso_technical_committees/iso_technical_committee.htm?commid=4514241 (accessed 02.01.2021).
7. More information can be found at https://www.iso.org/management-system-standards-list.html (accessed 21.01.2021).

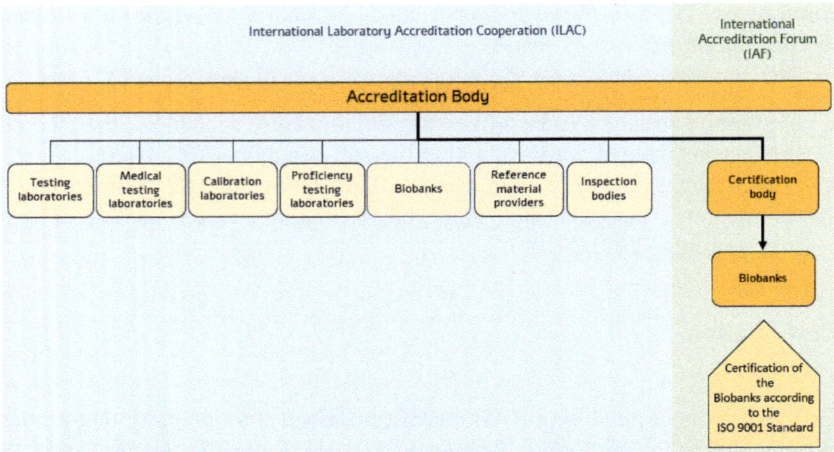

FIGURE 7.4 Diagram representing the entities involved in the certification of microbial biobanks according to the ISO 9001 standard, under the framework of the international quality infrastructure.

A management system may be defined as a set of inter-related elements of an organisation to establish policies and objectives and processes to meet those objectives (ISO, 2015a). In a microbial biobank, a management system supports the management of the interrelated parts of the biobank – equipment, personnel, premises, procedures, methods, policies, for instance – to enable it to achieve its objectives. These objectives may relate to several different topics such as the quality of the biological material preserved and supplied, biosafety, biosecurity, the competence of the personnel, operational efficiency and the microbial biobank overall performance.

MMS may be used with three main purposes: (1) for the establishment and development of a management system, (2) for assessment, by clients or other interested parties and (3) as a criterion for certification. Some benefits of implementing a management system include improvement of the performance, more efficient use of resources, improved risk management and increased customer satisfaction (Domingues et al., 2016).

There are two different types of MMS: A and B. Whilst type A conveys requirements (mandatory provisions) for the management system, type B provides recommendations. Type B MSS are often standards providing recommendations for the implementation of type A MSS.

Organisations may implement more than one MSS (with different aims): quality and environment, for instance, or safety and security. The possibility of easy integration of several management system standards is a long-awaited wish that has recently become reality.

The integration idea started in the early 1990s when the Technical Management Board required both the TC 176 and the TC 207 to work together to ensure alignment of their standards. However, TC 207 chose to go for the Plan-Do-Check-Act (PDCA) approach, whilst TC 176 had chosen the process approach. Those developments caused a lot of delays in making progress on alignment.

The two committees recognised however that they needed to work to (1) creating a common structure (main clause sequence) for their standards and (2) to make their texts as 'identical' as possible. This resulted in the production of the 'Joint Vision' for identical requirements MSS and the High-Level Structure (HLS). The alignment work was expanded by the Joint Technical Coordination Group (JTCG) from ISO resulting in huge progress. Nowadays HLS is mandatory for new MSS and for all the MSS in revision (ISO, 2006).

Standards for microbial biobanks

CCs were formerly considered as individual research initiatives hosted by universities and research institutes. However, the rise in the scientific and biotechnological importance of the biological material along with increased awareness by governments of the necessity of protecting the microbial materials in its three vectors – quality, origin and misuse – has resulted in a global understanding of the need to create stable and well-managed operations inside CC, assuring that only authentic microbial strains of known origin are preserved and supplied.

For at least 3 decades, many CC have been managing their operation under the recommendations of the World Federation for Culture Collections Guidelines (WFCC, 2010). The WFCC Guidelines convey recommendations to assist the affiliate CC in offering services outside their institution (service collections) and also to in-house or research collections.

In 2001, the Organisation for Economic Co-operation and Development (OECD) introduced a new concept for CC – the Biological Resource Centres (BRCs) – referring to the repositories and providers of *high-quality* biological material complying with international standards on quality matters. Also recommended is the creation of a BRC network – the Global Biological Resource Centres Network (GBRCN) – each member working under a national certification or accreditation scheme – supported by national governments – based on scientifically acceptable international criteria (OECD, 2001) and having conformity assessment programmes in place (OECD, 2004).

Six years later, a set of best practice guidelines (BPGs) for BRC – developed and endorsed by the OECD's Committee on Scientific and Technological Policy – were published (OECD, 2007). Even though these guidelines could not be used as a criterion for certification or accreditation, they would later provide the basis to establish a standard, conveying requirements for biobanks in all domains.

In 2008, AFNOR (2008), the French standardisation body, published a standard specifically developed to certify BRCs from human and microorganism domain. The standard was based on the OECD BPG and ISO9001. Brazil also developed a standard involved in microbial biobank accreditation: the NIT-DICLA-061 (INMETRO, 2012).

By the end of 2014, through its TC 276 – Biotechnology, ISO started to develop an international standard for biobanks in four domains – microorganisms, human, animal and plants.

At this point, the term biobank was adopted as the umbrella designation for all the BRCs holding human, microorganism, animal and/or plant resources.

The new ISO standard – published in 2018 – was based in the provisions selected from various standards and best practice documents including, as previously mentioned, the OECD BPG (Furuta and Schacter, 2015). Considering modern needs for the future of quality (Sampaio, 2014) and the fact that TC 276 counts, as its members, countries from five continents, the ongoing standardisation work in biotechnology represents a great global achievement resulting in a unified strategy to ensure quality in biobanking.

The ISO21710 *Biotechnology – Specification on data management and publication in microbial resource centers* was published in 2020 as a result of a project from China. It conveys requirements for data management and publication in microbial biobanks in order to enable consistent formatting, and a quality control workflow to improve the overall quality of data. It also provides recommendations for microbial biobanks to improve data sharing and integration of microbial material and associated data.

Fig. 7.5 displays a timeline presenting the published (and under development) main standards for microbial biobanks.

FIGURE 7.5 Timeline presenting the main standards developed for microbial biobanks.

ISO standards in microbial biobanks

The implementation of standards in microbial biobanks has been dominated by ISO9001 (ISO, 2015b). Some microbial biobanks have opted for ISO/IEC17025 and a few decided to undertake accreditation envisaging to provide proficiency testing.

The ISO9001 standard

ISO9001[8] is the most sought amongst the ISO standards. It bases on the seven quality management principles and the clause structure is built around the PDCA sequence. However, the process approach is still the standard's underlying concept whose understanding and application are improved by the explicit requirements for risk-based thinking. It has currently an improved applicability for services, serving better the microbial biobanks scope.

It is a generic standard that has been applied to BRCs for several years with many advantages, such as (a) helping in establish the biobanking processes, (b) helping qualify the resources, (c) supporting process control, analysis of the results and implementation of measures for process improvement and d) enhancing proactivity by promoting risk-based thinking.

ISO9001 standard provides requirements to manage the flow of information generated through the microbial biobanking activity and to control external providers of products, services and processes. ISO9001 provisions might also support the microbial biobank in establishing a business strategy.

Our interpretation of the ISO 9001 standard model for quality management, showing the overarched activities, is represented in Fig. 7.6.

The ISO/IEC17025 standard

The ISO/IEC17025 standard is not a management system standard although it requires the implementation of a management system. For this reason, ISO/IEC17025 follows a similar management model to that of ISO9001, which is presented in Fig. 7.7. Major differences are the inclusion of provisions for the verification and validation of methods, sampling requirements, calculation of the uncertainty of the testing results and provisions to ensure impartiality and confidentiality.

The ISO20387 standard

The decision of using a certain standard for conformity assessment purposes, such is the case of the accreditation of CABs (microbial biobanks, for

8. The ISO9001 certification is one of the indicators used to explore a broad vision of innovation - Global Innovation Index, available at: www.globalinnovationindex.org/home (accessed 21.01.2021).

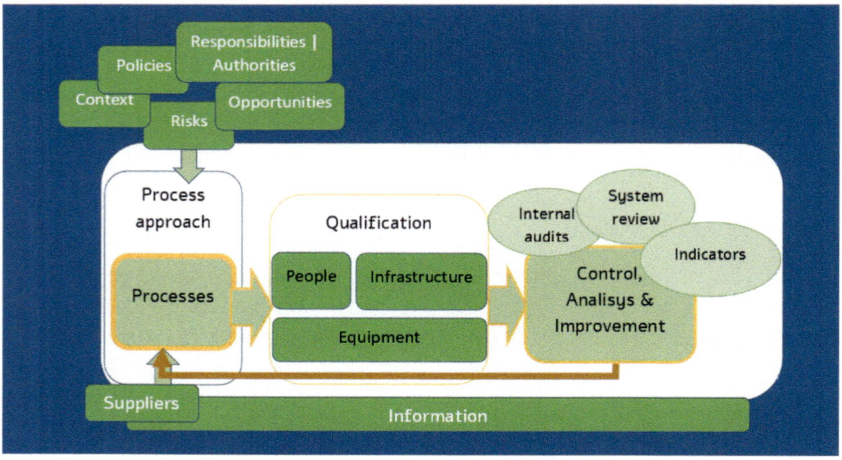

FIGURE 7.6 Diagram representing the activities covered by the ISO 9001:2015 standard and their relationship (management model).

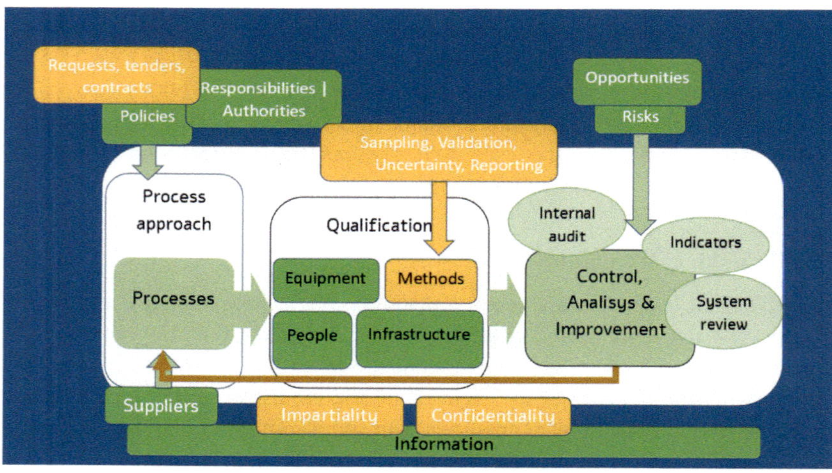

FIGURE 7.7 Diagram representing the activities covered by the ISO/IEC17025:2017 standard and their relationship (management model).

instance), is assigned to ILAC/IAF. Through resolution GA 22.19,[9] ILAC's General Assembly has decided that '*the standard applicable to biobanks for the purposes of accreditation will be ISO 20387 Biobanking – General requirements for biobanking, to be used as a standalone standard*'.

9. Available at https://ilac.org/publications-and-resources/ga-resolutions/ (accessed 21.01.2021).

The ISO20387 standard was developed by the working group 2 of the ISOTC 276, between the years of 2014 and 2018 envisaging biobank accreditation.

The ISO20387 structure and content were based on the 2005 version of ISO/IEC17025 which did not contain the common elements as they were not existing at that time. However, ISO20387 contains requirements on impartiality, confidentiality, complaints appeals and management system. The CASCO secretariat was consulted during the development process confirming appropriateness of the approach.

Quality management model in microbial biobanks

The microorganism preservation lifecycle

The products we use every day have a lifecycle. The lifecycle is the course of events in the products' life from development, use and discontinuation of use (European Commission, 2015). They are produced from raw materials, then are transported to the shops, market, bought and used by consumers and, eventually, eliminated (UNEP/SETAC, 2005). Equal to products, microorganisms used in biotechnology have a lifecycle as well: they are collected from nature, cultured, manipulated, transported, supplied, used and discarded (MCT, 2002). To avoid misunderstanding between the lifecycle in the biotechnology context and the biological lifecycle (that happens to microorganisms in nature), we will designate the former as the microorganism *'technological life-cycle'*. As may be seen in Fig. 7.8, it includes several stages from the microorganism collects from nature to its use and discard.

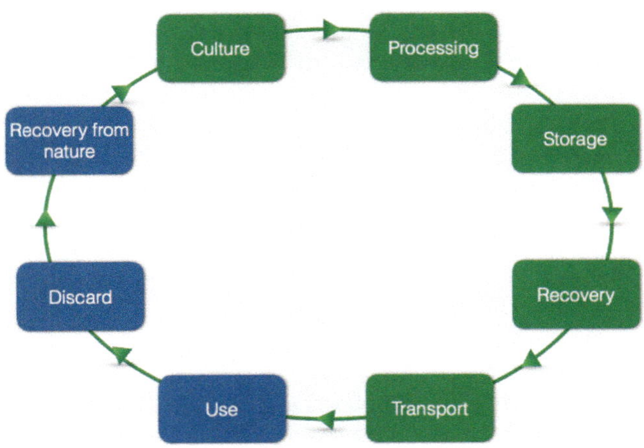

FIGURE 7.8 General description of the microorganism *technological lifecycle.*

Eight different stages can be distinguished throughout the microorganism *technological lifecycle*. During the first and second stages, the microorganism is taken from nature and set to grow. It is cultured by inoculation in a culture medium that will provide the necessary nutrients to support its growth. When appropriate cultures are obtained, it passes to the processing stage. During processing,[10] preservation techniques are applied to the microorganism preparing it to live (without changings) for the long-term storage. Examples of preservation techniques are deep-freeze at −80°C, liquid nitrogen cryopreservation, freeze-drying, amongst others. During this period, the microorganism is stored under specific conditions of temperature, relative humidity and/or other, according to the undertaken preservation technique. When the microorganism is intended for supply, it is submitted to a recovery technique whose purpose is to bring it 'back to life'. After recovery, the microorganism is shipped under specific transport conditions to the final user. After its use (in industry, research and teaching, for instance) the microorganism is discarded (eliminated). Within the microorganism *technological lifecycle*, five stages can be distinguished as accountable to the microbial biobanks: the microorganism culture, processing, storage, recovery and transport to the final user. We will designate these five stages as the microorganism *preservation lifecycle* (Fig. 7.9).

Microbial biobanking goals

Microbial biobanks provide two main types of services: (1) microorganism deposit (microorganism preservation for the long term) and (2) microorganism supply.

The diagram in Fig. 7.10 presents the activities[11] for the 'microorganism supply'. Adopting a process approach,[12] this set of interrelated activities (yellow boxes) might be managed as a process, with inputs, outputs and criteria for the output approval. The process input is the customer request. The process output is the microorganism delivery/supply and the criterion (for the output approval) is the microorganism authenticity (compliance with the supplier's requirements). If no other attribute is expected from the microorganism, the authenticity criteria are identity, purity, and viability.

Failure in this process may be corrected by a new (conform) supply.

---

10. Processing includes (1) the activities performed on microorganisms whose ultimate purpose is to temporarily inactivate its biological activity enabling their long-term storage and (2) the activities performed during recovery [Adapted from the FDA (2011) definition].
11. People collaborating within a process to carry out their daily activities. Some activities are prescribed and depend on the organisation, whilst others are not and react to external stimuli to determine their nature and execution (adapted from ISO, 2015a).
12. Way of managing an organisation as a system of processes. By controlling the interrelationships amongst these processes, the overall performance of the organisation can be enhanced (adapted from ISO, 2015b).

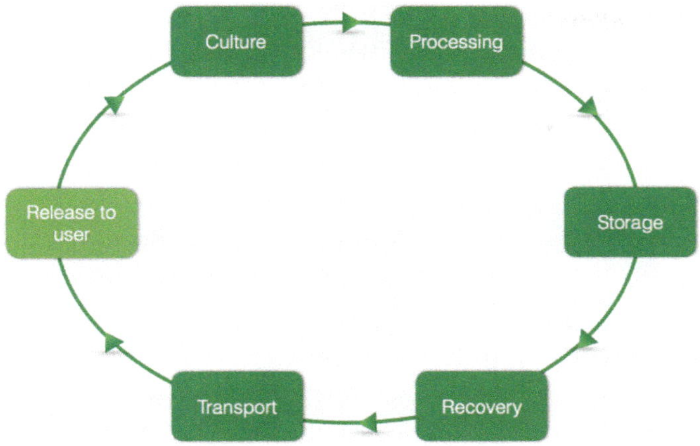

FIGURE 7.9 Microorganism *preservation lifecycle.*

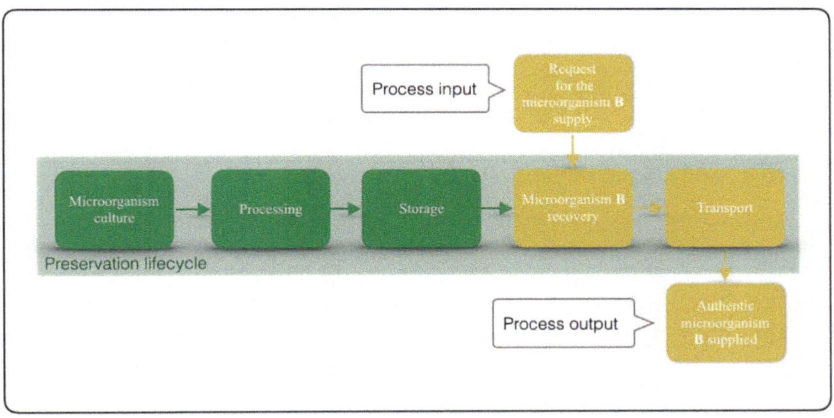

FIGURE 7.10 Sequence of activities for the microorganism supply (yellow boxes); identification of the process input and output.

The diagram in Fig. 7.11 represents the process ‘microorganism deposit’. The process input is the customer’s request for the deposit and the output is the maintenance of the microorganism authenticity during the established period for deposit; the customer may be supplied several times with the microorganism during this period; eventually, the microorganism might be returned to the customer, once finished the established period for deposit. The microbial biobank must ensure that during the deposit period the authenticity of the microorganism is preserved.

Failure in this process may be impossible to correct if the microorganism is unique and/or has unique characteristics.

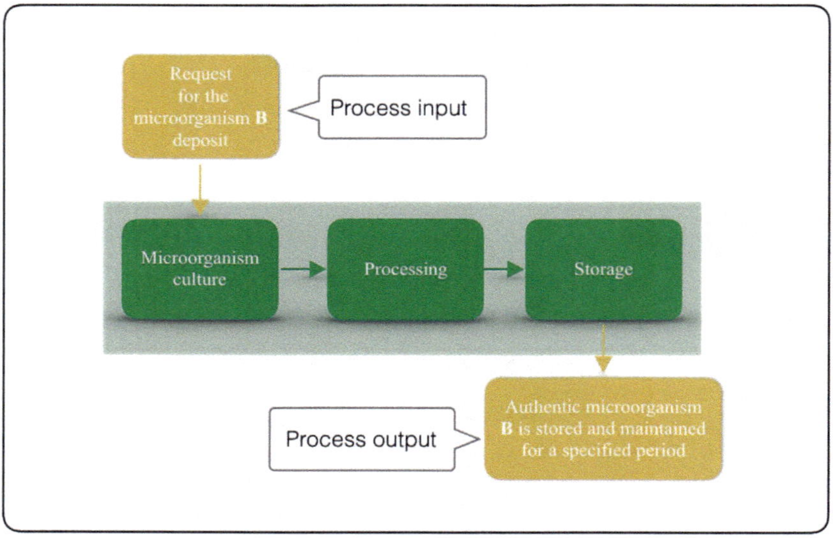

FIGURE 7.11 Sequence of activities involved in the request for the deposit of microorganisms, identification of the process input and output.

Customers' requirements for both services (supply and deposit) and the expected results from the corresponding processes are:

- authentic microorganisms preserved, for deposit process
- authentic microorganisms supplied, for supply process

Lack of consistency[13] in both cases may lead to irretrievable consequences affecting not only the microbial biobank but all the interested parties. One failure may lead to wrong investigation results, inconsistencies in industrial processes, loss of achieved knowledge, loss of unique strains, complaints, reworks, damage in the microbial biobanks' market image, waste of resources and amends resulting from accountability issues.

In sum, the biobanking quality[14] depends much on the consistent achievement of the following objectives:

a) Ensure that the processed microorganisms to preserve are authentic,
b) Ensure the supplied microorganisms are authentic,
c) Ensure the authenticity over time of the deposited microorganisms.

13. For this work, process consistency is defined as the degree to which a process leads to identical results over time.
14. The biobanks' ability to satisfy customers/users and the intended and unintended impact on relevant interested parties (adapted from ISO, 2015a).

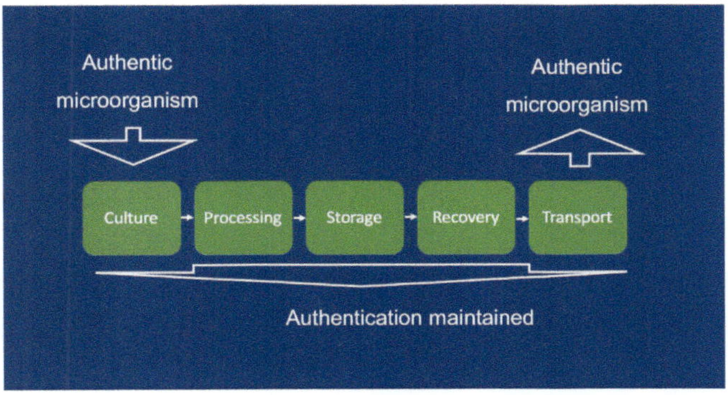

FIGURE 7.12 Objectives/targets for the (quality) management model in microbial biobanks.

To consistently achieve *a)* and *b),* proper testing making use of proper resources (competent[15] personnel and controlled[16] equipment) must be in place.

To consistently achieve *c)*, high process capability[17] throughout the *preservation lifecycle* must be ensured. Processes should operate in such a way that predictable outcomes (results complying with the requirements) are ensured.

The management model[18] in place must then focus on 'effectiveness'[19] and 'consistency' giving the microbial biobank and the interested parties confidence about the management system's ability to (1) preserve at all times the authenticity of the microorganisms and (2) supply authentic microorganisms.

The microbial biobank management-goals alongside the microorganism *preservation lifecycle*, are represented in Fig. 7.12.

Current management approach in microbial biobanks

The diagram in Fig. 7.13 presents the current management model for microbial biobanks to manage the quality of their activities throughout the microorganism lifecycle.

---

15. Competence is the ability to apply knowledge and skills to achieve intended results (ISO, 2018a,b).
16. Equipment managed in such a way as to ensure it provides measurement results with known error and uncertainty (obtained through calibration) and/or under periodic preventive maintenance.
17. Process capability is a measurable parameter related to process variability over time (e.g., number of defective units produced per million); higher process capability is related to lower variability.
18. For this work, management model is the set of methods (included in a management system) established to attain defined goals.
19. Process effectiveness is the extent to which planned activities are realised and planned results achieved (ISO, 2015a).

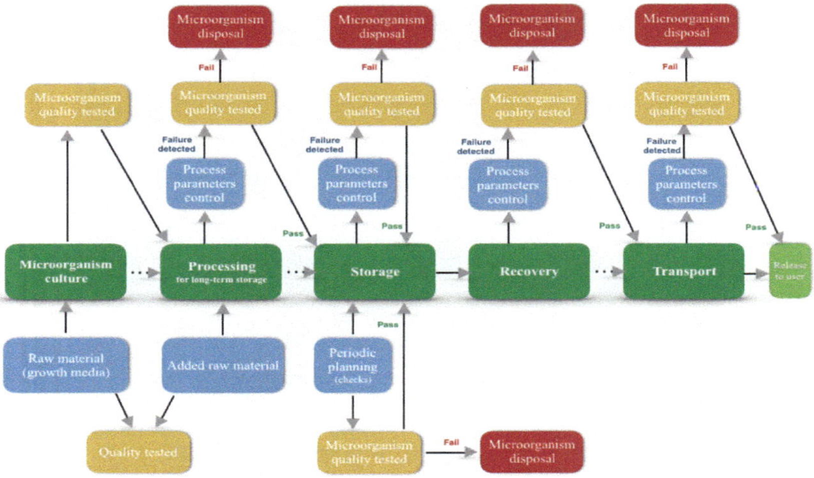

FIGURE 7.13 Diagram representing the current paradigm for quality management in microbial biobanks, based in the microorganism preservation lifecycle.

At stage one, a microorganism is accepted for deposit. It is cultured and tested for authenticity. In case of failure during the testing for authentication, a wrongly authenticated microorganism is at risk of being preserved (different identity, not pure, not viable or loss of a particular characteristic). The motive of failure in authentication might be related to several different causes such as the testing method, the equipment and the operator.

All raw materials entering the process are tested before use (inside or outside the biobank); raw material shall only be accepted if compliance with the requirements has been proven.

The second stage on the microorganism preservation lifecycle – processing – is undertaken to enable the microorganism long-term preservation.[20] The activities undertaken during the processing stage are well established so as to attain conformity and activities are fixed and the process factors[21] (inputs) are stringently set (no deviation is permitted for both). Several control points (measuring and monitoring activities) are implemented to early detect possible non-conformities. This enables the microbial biobank to timely re-establish conformity in case of any deviation, to avoid damage in microorganism

20. For this work, 'to preserve' means to maintain the biological material in such specific conditions that its authenticity is ensured. It includes any necessary processing activity and storage.

21. A process is a transformation of inputs (factors or process variables such as people, materials, methods, environment, machines and procedures) in outputs (such as performance or quality characteristics of a product). Factors can be either qualitative - the equipment A, B or C - or quantitative - a temperature of 10, 15 or 20°C, for example (Moen et al., 1991).

integrity. If any failure is detected, microorganisms could have been affected, so testing is made to identify possible loss of microorganism authenticity.

During stage 3 – storage[22] – the microorganisms under preservation are tested for authenticity at planned intervals. This monitoring (and measuring) procedure enables the microbial biobank to detect any consequence of undetected failures.

Stage 4 – microorganism recovery – begins with a requested for supply. As in stage 2, the recovery activities (procedures) are very well established, the process variables stringently set, and control points are implemented to detect any failure.

Before supply, microorganisms are tested for its authenticity and subsequently transported to the user. During transport, the procedures are fixed, the process parameters are stringently set and, if necessary, measured and monitored.

Red boxes represent microorganism loss, non-conforming product. In general, failure is detected by control activities represented by the blue boxes: each process variable is monitored and measured, and the results interpreted according to the criteria for acceptance. If some process variable is out of specification, correction and corrective actions are triggered and the processes may be readjusted as a consequence.

Red box non-conformities born mostly at the green level. They may arise from failure during culture, processing, storage, recovery and transport to the user. These failures are not predictable as they result from input variability (operator, equipment, materials and premises). By the time a failure is detected, any process parameter might already be out of specification, so microorganisms might be already damaged.

According to the current management model in microbial biobanks, represented in Fig. 7.12,

a) Consistency in preservation for the long term [objective (c)] is not ensured as microorganism might be lost at almost all stages.
b) Consistency in preserving and supplying authentic microorganisms [objectives a) and b)] might be achieved as long as valid results are ensured from testing; this might be achieved by verification and validation of methods/examinations, as required by ISO/IEC20387 and ISO/IEC17025.

Variability results in lack of consistency so, as microbial biobank goal is to achieve consistency, rather than focussing on 'variability detection', microbial biobanking management model should focus on 'variability control'.

22. For this work, 'to store' means to maintain/keep the preserved biological material for future use; storage conditions must be such as to ensure microorganism authenticity during the storage period (e.g., labelling, temperature, humidity, pressure, traceability).

Variability detection is in its essence related to a quality model known as quality by testing (QbT), whilst variability control is related to a model called quality by design (QbD).

Quality by testing

The microbial biobanks' current paradigm to achieve quality is a classic, empiric, reactive approach known as QbT according to which 'product' quality (final product and in-process product) is assured by the combination of fixed (and thus inflexible) manufacturing steps and extensive testing carried out through the manufacturing process.

The process through which the 'product' is built or transformed is well established and the activities, parameters and product quality characteristics are stringently settled and tightly controlled as they are used to ensure process consistency (Yu, 2008).

The quality of used raw materials/reagents is monitored by testing to verify if specifications (quality criteria) are met. The final product is tested for conformity with specifications before release, an operation known as verification[23] (Maropoulos and Ceglarek, 2010). Final product specifications are, most of the times, set out by observing data from a small number of batches believed to be acceptable and then establishing the acceptance criteria for future batches (Yu, 2008).

QbT is a reactive approach to quality management: when final product fails to comply with the specification, (1) the corresponding batch is discarded, (2) the root cause of the failure is identified and corrected and (3) steps on the process might be modified in order to obtain testing results within the predefined limits.

Some of the characteristics found in QbT approach are as follows (Yu, 2008):

- Root cause for failure is usually not well understood, which can lead to reoccurrence,
- Production can be halted until root causes of failure are understood and addressed or acceptance criteria revised,
- Limited characterisation of variability,
- Lack of process understanding,
- Changes in the process will usually require more testing and additional controls.

Rather than striving for failure identification and correction, microbial biobanks should strive to prevent failure opting for a proactive approach. Proactive approaches to quality are, in general, related to early stages of the

23. Verification is a quality control procedure; it is used to evaluate whether or not a product, service or system complies with requirements imposed at the start of a development phase.

product lifecycle, whilst reactive approaches are more related to the last stages of the product lifecycle (Kolaric, 1995).

Quality by design

QbD, first outlined by Joseph Moses Juran (1992), is a holistic, systematic, scientific, risk-based, proactive tool for quality assurance which enables to plan, set and reach quality goals. This approach is based on understanding and controlling product characteristics and manufacturing variables. The principles underpinning the QbD approach have been increasingly used to advance the product and process quality in many industries (Skrdla et al., 2009; Sangshetti et al., 2017). In pharmaceutical industry, it is being used with several purposes such as formula optimisation (Girotra et al., 2016), scale-up activities (Agrawal and Pandey, 2015) and monitoring operations (Largoni et al., 2015). It has also been considered as a strategy for analytical method development in robustness domain (Debrus et al., 2011; Rozet el al., 2013; Boussès el al., 2015; Mallik et al., 2015).

The implementation of QbD begins by establishing the project objectives (generally the targeted product quality profile) and continues by seeking product and process understanding through the acquisition of data/information in a knowledge-driven manner. The understanding of all operation is explored and maximised within three main stages: (1) identification of the raw materials' attributes and the critical process parameters[24] (CPPs) that might influence the product critical quality attributes[25] (CQAs), (2) determination of the functional relationships linking raw materials' attributes and process parameters to the product CQA and (3) the design and development of the manufacturing process (Verma et al., 2009).

Having its manufacturing process designed, the organisation is provided, on a regular basis, with information resulting from process measuring and monitoring activities. As a result, both product and manufacturing process can be updated in an ongoing basis to support new knowledge, contributing to a consistent quality (Zidan et al., 2007).

The consistency achieved through QbD comes from the design and control of the manufacturing process: product quality is assured by controlling product attributes and manufacturing variables. Final product testing is not part of the process consistency or process control; it is merely used to confirm the product's quality (Yu, 2008).

---

24. In microbial biobanking context, CPP are the process variables that have significant impact on the microorganism attributes.
25. In microbial biobanking context, the CQA are the microorganism attributes which grant it authenticity.

The proactive essence of QbD comes from identifying steps in the process where failure could result in the inability to meet quality targets, during the design phase.

In sum, QbD includes the following activities (Yu, 2008; Badawy et al., 2012):

1. Definition of the target product quality profile,[26]
2. Design of both product and manufacturing processes,
3. Identification of the raw material attributes, process parameters, sources of variability and the product CQA,
4. Identification of the potential CPP (building understanding of the impact of process parameters and raw materials on CQA),
5. Control of those critical sources of process variability (that might have an impact on product quality),
6. Establishment of a control[27] strategy,
7. Continuous control of processes using suitable control strategies and process improvement to produce consistent quality over time.

QbD makes use of several tools such as knowledge management, risk assessment, experimentation and design space. Risk assessment is performed to identify the main variables influencing the product quality attributes. Design of experiments studies is done to (1) determine the most significant factors (variables) affecting product quality and process parameters and (2) to obtain the exact relationship between the product CQA and the various factors (Kan et al., 2014). Afterwards, the design space is established. It is the multidimensional combination of input variables that have been demonstrated to provide quality assurance; in other words, the design space includes the critical factors and their acceptable ranges. The ideal design space will be one in which the process parameters have no impact on the process stability and product quality.

Under both approaches QbT and QbD, the role that the product specification (criteria for quality) plays is completely different. Whilst under QbT each batch must be tested to assure its quality and manufacturing consistency, under QbD batches may not be tested as process understanding and control provides sufficient evidence that the product will meet specification (if tested). This allows the final products' real-time release. Specifications for the final product are solely used for the confirmation of product quality, not to assure manufacturing consistency and process control (Yu, 2008).

To facilitate the understanding of differences between QbT and QbD, a diagram contrasting both approaches is presented in Fig. 7.14.

26. In the microbial biobanking context, the product profile includes the attributes that grant it authenticity.
27. For this work, 'control' means verification that the acceptance criteria are met (which may be accomplished by measuring), monitoring and interpreting the results.

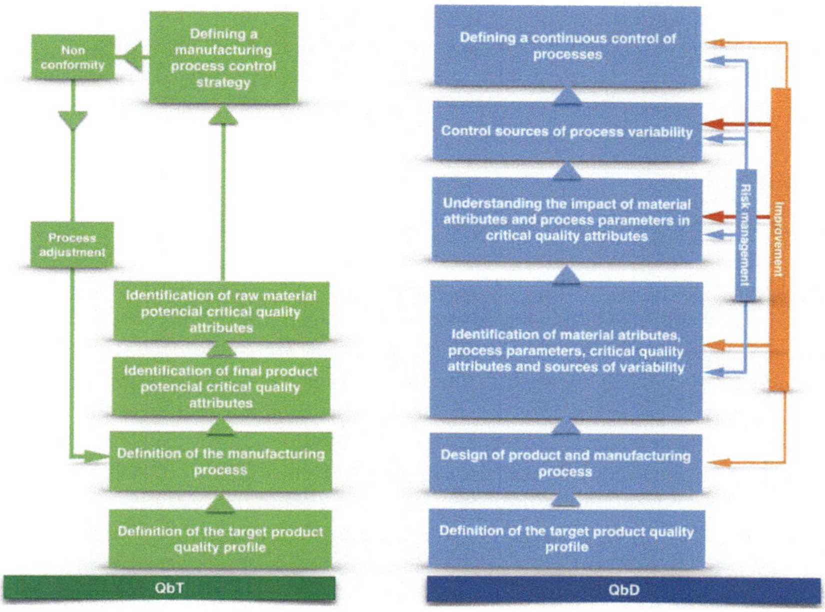

FIGURE 7.14 Comparison of the quality by testing and quality by design approaches.

Standards' centrepieces

The most commonly implemented standards in microbial biobanks have a set of common elements (many of them integrating management system standards); their rationale and application is briefly presented in the following chapters.

Impartiality

Impartiality is defined as *objectivity with regard to the outcome of a conformity assessment activity* (ISO, 2020b). It is key to build trust, confidence and ensure value in the conformity assessment activity. For this reason, impartiality is a mandatory element in accreditation standards.

Impartiality builds upon three main components: objectivity, independence and absence of conflict of interests. To ensure that conformity assessment activities are undertaken objectively, the microbial biobank must have in place clear procedures and clear criteria for decisions. Clear procedures and criteria will also ensure independence which is the *freedom of a person or organisation from the control or authority of another person or organisation* (ISO, 2020b). Independence presumes that effective separation between the microbial biobank and others who have an interest in the microbial biobank results exists. Regarding the conflicts of interests, microbial biobanks need first to identify existing and potential conflicts of interests and then manage them.

The microbial biobanks should raise awareness of their personnel about the responsibility and liability that comes with the biobank activities and decisions/attestations by promoting training and awareness sessions and bound their personnel to impartiality through a declaration of impartiality.

Confidentiality

To carry out their tasks effectively, many times microbial biobanks need to gain access to information from their customers/users and provide them confidence that confidential information will be protected and not disclosed. To do so, the biobank must identify which information must be maintained confidential and ensure that protection measures are implemented. Examples of such measures are the establishment of levels of access, giving access to confidential information only to those who need it, and to bound the personnel to confidentiality through a declaration of confidentiality.

Also, during the normal activity of the microbial biobank, external personnel including contractors, personnel of external bodies or persons acting on behalf of the biobank may access or obtain confidential information. For this reason, personnel from external entities also ought to be bound to confidentiality.

In case of having information about a person from a source other than the person as is the case of a regulator or a complainant, the microbial biobank should maintain confidential the identity of the source, unless otherwise agreed by the source.

Handling of appeals and complaints

An appeal is a request of the microbial biobank's user/customer to the microbial biobank for the reconsideration of a decision made by the microbial biobank relating to the object of conformity assessment provided by the user/customer.

Appeals ought to be processed appropriately as the effective resolution of the appeals strengthens the protection of the microbial biobank against errors or omissions. To ensure appeals are handled effectively, the biobank must consider establishing a method for receiving, acknowledging, investigating and undertaking actions for response (how to resolve and prevent reoccurrence, if applicable) ensuring the actions were effective and providing the appellant with the outcome. In the interests of transparency and objectivity, the process for handling appeals should be publicly available. In addition, to ensure an independent review, persons involved in the decision/result that might be challenged via an appeal process should not be involved in the appeal process.

A decision on the appeal is an internal process of the microbial biobank and should be recorded to facilitate transparent and consistent decision-making. Retaining records from appeals also allows the microbial biobank to identify trends and risks.

Handling of complaints

A complaint is any expression of dissatisfaction with the activities of the CAB other than the appeal (ISO, 2020b). All the complaints that are related to the microbial biobank conformity assessment activity ought to be resolved. The biobank should gather all necessary information to determine whether the complaint is substantiated and should follow a method similar to that established for the purpose of handling appeals.

Competence

Competence is key in order to provide confidence in the conduct and outcome of the conformity assessment activity. In the context of accreditation, competence is understood as applying to both microbial biobank and its personnel.

The microbial biobank is responsible for and must take actions to ensure the competent conduct of its operation. To do so, the microbial biobank must ensure that each activity is performed by a competent person and should monitor this competence over time.

Personnel knowledge and competence

The microbial biobank must ensure that personnel doing work under its control, regardless of their employment status, are competent. All personnel should have their competences assessed.

Educational and competence requirements should be determined for each function according to the existing qualification schemes and must be reviewed periodically. Some responsibilities or authority levels may require additional training such is the case of personnel involved in the management system or controlling the testing equipment. Training (whether practical or theoretical) should be provided, and the effectiveness of the training must be reviewed to ensure the necessary competence has been achieved.

Management system

The accreditation standards include provisions to implement a management system (MS) as part of their overall requirements. The reason is that management systems are a recognised tool to support the consistent fulfilment of requirements for CABs and their activities (ISO, 2020b). For this reason, MS requirements are mandatory for ISO/CASCO documents intended for CABs although compliance with the accreditation standard does not imply compliance with any MS standard such as ISO 9001.

In general, to implement the necessary MS provisions, the microbial biobank should start by (1) determining the processes needed for the MS and (2) undertaking their characterisation, sequence and interaction, should define the best method to monitor each process performance (ensuring effectiveness), determine the resources needed (including the documented information needed to support the operation) and the responsibilities and authorities for each of these processes. The microbial biobank should also determine the risks and opportunities for the processes, assess the risks and implement any changes needed to improve the processes and the overall MS performance.

Performance indicators

The sustainable success of a microbial biobank depends much on its ability to satisfy the needs and expectations of all its interested parties in a balanced way. Examples of interested parties are the users of the biological material the microbial biobank provides, governments, regulators, the personnel, suppliers, investors, owners or management board and the public.

Key Performance Indicators (KPIs), also known simply as indicators, are widely used to evaluate the performance of organisations. A KPI is a measurable value that demonstrates how effectively an organisation is achieving key business objectives. Microbial biobanks may use KPIs to evaluate their success at reaching the defined targets and monitor their KPIs progress over time — KPIs are constantly measured and monitored.

KPIs may (and should) be defined at multiple levels:

- *High-level KPIs* focusing on the overall performance of mBb,
- *Low-level KPIs* with focus on the different processes such as operational processes (biological material reception, preservation, storage and supply) or support processes (human resources, infrastructure maintenance, control of the equipment, marketing, internal audit, information management).

Microbial biobanks may select, monitor and analyse any information (data) on the performance of their activities. The nature and amount of such data depend on many factors such as the mBb culture and goals, external requirements and level of IT capability. It is important to ensure that the selected data is comprehensive, reliable, appropriately analysed and acted upon.

For microbial biobanks targeting certification or accreditation, adopting the following indicators may be appropriate:

Client feedback, to measure the effectiveness in meeting the client needs and expectations:

- Complaints, including volume, frequency, cause, adequacy of implemented corrective actions,
- Audit findings: analysing trends in positive and negative audit findings,
- Non-conformities: including volume, frequency, cause, adequacy of implemented corrective actions,
- Training, including trends on training areas, personnel, competence evaluation;

- Biological material reception, preservation and supply, including biological material not accepted, biological material loss, environmental conditions monitoring, quality-check results,
- Equipment failure,
- Mitigated risks.

Risk management

Risk management is the identification, assessment and prioritisation of risks followed by coordinated and application of resources to minimise, monitor and control likelihood or impact of unfortunate events. The risks must be identified, assessed, prioritised and mitigated. Concerning the positive risks (opportunities), its identification should be followed by a coordinated application of resources to maximise the realisation of the opportunities.

At operational level, the microbial biobank needs to evaluate the impact of potential failures on testing results as they might affect the microorganism authenticity, the safety of the personnel and the process effectiveness. Processes must be changed accordingly to reduce or eliminate potential risks.

Internal audits

Internal audits are an effective[28] and reliable tool to provide information on which the microbial biobank can act to improve its performance.

Internal audits might be conducted against a range of audit criteria,[29] separately or in combination, including requirements defined by the mBb, policies and provisions defined by interested parties, contractual obligations, statutory and regulatory requirements and standards (ISO, 2018a,b).

The microbial biobank should define an audit programme. The audit programme generally includes the arrangements for a set of one or more audits, the time frame for execution and the purpose. The audit programme should include, at least, the objectives of the audit, the audit criteria, the resources needed, the scheduling, the audit team, the methods and processes for handling confidentiality, information security, health and safety (ISO/IAF, 2016).

Although the focus of a management system audit is to determine if the management system has been developed, is effectively implemented and is being maintained (ISO/IAF, 2016), with emphasis on conformity instead of documenting non-conformities,[30] when the audit evidence indicates a non-conformity, it must be documented and a plan for corrective actions should be developed.

---

28. Effectiveness is defined as the extent to which planned activities are realised and planned results achieved (ISO, 2018a,b).
29. Audit criteria may be defined as the set of requirements used as a reference against which objective evidence is compared (ISO, 2018a,b).
30. Non-fulfilment of a requirement (ISO, 2015a).

To be well document, the description of the nonconformity must address the following:

- the statement of non-conformity (precise, self-explanatory, unambiguous and concise),
- the requirement that is not being fulfilled (the exact text in the source – audit criteria),
- the audit evidence that supports the auditor findings (sufficiently detailed).

Correction and corrective actions

Correction and corrective actions are the microbial biobank response to identified nonconformities whether they are identified as a result of an audit or during daily work. When nonconformities are well addressed, organisations will increase the likelihood of achieving customer and interested parties' satisfaction.

In addressing a non-conformity, the following actions must be taken:

- Correction: action(s) implemented to eliminate the detected nonconformity,
- Cause analysis: action taken to determine the cause of the problem,
- Corrective action: action(s) implemented to eliminate the cause of the problem to prevent recurrence,
- Evaluation of the effectiveness of the corrective action.

Cause analysis is key in eliminating the nonconformity. Corrective actions will be effective if the true cause is identified. Many methods may be employed in determining the cause of a non-conformity such as *root-cause analysis*, *Ishikawa diagram* and *five-whys*.

The actions taken to resolve and prevent recurrence of nonconformities should be recorded for future reference. Records should be clear and concise and provide a thorough description of the actions taken as well as the date of completion. Evidence supporting the claim that the corrective action has been fully and effectively implemented should also be recorded.

Final remarks

CCs sustain the priceless microbial diversity that is our heritage and our legacy to future generations. They provide researchers with valuable assets to improve food security, environmental remediation, agriculture crops, national defence and their therapeutical arsenal. Along with other biological collections, they provide the specimens needed to understand how infectious diseases emerge and how they might be mitigated before reaching the destructive level of the modern-day COVID-19 pandemic.

With the improved ability to store, access and use microorganisms, that was achieved with the new methods of automation, preservation, information extraction, data integration and related technologies, standardisation gains more significance, especially if it grants CCs the access to harmonisation and global recognition of compliance and the access to the international quality infrastructure.

We hope that the content presented here may contribute to clarify (1) the framework for the establishment of the international quality infrastructure, the role it plays at a global level, (2) the benefits it might bring to enhance CC cooperation, how they communicate with users and respond to their needs (3) and some of the management requirements provided by standards applicable to microbial biobanks. Most of all, the content may aid CCs to advance in line with the legislative frameworks.

Acknowledgements

This research had partial financial support of Portuguese national funds through the FCT (Foundation for Science and Technology) within the framework of the CEB project UIDB/04469/2020. It also had the partial support of the European Union's Horizon 2020 research and innovation programme under grant agreement No 871129 - IS_MIRRI21 Project.

References

Agrawal, A.M., Pandey, P., 2015. Scale up of pan coating process using quality by design principles. J. Pharmaceut. Sci. 104 (11), 3589–3611.

Allen, R., Sriram, R., 2000. The role of standards in innovation. Technol. Forecast. Soc. Change 64, 171–181.

AFNOR, 2008. NF S96-900 - Qualité des Collections de Ressources Biologiques (CRB) — Système de management d'un CRB et qualité des ressources biologiques d'origine humaine et microbienne. Association Française de Normalisation, Paris, France.

Badawy, S.I.F., Narang, A.S., LaMarche, K., Subramanian, G., Varia, A., 2012. Mechanistic basis for the effects of process parameters on quality attributes in high shear wet granulation. Int. J. Pharm. 439, 324–333.

Blind, K., Jungmittag, A., Mangelsdorf, A., 2011. The Economic Benefits of Standardization. DIN German Institute for Standardization, Berlin.

Boussès, C., Ferey, L., Vedrines, E., Gaudin, K., 2015. Using an innovative combination of quality-by-design and green analytical chemistry approaches for the development of a stability indicating UHPLC method in pharmaceutical products. J. Pharmaceut. Biomed. Anal. 115 (10), 114–122.

CEN-CENLEC, 2013. Standards and Your Business. European Committee for Standardization/European Committee for Electrotechnical Standardization, Brussels — Belgium. http://www.cencenelec.eu/News/Publications/Publications/Standards-and-your-business_2013-09.pdf. (Accessed 21 January 2021).

Debrus, B., Lebrun, P., Ceccato, A., Caliaro, G., Rozet, E., Nistor, I., Oprean, R., Rupérez, F.J., Barbas, C., Boulanger, B., Hubert, P., 2011. Application of new methodologies based on design of experiments, independent component analysis and design space for robust optimization in liquid chromatography. Anal. Chim. Acta 691, 33–42.

Domingues, P., Sampaio, P., Arezes, P.M., 2016. Integrated management systems assessment: a maturity model proposal. J. Clean. Prod. 124, 164–174. https://doi.org/10.1016/j.jclepro.2016.02.103. (Accessed 21 January 2021).

Ernst, D., Lee, H., Kwak, J., 2014. Standards, innovation, and latecomer economic development: conceptual issues and policy challenges. Telecommun. Pol. 38 (10), 853–862.

European Commission, 2015. EU Guidelines for Good Manufacturing Practice for Medicinal Products for Human and Veterinary Use. Annex 15: Qualification and Validation. http://ec.europa.eu/health/documents/eudralex/vol-4/index_en.htm. (Accessed 21 January 2021).

FDA, 2011. Process Validation: General Principles and Practices Current Good Manufacturing Practices, Rev 1, Guidance for Industry. US Food and Drug Administration, Rockville (MD) US.

Furuta, K., Schacter, B., 2015. Report on status of ISO276/WG2 on biobanks and bioresources: international standards for biobanking. Biopreserv. Biobanking 13 (6), 452–453.

Gasser, U., Palfrey, J., 2013. Fostering innovation and trade in the Global Information Society: the different facets and roles of interoperability. Law, Pol. Econ. Tech. Stand. J. 1 (1), 13.

Girotra, P., Kumar, S.S., Kumar, G., 2016. Development of zolmitriptan loaded PLGA/poloxamer nanoparticles for migraine using quality by design approach. Int. J. Biol. Macromol. 85, 92–101.

ILAC, 2015. Rules for the Use of the ILAC MRA Mark, ILAC-R7:05/2015. Silverwater, Australia. https://ilac.org/publications-and-resources/ilac-rules-series/. (Accessed 21 January 2021).

ILAC, 2019. Structure of the ILAC Mutual Recognition Arrangement and Procedure for Expansion of the Scope of the ILAC Arrangement, ILAC-R6:05/2019. ILAC, Silverwater, Australia. https://ilac.org/publications-and-resources/ilac-rules-series/. (Accessed 21 January 2021).

INMETRO, 2012. NIT-DICLA-061, Rev.2, Requisitos sobre a acreditação dos laboratórios de ensaio e dos produtores de materiais de referência dos centros de recursos biológicos. Associação Brasileira de Normas Técnicas, Brasil.

ISO, 2006. The Future of ISO 9000 and ISO 14000. ISO Management Systems – January-February. International Organization for Standardization, Geneva, Switzerland. http://www.qualityintegrationsolutions.com/pdfs/future.pdf. (Accessed 21 January 2021).

ISO, 2015a. ISO9000:2015 - Quality Management Systems – Fundamentals and Vocabulary. International Organization for Standardization, Geneva, Switzerland.

ISO, 2015b. ISO9001:2015 - Quality Management Systems – Requirements. International Organization for Standardization, Geneva, Switzerland.

ISO, 2018a. ISO19011 - Guidelines for Auditing Management Systems. International Organization for Standardization, Geneva, Switzerland.

ISO, 2018b. ISO20387 - General Requirements for Biobanking. International Organization for Standardization, Geneva, Switzerland.

ISO, 2020a. ISO Committee on Conformity Assessment (CASCO) Structure, Process and Rules of Operation, QS-CAS-PROC01. International Organization for Standardization, Geneva, Switzerland.

ISO, 2020b. CAS/PROC/33 Common Elements in ISO/CASCO Standards. International Organization for Standardization, Geneva, Switzerland.

ISO/IAF, 2016. ISO9001 Auditing Practices Group Guidance on: Nonconformity – Documenting. ISO International Organization for Standardization, Geneva, Switzerland.

ISO/IEC, 2004. ISO/IEC Guide 2 - Standardization and Related Activities - General Vocabulary, eighth ed. International Organization for Standardization, Geneva, Switzerland.

Juran, J., 1992. Juran on Quality by Design: The New Steps for Planning Quality into Goods and Services. The Free Press, USA.

Kan, S., Lu, J., Liu, J., Wang, J., Zhao, Y., 2014. A quality by design (QbD) case study on enteric-coated pellets: screening of critical variables and establishment of design space at laboratory scale. Asian J. Pharm. Sci. 9, 268–278.

Kolaric, W., 1995. Creating Quality. McGraw-Hill International Editions, Singapore.

Largoni, M., Facco, P., Bernini, D., Bezzo, F., Barolo, M., 2015. Quality-by-Design approach to monitor the operation of a batch bioreactor in an industrial avian vaccine manufacturing process. J. Biotechnol. 211, 87–96.

Mallik, R., Raman, S., Liang, X., Grobin, A., Choudhury, D., 2015. Development and validation of a rapid ultra-high performance liquid chromatography method for the assay of benzalkonium chloride using a quality-by-design approach. J. Chromatogr. A 1413, 22–32.

Maropoulos, P., Ceglarek, D., 2010. Design verification and validation in product lifecycle. CIRP Ann. - Manuf. Technol. 59, 740–759.

MCT, 2002. Sistema de Avaliação da Conformidade de Material Biológico. Ministério da Ciência e Tecnologia, Brasília, Brasil. https://www.fcav.unesp.br/Home/servicos/CREBIO1852/sistema-de-avaliacao-da-conformidade-de-material-biologico.pdf. (Accessed 21 January 2021).

Moen, R., Nolan, T., Provost, L., 1991. Improving Quality through Planned Experimentation. MacGrow-Hill Book Co., Singapore. ISBN 0-07-100837-3.

NASEM, 2020. Biological Collections: Ensuring Critical Research and Education for the 21st Century. National Academies of Sciences, Engineering, and Medicine, The National Academies Press, Washington, DC. https://doi.org/10.17226/25592. (Accessed 21 January 2021).

OECD, 2001. Biological Resource Centres - Underpinning the Future of Life Sciences and Biotechnology. Paris. ISBN 92-64-18690-5 93 2001 04 1 P.

OECD, 2004. OECD Global Forum Knowledge Economy: Biotechnology. Guidance for the Operations of Biological Research Centres. Certification and quality criteria for BRCS, Paris.

OECD, 2007. OECD Best Practice Guidelines for Biological Resource Centres. OECD, Paris.

Raven, M., Blind, K., 2017. The characteristics and impacts of scientific publications in biotechnology research referenced in standards. Technol. Forecast. Soc. Change 115, 167–179. https://doi.org/10.1016/j.techfore.2016.10.001. (Accessed 21 January 2021).

Rozet, E., Lebrun, P., Debrus, B., Boulanger, B., Hubert, P., 2013. Design Spaces for analytical methods. Trends Anal. Chem. 42, 157–167.

Sampaio, P., 2014. Paulo Sampaio About the Future of Quality. https://www.youtube.com/watch?feature=player_embedded&v=JIQW0Igb14Y. (Accessed 21 January 2021).

Sangshetti, J., Deshpande, M., Zaheer, Z., Shinde, D., Arote, R., 2017. Quality by design approach: regulatory need. Arab. J. Chem. 10 (2), S3412–S3425. https://doi.org/10.1016/j.arabjc.2014.01.025. (Accessed 21 January 2021).

Santos, I., Lima, N., 2001. Criteria followed in the establishment of a filamentous fungal culture collection – Micoteca da Universidade do Minho. World J. Microbiol. Biotechnol. 17, 215–220.

Skrdla, P.J., Wang, T., Antonucci, V., Dowling, T., Ge, Z., Ellison, D., Curran, J., Mohan, G., Wyvratt, J., 2009. Use of a Quality-by-Design approach to justify removal of the HPLC weight % assay from routine API stability testing protocols. J. Pharmaceut. Biomed. Anal. 50, 794–796.

Tassey, G., 2000. Standardization in technology-based markets. Res. Pol. 29, 587–602.

UNEP/SETAC, 2005. Life Cycle Initiative, Life Cycle Approaches – the Road from Analysis to Practice.

Verma, S., Lan, Y., Gokhale, R., Burgess, D.J., 2009. Quality by design approach to understand the process of nanosuspension preparation. Int. J. Pharm. 377, 185–198.

WFCC, 2010. WFCC Guidelines, third ed. World Federation of Culture Collections. ISBN 92 9109 043 3. http://www.wfcc.info/#. (Accessed 21 January 2021).

WTO, November 29, 2012. Agreement on Technical Barriers to Trade, Committee on Technical Barriers to Trade. World Trade Organisation. http://www.wto.org/english/tratop_e/tbt_e/tbt_e.htm. (Accessed 21 January 2021).

Yu, L.X., 2008. Pharmaceutical quality by design: product and process development, understanding, and control. Pharmaceut. Res. 25 (4), 781−791.

Zidan, A.S., Sammour, O.A., Hammad, M.A., Megrab, N.A., Habib, M.J., Khan, M.A., 2007. Quality by design approach to understand the process of nanosuspension preparation. Int. J. Pharm. 332, 55−63.

Further reading

ISO/IEC, 2017. ISO/IEC17025:2017 - General Requirements for the Competence of Testing and Calibration Laboratories. International Organization for Standardization, Geneva, Switzerland. http://imed.ir/userfiles/files/11/ISO-IEC%2017025-2017.pdf.

Chapter 8

Microbial preservation for biotechnological development

Supawadee Ingsriswang[1], Lily Eurwilaichitr[2], Suwanee Chunhametha[1], Piyanun Harnpicharnchai[1], Somsak Likhitrattanapisal[1] and Supattra Kitikhun[1]

[1]*Thailand Bioresource Research Center (TBRC), National Center for Genetic Engineering and Biotechnology (BIOTEC), National Science and Technology Development Agency (NSTDA), Pathum Thani, Thailand;* [2]*Internaional Collaboration Office (INCO), National Sceince and Technology Development Agency, (NSTDA), Pathum Thani, Thailand*

Chapter outline

Importance of Microbiology Teaching and Microbial Resource Management for Sustainable Futures
https://doi.org/10.1016/B978-0-12-818272-7.00008-0

Introduction

Microorganisms have become increasingly important in our modern industry, including bioenergy and biofuel, biotechnological industry, biomedical and therapeutics production and wastewater treatment (Stahl and Wagner, 2006; Prakash et al., 2013). In the field of research and development, the availability of reference microbial strains is essential for analysis and testing, especially when the new use of old strains is discovered. Therefore, it is imperative to preserve and maintain microbial resources to support the needs of users, both for specific uses of laboratories or research groups and for wider utilisation. A very simple yet prominent example of the importance of preserving microbial resources is type strains and type species that have been well preserved by biological resource centres (BRCs) around the world, as evident by their isolation, characterisation, description and continuing availability for the benefit of researchers until the present day (Clark and Loegering, 1967).

To utilise microorganisms on a large production scale, cell propagation and large-scale production are essentially employed. Consequently, it is important to ensure cell stability as well as product yield level. In fact, one of the BRC's most important roles is to preserve microbial cultures without alteration of genetic or phenotypic characteristics of the microbial strain. This allows the preserved strain to be used for further screening, characterisation, genetic improvement and production of valuable products for the scientific community in the future (OECD, 2001). Therefore, the availability of the microbial strains with a genetically and physiologically unchanged state should be ensured even after decades of storage. The importance of maintaining microorganisms has also been realised by accredited international organisations. For example, the Organization for Economic Co-operation and Development (OECD) has published the 'OECD Best Practice Guidelines for Biological Resource Centres' (OECD, 2007), which provides useful guidance for BRCs to properly preserve and maintain microorganisms in order to effectively support research and development communities. Recently, a more specific guideline for BRCs, namely ISO 20387:2018, Biotechnology – Biobanking – General Requirements for Biobanking, has been published by the International Organization for Standardization (ISO) to systematically assess quality management of BRCs (ISO, 2018).

Nowadays, BRCs are required to preserve and maintain a large number of microbial strains. With the renewed interest in microbial cultivation in recent years, the number of isolated microbial strains has exploded, and preservation of these organisms is becoming difficult in both techniques and equipment. One of the reasons is the recent trend of –omics studies, especially the culturomics studies to isolate and culture the previously 'unculturable' microbes. As microorganisms isolated from the natural environment are high in biodiversity, preservation methods for various types of microorganisms are, therefore, usually different for various groups of microorganisms. It is speculated that large numbers of microorganisms have not been preserved and are subsequently lost due to the unavailability of suitable conditions for preservation (Fritze, 2008; Dugan and Tang, 2000). Consequently, novel preservation techniques have continuously been developed to prevent microbial viability loss and productive degeneration. In general, several efficient preservation techniques, including preservation with glycerol stock, sub-culturing, lyophilisation, cryopreservation and preservation in liquid nitrogen, are currently employed. These preservation techniques have advantages and disadvantages, and their suitability for preserving a particular microorganism varies from species to species. There are several factors of preservation methods that are critical for viability and longevity of microorganisms, including the composition of the medium, type of cryoprotectants (i.e., the essential antifreezes to protect life in the frozen state (Mazur, 2004; Woke, 2007)), speed of cooling and thawing, growth stage and cell types and sizes (Chian, 2010). It is not clear which mechanisms lead to the induction of changes in microbial cells during preservation, and this outcome, therefore, demands more investigation.

Another important issue of microbial preservation is quality control to monitor and reassure the genotypic and phenotypic stability of cultures. Several approaches, i.e., phenotypic and genotypic characterisation, have been employed. It is necessary that to verify the stability and viability of microorganisms in long-term preservation, several methods are needed to reassure the reliable and accurate outcome.

For uninterrupted use of microbial bioresources, associated data on each microbial resource as well as data from their long-term preservation is crucial. Strategies and tools for long-term data preservation are thus discussed in this chapter. In addition, the latest approaches in microbial cell preservation for long-term storage, as well as other important issues related to cell stability, viability and functionality, are also overviewed.

Microbial isolation: from sources to preservation

Microorganisms preserved in the BRCs are often deposited into the collection after being isolated from various sources, including humans, animals, plants and environments. As such, isolation of targeted microorganisms is the first step towards obtaining desirable specimens for further uses in industry or research. A vast variety of methods have been developed and employed for microbial isolation. Some methods have been developed to target a broad range of microorganisms, whilst others aim at isolating specific groups or species of microorganisms (e.g., actinomycetes, Kurtböke, 2017). An example of common isolation methods is the direct isolation approach, in which samples (such as fungal mycelium or spores) are directly placed on nutrient-containing agar (Abadias et al., 2001; Kaewkla and Franco, 2013; Ke et al., 2013; Hamzah et al., 2018). Another method is the dilution plating, in which microbial cells are diluted and are allowed to grow into a single colony. It has also been applied as a standard method to enumerate microorganisms in different types of samples (Baur and Ensminger, 1977). Airborne microorganisms can be isolated by allowing spores or cells to fall on agar plates by gravity (Napoli et al., 2012) or by using commercially available air samplers/air impactors to draw air through an inlet or the impact (Juozaitis et al., 1994).

For isolation of specific groups of microorganisms, selective growth conditions are usually employed. For instance, oxygen availability is used to select anaerobic bacteria, as they can only be grown in oxygen-deficit condition (often in an anaerobic jar or chamber) (Mirza and Rodrigues, 2012). Other selective conditions include growth media (Hamaki et al., 2005; Kato et al., 2018), osmolarity (Gunde-Cimerman et al., 2000; Butinar et al., 2005), temperature (Pernicova et al., 2020; Satoh et al., 2011; Ahirwar et al., 2017; Kimura and Horikoshi, 1988) and pH (Kimura and Horikoshi, 1988; Ňancucheo et al., 2016). For isolation of some fungi, especially mushrooms, spores are subjected to spore printing/shooting process in which they are dropped by gravity to the surface underneath (Gupta et al., 2011; Choi et al., 1999) or attached to the surface above by electrostatic attraction (Lakkireddy and Kües, 2017). In addition, the baiting method uses specific substrates to attract or capture targeted microorganisms. Examples of this method are the use of fungal mycelium to attract hyphae-adhering bacteria from soil (Ballhausen et al., 2015) and the use of pieces of wood submerged in water to attract aquatic fungi (Sri-indrasutdhi et al., 2010).

With the recent focus on microbiome analysis, advances in technology have also allowed effective isolation of a large number of microorganisms from various niches and environments in a relatively short time. An example of such technology is the microfluidics technique, which utilises isolating assembly and culturing fluids as a microbial habitat in artificial ecosystems (Nagy et al., 2018; Wu and Dekker, 2016). With this technique, a large number of individual cells can be isolated and allowed to grow separately. To study microbial communities, Jiang and co-workers created nanolitre droplets in microfluidic streak-plated platforms to isolate soil communities that can degrade polycyclic aromatic hydrocarbons (Jiang et al., 2016). The study showed that rare species could be uncovered by this technique.

Preservation techniques for microorganisms and microbial communities

Preservation of microbial strains is undoubtedly one of the key factors contributing to the utilisation of microorganisms in industry, research and medicine. As different microbial strains behave differently under preservation conditions, there is no universal preservation method that is applicable to every microorganism.

Standard methods for microbial culture preservation

Freezing and freeze-drying (lyophilisation) are one of the most reliable methods for the long-term preservation of microorganisms (Smith, 2012). In addition, liquid drying is recommended for microorganisms sensitive to freezing and freeze-drying (Malik, 1990). These methods have been favoured by culture collections/BRCs worldwide as evident by the fact that microorganisms normally distributed in the form of freeze- or liquid-dried ampoules, thawed cryotubes, frozen cryotubes and fresh cultures prepared from frozen cultures. These long-term preservation methods ease the BRC's distribution function by reducing the number of passages or subcultures and minimising maintenance cost and time. Additionally, the long-term preservation methods have also been widely used by research and educational organisations where microorganisms are required to be kept long term for research or teaching purposes.

Freezing (cryopreservation)

Cryopreservation methods can be used to preserve a vast majority of microorganisms. It is a technically simple method where cells or spores mixed with cryoprotectants can be preserved long term below −70°C or in liquid nitrogen. Although the method is basic, several factors, such as growth phase and rate, growth medium composition, cryoprotectant type, cooling rate and thawing rate, must be taken into consideration. Cryoprotectants contribute greatly to the success of cryopreservation. Several chemical compounds have been used successfully as cryoprotectants to protect microorganisms at low temperature including dimethyl sulfoxide (DMSO), glycerol, blood serum or serum albumin, skimmed milk, peptone, yeast extract, saccharose, glucose, methanol, polyvinylpyrrolidone, sorbitol and malt extract (Hubalek, 2003). Cryoprotectants suitable for most microorganisms appear to be 10%−15% glycerol and 5% DMSO (Prakash et al., 2013; Feltham et al., 1978). Another important factor affecting cryopreservation is the cooling rate. Although microorganisms suspended in cryoprotectants may be transferred directly to deep freezers (Nagel and Kunz, 1972), lowering the temperature by 1°C/min before transferring the microorganisms to deep freezers or nitrogen storage tanks is recommended. This process can be controlled by programmable machines in which liquid nitrogen is used as a cryogen in order to precisely lower the temperatures. A cooling rate of 1°C per minute to −40°C followed by a cooling rate of 10°C per minute to −90°C is recommended for freezing of fungi by ATCC (ATCC, 2020). Alternatively, cryotubes can be temporarily stored in a simpler apparatus or box, such as Thermo Scientific 'Mister Frosty' box, to lower the temperatures (CABRI, 1998; Homolka, 2013; Balfour-Cunningham et al., 2017) prior to being transferred to permanent storage containers.

Preservation methodology may vary according to types or species of microorganisms. Bacteria, yeasts and yeast-like fungi can be easily prepared and preserved by immobilising cells on glass beads and stored under −70°C

(Espinel-Ingroff et al., 2004; Feltham et al., 1978; Giampaglia et al., 2009; Gorman and Adley, 2004; Winters and Winn, 2010; Nagel and Kunz, 1972). For filamentous fungi, mycelium may be prepared by scraping with a pipette and suspending the fragments in cryoprotectants or cutting out agar plugs from colonies grown on agar media (ATCC, 2020). Alternatively, mycelium may be grown on carriers such as wheat grain (Zaghi Junior et al., 2020) and perlite (Homolka, 2013). Some species of pathogenic bacteria have been reported to survive when impregnated on cotton tip swabs and stored at −80°C without adding any cryoprotectant (Villa et al., 2020).

Preservation conditions need to be adjusted especially for fastidious microorganisms. For example, certain fungal species that are fastidious to the drying process can be successfully preserved for at least 5 years by using 10% glycerol as a protectant. In addition, the freezing rate for these fastidious strains should be controlled at 1°C per minute to −35°C and then subjected to an uncontrolled accelerated rate to −165°C to −196°C. To revive the cells after cryopreservation, rapid thawing at 38°C–40°C is recommended (Hwang, 1966). Survival rates of some species of microalgae (*Chlorella vulgaris*, *Isochrysis galbana* and *Dunaliella salina*) were shown to improve by using DMSO as a protectant and slow freezing rate of 1.5°C per minute, and the level of fatty acids was not seriously affected by the freezing conditions (Guermazi et al., 2009). Marine thraustochytrids were able to survive for at least 1 year when glycerol and trehalose were used as protectant agents and the strains were stored at −80°C or under vapour phase of nitrogen (Unagul et al., 2017).

To successfully preserve microorganisms using the freezing method, maintenance of storage environments also needs to be controlled. Freeze–thaw cycles need to be minimised as it affects survival rates of the preserved microorganisms (Postgate and Hunter, 1961). This risk may be prevented by separating working stock cultures from the seed/backup stock cultures and having power generators or nitrogen back-up in place to ensure that the storage temperature can be maintained at all times. Spare freezers are required in case of freezer malfunction. However, this measure may only be possible economically for well-established BRCs.

Freeze-drying

Freeze-drying can be employed to successfully preserve most bacteria, yeasts and sporulating filamentous fungi. The method involves freezing of microorganisms suspended in protective media followed by vacuum drying. Although this preservation process required several steps and is more complicated than cryopreservation, storage of freeze-dried microorganisms and shipment of freeze-dried products to users are easy and economical, making it favourable to many culture collections and BRCs worldwide.

The processes of freeze-drying and specifications of vials/ampoules used are largely adapted to types and models of freeze dryers. For the freezing process, cells suspended in protective agents may be frozen prior to loading to the freeze dryer or frozen within the freeze dryer chamber. The drying process involves two steps, namely primary drying and secondary drying. During the primary drying, water from the frozen samples is removed by sublimation. Secondary drying aims to remove bound water in the solid matrix. After the drying process, vials are plugged with stoppers under a vacuum condition or flooding with nitrogen, whilst ampoules are sealed by flame.

Several factors affecting the success of the freeze-drying process are growth media, initial cell concentration, protective agents, freezing temperatures, storage temperature and rehydration media. A protective agent is one of the most important factors contributing to the long-term survival of the freeze-dried microorganisms as they are required to go through two harsh processes during the freeze-drying process, i.e., freezing and vacuum drying. Recommended protectants are a mixture of proteins and sugars. In particular, skim milk was found to be a good protectant not only for the freezing process (Dahmen, 1983; Kim and Kubica, 1973) but also for the drying process. It provided even better results when mixed with sugars such as lactose (Abadias et al., 2001) or sucrose (Ming et al., 2009), as well as other substances such as gum acacia (Desmond et al., 2002). Storage condition also plays an important role in long-term storage of freeze-dried microorganisms. After the freeze-drying process, freeze-dried microorganisms should be kept at a low temperature, whilst excess oxygen and light should be avoided. It was found that storage temperature greatly affects the shelf life of the freeze-dried bacteria (Achour et al., 2001). Viability could decrease 3 logs after storage at room temperature (25°C) compared to 0.5 log at 4°C. A decrease in survival rates was also observed when the freeze-dried microorganisms were exposed to oxygen compared with those flooded with nitrogen or kept under a vacuum condition (Bozoğlu et al., 1987).

Similar to cryopreservation, several factors affecting the freeze-drying process may need to be adjusted according to type, species or strains of microorganisms. For example, sucrose is the best protectant for *Pseudomonas chlororaphis*. Its survival rate is similar to that of using sucrose mixed with skim milk as a protectant. The highest survival rate was observed when initial cell concentration reached 10^9 to 10^{10} cells/mL. Cells subjected to heat treatment and carbon starvation survive better than those cultivated under optimal growth conditions before the freeze-drying process (Palmfeldt et al., 2003). It was also found that pre-treatment of *Lactobacillus salivarius* by adjusting growth temperature and pH of growth medium greatly affected the survival rate. Immediate freezing of cell suspension at −80°C showed a better survival rate than those stored at 4°C for 2 h before freezing at −80°C (Ming et al., 2009). Heat- or cold-shock treatment increased tolerance of *Lactobacillus lactis* during the freeze-drying process (Broadbent and Lin, 1999). *Candida sake* survived best when using 10% skim milk and 10% lactose as a protectant and 10% skim milk as a rehydration medium. However, biocontrol effect of *C. sake* was higher when using 1% peptone as a rehydration medium (Abadias et al., 2001). For *Lactobacillus brevis and Oenococcus oeni*, yeast extract and sodium glutamate were better protectants compared to sugars. For these species, fast freezing (−65°C) is recommended rather than slow freezing (−20°C) (Zhao and Zhang, 2005).

Liquid drying

Liquid drying method differs from the freeze-drying method since freezing of microbial samples is not required prior to vacuum drying. This method has been proven to be effective for microorganisms sensitive to a freezing condition (Malik, 1990; Malik and Hoffmann, 1993). During the process, the microbial suspension is vacuum dried by evaporation. Although this method requires less steps than that of freeze-drying, the methodology and factors to be concerned are similar to that of the freeze-drying method. The processes of liquid drying and specifications of vials/ampoules used are largely adapted to types and models of vacuum dryers. Although information or publications regarding this method is far less than the freeze-drying method, it has been adopted at some BRCs such as NITE Biological Resource Center (NBRC) and Thailand Bioresource Research Center (TBRC).

Simple methods for microbial culture preservation

Various simple methods for microbial preservation have been developed and used by researchers. Most methods aim at refraining from using complicated or expensive equipment in order to lower preservation maintenance costs. Some methods have been adapted to fit the field use.

Drying

Drying is the most common methodology adopted for the simple preservation of microorganisms, as it does not require storage at extremely low temperatures. For example, actinomycetes are allowed to grow until spores are produced on filter paper discs. The discs were then dried and stored at low temperature. The microorganisms are able to maintain their viability after 18 months of storage at −20°C (Kulkarni and Chitte, 2015). Fungal mycelium could also be preserved using the filter paper method, with which the fungal mycelium is allowed to grow fully on filter paper, dried and then stored at low temperature (Fong et al., 2000).

Gelatin disc method is also a preferable choice for laboratories and research institutes as it only requires simple apparatus. Droplets of bacterial cells mixed with medium containing gelatin were dried in a desiccator. The discs may be stored in a desiccator or transferred to tubes and kept at low temperature. Several species of bacteria can be kept for many years with this method (Obara et al., 1981; Stamp, 1947). Other substances such as acacia gum and pullulan were also used as a protective medium. Microorganisms immobilised in these substances could survive even when stored at ambient conditions (Krumnow et al., 2009; Morkus et al., 2019).

Preservation of microorganisms under paraffin oil

Paraffin oil can be applied for the preservation of bacteria, fungi and yeasts (Hartsell, 1953, 1956). The microorganisms are simply grown on suitable culture media prior to being layered with sterile paraffin oil. Storage can be done at room or low temperature. Survival rates of the preserved microorganisms largely differ from microbial species to species. Some species do not survive after a few weeks of preservation, whilst several species could survive for more than 10 years (Hartsell, 1953, 1956).

Preservation of microorganisms in sterile water

Plant- and human-pathogenic bacterial cells suspended in water or phosphate-buffered saline could be stored in the dark at room temperature and survived for up to 16 years (Liao and Shollenberger, 2003). Similar successful results have been reported when plant pathogenic bacteria suspended in water and stored at 6°C could survive after 24 years of storage (Iacobellis and Devay, 1986). The survival rate was also impressive for some species of fungi that survived after 24 years of storage in water at room temperature (Capriles et al., 1989).

Preservation of microbial community

The more challenging matter for microbiologists recently has been the preservation of microbial community which is more complicated than the preservation of individual microbial strain mainly because most of the preservation methods are suitable for specific groups of microorganisms rather than universally applicable to every microorganism. Preservation and viability testing of the microbial community may not be straight forward compared to single or pure cultures. As the component of the community may be complex and unculturable, successful preservation of microbial community may be indicated by the continuation of its properties or outcomes of processes, rather than the number of viable microbial cells. For example, anammox bacteria used in nitrogen recycling can be preserved by freeze-drying the washed sludge. After the freeze-drying, changes in nitrogen concentration and the removal of total nitrogen in its culture medium can be measured as an indicator of stability of the preserved microbial community (Zhang et al., 2020). Recently, much attention has been focussed on the use of microbial communities in human (human microbiome), especially human stools, for the treatment of diseases or fecal microbiota transplantation therapies. Cryopreservation has been used by stool banks to preserve the fecal microbiome (Terveer et al., 2017). Although the effectiveness of fresh and frozen stools was similar (Lee et al., 2016), cryopreservation helps keep the stools for treatment tracking. Using this approach, stools are kept by cryopreservation at −80°C by OpenBiome (www.openbiome.org) for up to 2 years (Box 8.1).

BOX 8.1 Examples of microbial culture preservation

Preservation techniques used at Thailand Bioresource Research Center (TBRC)

At TBRC, bacteria, yeasts and sporulating fungi are cryopreserved at −80°C (Fig. S1). First, the microorganisms are cultured on suitable agar media until late log phase is reached. Sporulating fungi and sporulating actinomycetes are cultured on suitable sporulating agar media until spores are mature. Then, the microbial culture is mixed well with a suitable cryoprotectant before being transferred to 2 mL cryotubes. The content is left at room temperature for 1 h before transferring to a −80°C freezer.

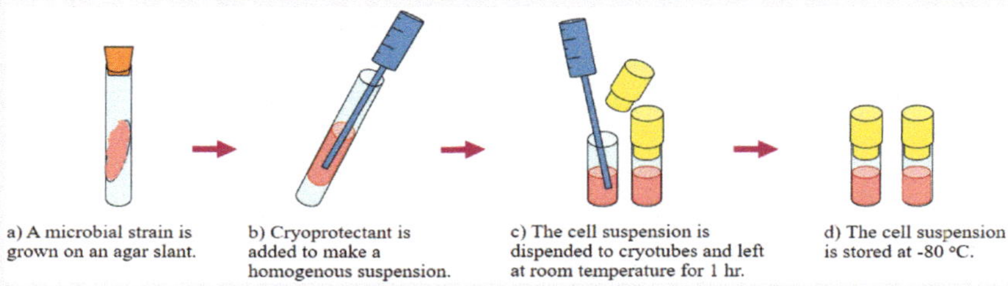

FIGURE S1 The freezing method used for preservation of bacteria, yeasts and sporulating fungi at TBRC.

Non-sporulating fungi are cultured on a suitable agar plate until fully grown. The adhering colony on agar is cut to pieces (approximately 0.5 × 0.5 cm agar block), and 5−6 agar blocks are then transferred to a 2 mL cryotube filled with 1 mL cryoprotectant. Sporulating filamentous fungi and sporulating actinomycetes can also be preserved using this method. The cryotubes are left at room temperature for 1−2 h then stored at 4°C for 48 h. Then the cryotubes are transferred to Thermo Scientific 'Mister Frosty' box and stored at −80°C for 4 h before quickly transferring to a −80°C freezer and a liquid nitrogen tank (Fig. S2).

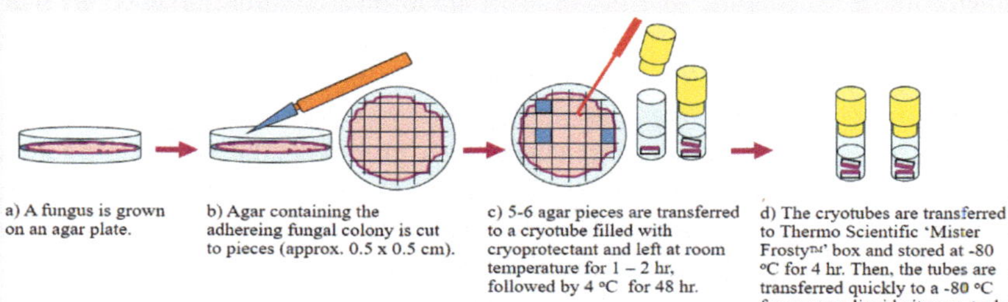

FIGURE S2 Preservation of non-sporulating fungi by the freezing method at TBRC.

BOX 8.1 Examples of microbial culture preservation—cont'd

For liquid drying method, the fully grown microbial culture is mixed with a suitable protectant to produce a dense (10^9–10^{10} cells/mL) homogeneous suspension. Then, 0.1 mL of the cell suspension is transferred to an ampoule. The projecting part of the cotton plug is removed with scissors and the remaining cotton plug is pushed halfway down the ampoule. The ampoule is constricted approximately 1 cm above the cotton plug to produce a short capillary section of about 1 mm in diameter. The cell suspension is vacuum dried for 4 h before the ampoule is sealed under a vacuum condition. The vacuum condition in the ampoules is tested using a high-frequency spark tester. The ampoule is kept in dark in a 4°C refrigerator (Fig. S3).

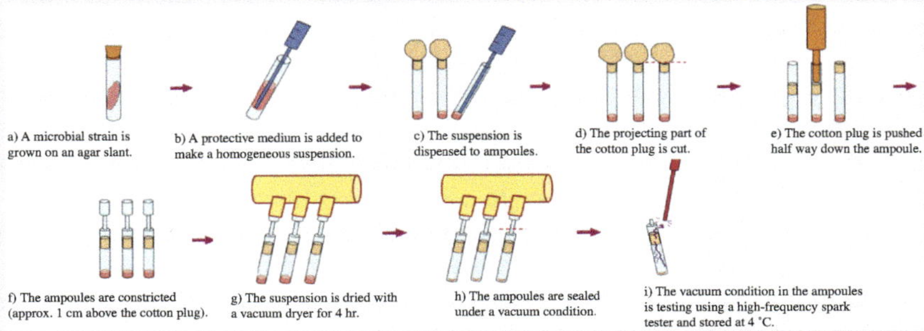

FIGURE S3 Preservation of microorganisms by liquid-drying method at TBRC.

Microbial stock management and quality control in BRCs

Microbial stock management

It is apparent that selecting suitable preservation methods enables BRCs and research organisations to maintain their service and research functions. Long-term preservation methods help to minimise the number of passages and maintenance costs. The stock of microorganisms needs to be managed and monitored to guarantee the availability of microorganisms once requested by customers or researchers. Fig. 8.1 illustrates an example of microbial stock management applicable to any laboratories or BRCs that may obtain a microorganism by deposition from a researcher, isolation from a specimen or purchase from other BRCs. After obtaining the microbial strain (original culture), it is recommended that three sets of microbial stocks (designated working stock, seed stock and backup stock) should be prepared. The number of tubes for each stock can be adjusted according to the frequency of use. The working stock is mainly used for distribution or routine uses. The seed stock is used for the preparation of working stocks. It is recommended to keep at least one tube of each generation for quality control and tracking purposes. The backup stock maintained at a different location is recommended for important microorganisms. Alternatively, important microorganisms can be deposited into BRCs. Microorganisms obtained from BRCs may not need a backup storage as it may be re-ordered at any time.

Quality control

With an ever-increasing number of microbial strains deposited and preserved in culture collection/BRCs, quality control (QC) of these microbial strains becomes more challenging, especially as the BRCs are responsible for assuring the quality of the product through all operations of microbial preservation so that these strains are not lost, and the cultures are always available for dispatch services.

Quality control of bioresources is important for maintaining good services compliance to rules and standards such as ISO 17034:2016. QC depends on effective maintenance of stock cultures to guarantee that the product retains the highest standard and fulfils product claims as defined in their collection catalogues before it reaches consumers. In addition, to ensure the trust of consumers, guidelines provided by umbrella organisations, such as the World Federation of Culture Collections (WFCC) and Common Access to Biological Resources and Information (CABRI), recommend that deposits in culture collections should be maintained by at least two different methods, at least one of which should be a metabolically inactive method, to minimise the risk of genetic and phenotypic change (http://www.wfcc.info/guidelines/). If two distinct methods are not applicable, two separated preserved stocks should at least be maintained at separate locations. In addition, regular audits must always be planned and carried out to ensure the continuing quality of the preserved biological resources.

Quality control of bioresource preservation can be achieved in the following ways.

Proper propagation, preservation and maintenance protocol

Thorough and systematic protocol with regards to propagation, preservation and maintenance will lead to effective long-term preservation and proper distribution of diverse microbial strains. It also allows the preserved cultures to remain as close as possible to the original culture and minimises the chance of microbial strains getting lost. One of the most important elements of maintaining a collection of microbial cultures is ensuring the genetic stability of the microbial strains. To this end, continuous sub-culturing will generally generate genetic drift and mutations to some degree that will render the cultures unsuitable for many uses. Regulatory guidance differs between advisory bodies on how many passages are allowed. However, it is generally agreed to limit the number of passages to a minimum. Minimisation of the number of passages will also minimise phenotypic variations and reduce the chance of contamination. It is also important to ensure that working cultures are not used as replacements for depleted stock or master cultures.

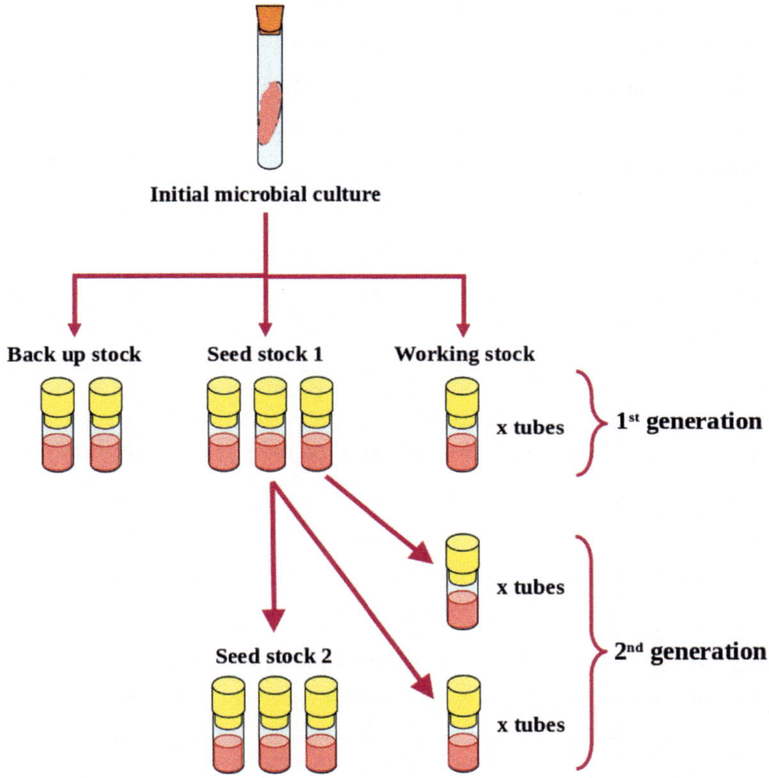

FIGURE 8.1 Preparation of a microbial stock. The first generation of a microbial stock contains a working stock, a seed stock and, if possible, a backup stock. Once the working stock of the first generation is completed, new batches of the working stock can be prepared from the seed stock. Once the seed stock of the first generation is almost completed, a new batch of the seed stock can be prepared from the first generation of the seed stock.

Accurate and reliable documentation system

Traceability of microbial strains in the collection can be achieved by an accurate and reliable documentation system that provides information about the preserved strain as much as possible including the identity, purity and viability of each microorganism. Documentation also includes the number of stocks (e.g., numbers of vials or ampoules) for each microbial strain. If the number of the preserved stock becomes low, a plan for a new preservation batch for the strain must be made. In addition, accurately documented procedures will ensure that each working culture can be traced back to the master culture. Documentation also includes detail of the inventory control, lead times and restocking practices, as well as proprietary systems for managing strain stocks that allow fair sharing of benefits amongst stakeholders.

Currently, various software have been developed to specifically allow accurate and efficient documentation of microbial culture collections. Examples of these tools include BioloMICS developed by BioAware and iCollect developed by TBRC, which allow the user to create a standard or customised database for straightforward archiving and retrieval of data. These tools are also advantageous for minimising human errors associated with the documentation as the cross-check system can be installed within the tool.

Reliable polyphasic strain identification and verification

Strain identification and verification are often needed for post-preservation quality check prior to purity and viability assessment of microbial strains stored in the collection. The main methods of bacterial identification and verification can be broadly classified as phenotypic- or genotypic-based techniques to differentiate a particular species or sub-species of microorganisms from the others. Phenotypic-based techniques include morphology characterisation (often microscopic), primary and secondary metabolite production, drug resistance characteristics and serotyping. Genotypic-based techniques are either pattern-generating techniques or sequencing techniques. The pattern-generating techniques typically produce a series of DNA or RNA fragments that form a unique pattern for each organism. Sequencing techniques take advantage of unique nucleotide sequences to characterise microorganisms at the species level. Sequencing of a targeted gene or whole genome has now emerged as a gold standard for characterisation of microbial strain.

Phenotypic identification and verification are relatively inexpensive, but these approaches depend on the metabolic activities that are taking place in the cell. Genotypic techniques have the advantage over phenotypic methods in that they are independent of the physiological state of an organism and are not influenced by the growth state of the microorganism. In addition, it usually offers a higher resolution for microbial identification than the phenotypic techniques. However, they often require expensive reagents and equipment.

It is imperative that microorganisms in culture collections should undergo several methods of identification and verification as a single method of identification is usually not sufficient. Both the number of tests needed and the actual tests used for identification often vary from one group of organisms to another to achieve accurate results. This will lead to the most reliable quality control of the preserved microbial strains.

Validation and evaluation of long-term preservation

Biological resources must be maintained without change to ensure reproducibility and sustainability. When maintaining culture collection for long-term storage, best practice guidelines demand that the preserved strains must be validated, and culture viability should be evaluated periodically to monitor possible phenotypic and genetic changes during the preservation. This verifies the cultures are stable in the storage conditions. Regular in-storage maintenance checks are thus necessary for the quality control of microorganisms deposited in BRCs.

Identification and verification of deposited cultures

Phenotypic techniques

Morphological characterisation Colony morphology characteristics of an individual colony of microorganisms growing on nutrient agar according to properties include colony size, shape, colour, texture, surface, margin (edge), elevation and optical appearances (Brehm-Stecher and Johnson, 2004; Jackman, 2011; van Teeseling et al., 2017). Morphology information can often allow identification of a certain microorganism to the genus level or to distinguish that it belongs to one or another group. Additionally, morphology characterisation can be used to assess culture purity as the presence of more than one colony type is often, but not always, indicative of a mixed culture. One of the most basic morphology assessments is the gram staining used as a technique to distinguish gram-positive and gram-negative bacteria. Other morphology assessments include the sediment formation and pellicle formation can be observed in liquid cultures. In addition, many microbial species exhibit notable characteristics. For example, colonies of several *Porphyromonas* species display auto-fluorescent under long-wavelength ultraviolet light (Bisen et al., 2012). Members of the families Mycobacteriaceae and Nocardiaceae belong to a few groups of microorganisms that are acid fast (Schumann, 2011). These characteristics are thus useful for the primary evaluation of microbial identity.

In recent years, automated systems have been employed especially for the detection and enumeration of microbial colonies. Using a microscope equipped with a camera, automated image analysis of cell morphological characteristics can be achieved, especially for the investigation of dynamic morphological changes over time. This helps reduce the time of the investigation and the images can be easily stored for further analysis. However, when the automated system is used, considerable expertise and experience in morphology analysis is still required in discerning morphology information for microbial genus or species analysis.

Primary and secondary metabolite characterisation Microorganisms produce numerous primary and secondary metabolites involved in physiological processes. Since the primary and secondary metabolite characteristics of each microorganism can give suggestive information for that particular microorganism, an array of biochemical tests can be used to confirm the microbial identification. Examples of these tests include catalase test to detect the presence of the catalase enzyme (Iwase et al., 2013), oxidase test to differentiate certain bacteria capable of producing cytochrome *c* oxidases (STEEL, 1961) and coagulase test to detect *Staphylococcus* sp. (especially *S. aureus*) that produce coagulase (Bascomb and Manafi, 1998) High-performance liquid chromatography or thin layer chromatography can also be employed for quantification or detection of many metabolites secreted by microorganisms (Frisvad and Thrane, 1987; Choma and Grzelak, 2011). It is worth noting that the detection of primary and secondary metabolite production also depends on culture conditions and sample preparation process that allows enough levels of the metabolites to be obtained.

Drug susceptibility test Drug susceptibility tests (DSTs) can be done to verify certain microbial strains with susceptibility or resistance to particular antibiotics. The test can usually be investigated by either broth dilution or disk diffusion technique. DST is also useful for evaluation of the ability of a particular microbial strain to retain the antimicrobial resistance after preservation. Recently, the broth dilution method has also been combined with biochemical assay associated with microbial growth to allow faster detection of the drug-resistant strain. For example, the broth dilution method for certain *Mycobacterium tuberculosis* strains can be coupled with a colorimetric assay for MTT [3-(4.5-dimethylthiazol-2-yl)-2.5-diphenyltetrazolium bromide] to investigate growth of that strain in the presence of antibiotics (Caviedes et al., 2000).

Serotyping Serotyping is the analysis of specific interactions between certain microorganism's cell surface antigens and antibodies. Variations within a microbial species allow the epidemiologic classification of organisms to the subspecies level. In addition, the protein and polysaccharides that make up a bacterium are sometimes characteristic enough to be considered identifying markers (Baron, 1996; Henriksen, 1978). The most useful of these are the molecules that constitute cell surface structures including the β-glucans, glycocalyx, flagella and pili (Janda et al., 1996; Wang et al., 2007). Examples include some species of *Streptococcus* that contain a unique carbohydrate molecule as part of the cell wall. Using techniques that rely on the specificity of interaction of antibodies and antigens, these carbohydrates, as well as any distinct protein or polysaccharide, can be detected.

Genotypic techniques

PCR-based fingerprinting Since the genetic makeup of microorganisms differs from one microbial species to another species, PCR of a targeted gene provides a possible basis for defining and validating a microbial species. PCR-based restriction fragment length polymorphism (RFLP), also known as cleaved amplified polymorphic sequence, involves the use of specific primers to amplify targeted genetic loci. The obtained amplicons are then subjected to restriction endonuclease analysis with different enzymes to generate a unique fingerprint, which is then compared to the fingerprints in the database or to other in-house samples to determine the microbial species (Lu et al., 2000; Vaneechoutte et al., 1992).

An example of a PCR-RFLP method is toxinotyping used for differentiation of toxin-producing pathogens, especially *Clostridium difficile*. The RFLP is focussed on pathogenicity locus (PaLoc), a region coding for toxins A and B on the genome. The technique is also a good basis for *C. difficile* phylogenetic studies (Rupnik and Janezic, 2016).

Another version of PCR-RFLP is PCR-ribotyping. Ribotyping involves the comparative analysis of discrete-sized genome fragments generated by restriction digestion of the rRNA operons, usually at the intergenic spacer region. This technique has been successfully employed for the characterisation of different microbial species such as *Salmonella* sp., *Pseudomonas* sp. and *Clostridium* sp. (Bouchet et al., 2008; O'Neill et al., 1996).

16S rDNA and ITS gene sequencing Gene sequencing is currently a useful tool for the identification and verification of microorganisms in culture collections. As the 16S rDNA gene, encoding the small subunit ribosomal RNA (rRNA) molecules of ribosomes, is highly conserved and universal amongst bacterial species, sequencing of the 16S rDNA gene is now often considered as a gold standard for assessment of bacterial species. In general, the 16S rDNA sequence is composed of both variable and conserved regions and the variation in the sequence between closely related species allows for differentiation amongst most bacterial species. High-quality, full-length 16S gene sequencing the comparison of these sequences can allow for identification at least to the genus level, and oftentimes to the species level (Janda and Abbott, 2007; Woo et al., 2008). For fungal and higher eukaryote species, the internal transcribed spacer (ITS), the non-coding spacer DNA situated between the small subunit rRNA and large subunit rRNA genes, is often employed for species identification and verification (Yang et al., 2018). Due to its applicability to broad taxonomic groups, the ITS region has also been recommended as the universal barcode sequence (Schoch et al., 2012) that can be used for identification and verification of fungal species.

Alignment of the 16S rDNA or ITS sequence of strain to the type strain of the species/subspecies of interest can be done to validate the correct taxonomic classification. Sequence alignment against several databases, including public databases such as GenBank, the Ribosomal Database Project and GreenGenes, as well as commercial databases such as the MicroSEQ, can ensure the identity of a particular microbial strain.

Validation and evaluation of long-term stock cultures

Viability count

The calculation of viable cells is very crucial in determining survival cells after preservation. During long-term storage of stock stains, it is most preferable that cell viability remains high. Microbial cell viability can be evaluated by counting either most probable numbers or colony forming units (CFUs) indicative of capability for growth and division. For most bacteria and yeast, the number of viable cells can be estimated by counting the CFUs on agar plates after serial dilutions (usually 1:10) of the microbial suspension in nutrient-containing medium or 0.1% peptone water has been applied onto suitable agar medium and appropriated times are allowed for microbial growth. The numbers of microbial colonies, indicative of the CFUs, can be counted on plates showing 30–300 colonies and the number of cells in the original (non-diluted) suspension can then be calculated. For sporulating fungi, the survival rate can be calculated from viable fungal propagules seen under a microscope after allowing for germination. If germination rates are too low, CFUs can be scored after allowing serial dilutions of the fungal suspension to grow on a suitable agar medium. Additionally, for non-sporulating fungi, multiple agar plugs containing the fungus can be allowed to grow on a suitable agar medium, and survival rate can be investigated from the plugs exhibiting radial growth on the culture plate.

As the manual counting of microbial CFUs can be considered time consuming, tedious and error prone, there are currently many types of automated colony counters available commercially to help increase speed and reduce manual labour. The equipment usually contains a digital camera, a detection system and computer software to allow reliable and accurate colony enumeration. In addition, images of colonies are provided and stored for further use. Thus, automating colony counts can bring precision, speed, accuracy and higher throughput to microbial viability count.

Besides culturability, viability of cells and cell density can also be assessed by measurement of certain cellular functions such as cell membrane integrity, respiration, generation of ATP and heat flow. For example, fluorescent dyes such as SYBRGreen can penetrate all cells and bind selectively to double-stranded (ds) DNA. On the other hand, propidium iodide only penetrates damaged membranes due to its positive charge (Joux and Lebaron, 2000). Employing SYBRGreen and propidium iodide results in green or red fluorescence, respectively. Using this method, viable microbial cells containing intact cell walls can then be discriminated from dead cells and can be counted rapidly by flow cytometry (Berney et al., 2007). Other dyes that can be utilised for the determination of cell viability include fluorescein diacetate, rhodamine dyes, cyanine dyes, propidium monoazide, ethidium monoazide, tetrazolium salts and resazurin (Mosmann, 1983; Kumar and Ghosh, 2019).

Automated biochemical detection system

Biochemical phenotypes of a microbial strain should remain unchanged during storage in the culture collections. To help detect the change in biochemical phenotypes of microbial strains during the long-term storage, automating microbial detection can be employed to help minimise human labour from much of the QC process. Automation ensures faster, more accurate results and frees personnel from redundant tasks to focus on other responsibilities. A number of automated or semi-automated systems are offered commercially for rapid authentication of microorganisms based on their biochemical profiles. These test kits include RapID panels (Thermo Fisher Scientific), API strip systems (bioMérieux) and VITEK 2 cards (bioMérieux) (O'Hara C, 2005; Crowley et al., 2012). The biochemical test panels from the kits allow multiple tests simultaneously and help reduce the time of analysis.

Single-nucleotide polymorphism analysis

Genome-wide single-nucleotide polymorphism (SNP) analysis can be performed to assess the genetic mutations occurring during the long-term preservation of microbial strains. SNP analysis can be performed by comparing a set of DNA sequence reads from a particular strain with the high-quality reference genome from a strain of the same species to identify the numbers and locations of SNPs that differentiate the reference and the query genomes. In theory, the difference of a single base pair can be detected. It is most likely that when genomic variation occurs during the storage, genomic SNPs can be found in both non-coding sequences (non-CDSs) and coding sequences (CDSs) (Sakurai and Kawasaki, 2018). Different microbial species may behave differently with regards to the risk of genomic variation in the same storage conditions. Thus, the SNP analysis can also help verify the extent of the risk inserted to each species.

Whole-genome sequencing with next generation sequencing and third-generation sequencing techniques

Discrimination of different species or subspecies of microorganisms and validation of a particular microbial strain in culture collections can be ultimately performed by whole-genome sequencing. The small number of sequence variations within the genome allows the resolution of whole-genome sequencing at strain level. Beginning in 2004, next generation sequencing (NGS) technologies, also known as high-throughput sequencing, have demonstrated the capacity to sequence DNA at rapid speed. The advantage offered by the NGS is such that the whole genome of an organism can be read at once within hours (Schuster, 2008; Behjati and Tarpey, 2013). Usually, this is accomplished by fragmenting the genome into small pieces, randomly sampling for a fragment and sequencing it using one of a variety of technologies. The platforms for the NGS include Illumina sequencing (using synthesis and incorporation of fluorescent base approach), Roche 454 sequencing (using pyrosequencing approach) and Ion Torrent: Proton/PGM sequencing (using ion semiconductor approach). Generally, read lengths of 50−700 bp are obtained. The obtained sequencing reads are then assembled into contigs to reveal the unique genome sequence of an organism.

Currently, 'third-generation sequencing' (TGS) is state-of-the-art sequencing methods being applied for microbial identification and verification. To circumvent the short sequence reads offered by NGS, the TGS uses technology that allows for longer DNA reads (up to 10,000−100,000 bp) suitable for genome sequencing and genome resequencing. However, TGS data still have high error rates and attempts have been made to increase the accuracy of the results. There are several companies currently exploring TGS technology development, including Pacific Biosciences, Oxford Nanopore Technology, Quantapore and Stratos (Bleidorn, 2016; van Dijk et al., 2018). These companies exploit different approaches to sequencing single DNA molecules. PacBio utilises the sequencing platform of single molecule real time sequencing based on the properties of zero-mode waveguides. Signals are read as fluorescent light emission from each nucleotide incorporated by the immobilised DNA polymerase. Oxford Nanopore's technology involves passing a DNA molecule through a nanoscale pore structure and then measuring changes in the electrical field surrounding the pore. On the other hand, Quantapore has a different proprietary nanopore approach. Stratos Genomics spaces out the DNA bases with polymeric inserts, 'Xpandomers', to circumvent the signal to noise challenge of nanopore ssDNA reading.

With advances in sequencing technology coupled with the capability for high throughput analysis, whole-genome sequencing with the NGS and TGS will be more widely applicable for routine uses in the validation of microbial strains in culture collections. In addition, with highly detailed information the whole-cell sequencing can generate, it is still considered a gold standard microbial typing technique.

Dereplication of stock cultures

Similar microorganisms preserved in BRCs can be grouped together by dereplication processes. Most importantly, a large number of microbial strains in BRCs can pose a burden for maintenance and quality control of the resources. It is recommended that when long-term storage of cultures is limited or difficult, dereplication of conspecific strains (belonging to the same species) or genetically identical clones in the BRCs can be implemented to control or reduce the downstream cost, labour and time of managing the collection. Filtering out conspecifics can also facilitate a cost-effective screening of these microorganisms for further uses; i.e., the number of strains needed to be screened can decrease dramatically after grouping conspecific strains together.

The modern dereplication techniques commonly employed by BRCs include the following:

Fatty acid methyl esters method

To verify and dereplicate microorganisms in BRCs, several methods can be employed. These methods should be applicable to most or all microbial strains, have a high taxonomic resolution and produce robust results. Mass spectrometry has recently become an attractive method for efficient verification and dereplication of microbial strains as a labour- and cost-efficient alternative to the traditional methods using 16S rRNA sequence analysis or repetitive element (rep) primed PCR. The coupling of mass spectrometry to gas chromatography (GC) for the analysis of fatty acid methyl esters (FAMEs) was the first application of mass spectrometry for bacterial identification (Slabbinck et al., 2008). A gas chromatograph is used to analyse whole-cell fatty acids of microbial cells. With this technique, fatty acids from phospholipids, triacylglycerols, sterols and various other lipid structures are hydrolysed and then labelled with a methyl group at the carboxyl group of fatty acids which forms a methyl ester. The GC system is used to measure the presence of these fatty acids within the cell and the fatty acid profile obtained is unique for each microorganism. Thus, the fatty acid profile is used as a fingerprint to identify and verify each organism based on comparison to a library of reference profiles (Sherlock Analysis Software, MIDI Systems, Inc.).

Similar to other types of mass spectrometry, the success of FAME depends on the reference library containing fatty acid profiles from numerous organisms. The Sherlock Analysis Software is currently able to accurately identify 1700 species of bacteria and yeast, many to the subspecies or strain level.

Matrix-assisted laser desorption/ionisation time-of-flight mass spectrometry method

In the last 2 decades, Matrix-Assisted Laser Desorption/Ionisation Time-Of-Flight Mass Spectrometry (MALDI-TOF MS) has become a popular proteotypic-based method for the robust, rapid and labour- and cost-effective characterisation of microorganisms. The method is based on comparing mass spectra of protein ions, mostly abundant ribosomal proteins and other housekeeping proteins, uniquely generated by a particular species of organism. The unique spectra, referred to as main spectra profile, are used as a fingerprint to identify and verify a particular microbial species. Researchers have also reported the use of MALDI-TOF MS profiling for characterisation at the subspecies level, indicating a high taxonomic resolution for some taxa that potentially exceeds that of 16S rRNA sequencing.

MALDI-TOF MS has been proposed as a promising alternative for the dereplication of recurrent bacterial isolates (Ghyselinck et al., 2011). To achieve effective dereplication, the process involves the generating mass spectra of small proteins/peptides from whole-cell or total cell extract (usually obtained by formic acid extraction) and assigning the sequence to a taxonomic bin (proteotic phylotype) based on known references either by classification or identification of the closest strain using an additional statistical analysis step to group measured mass spectra and identify unique isolates (Clark et al., 2013).

Similar to the 16S rRNA gene sequencing, the key factor for the success of dereplication by MALDI-TOF MS is its availability of the reference databases. The databases can be commercial or in-house. Several successful commercial platforms, such as Biotyper (Bruker Daltonics), SERAMIS (Shimadzu) and Vitek MS (BioMérieux), are mainly used to identify microorganisms. For example, Bruker MALDI Biotyper library contains spectra of more than 7000 bacterial and 1300 filamentous fungi isolates (as of April 2020). However, these systems, whose reference databases cover only a small fraction of the vast range of microbial diversity, as a large fraction of entries in the existing commercial libraries are from microorganisms with clinical relevance. To improve dereplication when using whole-cell or protein extracts for MALDI-TOF MS, two main bioinformatic strategies are commonly used. Additional reference strains can be used as the first strategy to expand the commercial databases, especially those from environments or industries. In the context of expanding a spectral library, it is worth noting that the confidence level of species or strain identification can be achieved when several reference strains of one species are available in the database. The second strategy involves using statistical analysis to group mass spectra. Using additional statistical analysis steps or software such as the principal component analysis can therefore improve the approach in creating a non-redundant set of isolates. Other software improvements have also been made to facilitate dereplication. Examples include, a user-friendly spectral data analysis tool named SPeDE for the dereplication of MALDI-TOF mass spectra by analysing a mix of global and local peak comparisons to identify sets of spectra at the level of the species or strain (Dumolin et al., 2019).

Data preservation in BRCs

With regards to long-term microbial preservation in bioresource centres, it is essential to ensure continued access of not only high-quality microbial cultures but also their associated information. Therefore, planning to deal with data management and preservation is absolutely significant for microbial resource centres that want to thrive in the rapidly evolving technology world. With the adoption of large-scale multi-omics technologies and high-throughput analytical facilities for investigating genomic structure and metabolic potential of microbial strains, each analysis of microbial culture can generate massively heterogeneous data at ever faster rates. Moreover, the technological and digital advances also create challenges in data preservation, such as the dependencies on specific proprietary hardware and software and the lack of backward compatibility. The data preserved can be swiftly inaccessible and obsoleted over a short period of time. The multi-dimensional data from different analysis platforms and versions also pose greater complexity and challenges of data preservation in order to achieve biological meaningful data integration for current and future uses. In this context, the BRCs have to cover the provision of organisational capacity building and digital infrastructure development to ensure sustainable scientific data preservation. The Open Archival Information System (OAIS) is an international standard (ISO 14721) providing the concept and framework for long-term digital information preservation (Lee, 2010). The OAIS can be used as a reference model for organising people and computing resources in an archival system (Deegan and Tanner, 2006). This section focusses on technical elements for BRCs' long-term digital data preservation.

Data integrity

Data can become degraded with the reduction in data quality. Data preservation, thus, plays a role in the ongoing maintenance of data integrity by preserving the validity, consistency and accuracy of data. Data preservation aims to prevent and recover data from unintentional changes over time. Poor data integrity practices can seriously undermine the quality of records and the quality of further derived information. Therefore, the data management in BRC should comply with a good data management guideline for biological data, such as FAIR (Findability, Accessibility, Interoperability and Reusability) (Wilkinson et al., 2016), ALCOA (Attributable, Legible, Contemporaneous, Original, Accurate) (Rattan, 2018) and Good Data Management Practice (GCDMP) (World Health Organization, 2015).

Some BRCs adopted the Laboratory Integrated Management System (LIMS) software to pipeline the data management workflow and systematise the data passage throughout the data life cycle. LIMS is a software suite specifically designed to help ensure the traceability and consistency of the data generated in the BRC. However, in many cases, the data repository is stored and managed on stand-alone computerised systems. The administrator and users must implement an appropriate grant permission system to ensure that any data records and changes are validated and audited with verifiable time/date stamps and user accounts.

Data in BRC's archived repositories will periodically need to be assessed to examine the obsolescence of those records due to the effect of changes such as taxonomic and classification system, name of geographic location or institutes, correction in genome assemblies, etc. BRC should implement monitoring protocols to detect any information update and apply traceable adjustment accordingly. The use of globally unique identifiers, such as DOI for digital records or journal articles, ISBN for books, ORCID for researchers, are encouraged to facilitate data interoperability and data reusability (Sansone et al., 2012).

These practices do not only help BRCs and scientific communities, but also benefit data owners as it allows easy research workflow and reproducibility, and thus indirectly increasing citation rates (Griffin et al., 2018).

Integrity violations may also be caused by hardware or software malfunctions, malicious intrusions and inadvertent user errors (Sivathanu et al., 2005). Data loss and corruption can occur due to hardware and software malfunctions such as power outages, disk errors and system bugs. Unauthorised modification of information can be executed by external and internal malicious software (e.g., virus code) and malicious users. The inadvertent deletion of information or software from a storage can result in data corruption, system malfunction and integrity violation. Common techniques to prevent the violation and ensure the integrity of the storage include data replication or mirroring and checksum verification.

Metadata and standards

Useful biological data is necessarily accompanied by meaningful metadata. The definition of metadata is the data that provide the contextual information about the obtained data (National Information Standards Organization, 2001). The descriptive metadata of biological data describes the information related to the data such as collector name, date of collection, date of experiment, sequencing platform, experiment conditions, etc. This information is crucial for future use and reference. Besides the descriptive metadata, there is structural metadata that tells how the digital objects in the data are structured. Another type of metadata which is important to BRC data preservation is administrative metadata which provides information for the management of digital objects and information resources, such as data permission, intellectual property rights and resource type, in order to secure the integrity and enable their accessibility.

Data and metadata can be recorded in unstructured formats, with which one may find it easy for data deposition and recording, but greatly hinder data reuse and sharing (Goodrich et al., 2014). The movement towards 'open data' culture, therefore, begets the need for data standards in which the data and metadata structure, common glossary of terms and minimal specification are systemically defined. For example, the Genomic Standard Consortium develops the MIxS standard (http://gensc.org/gc_wiki/index.php/MIxS) that includes the technology-specific checklists for genome sequences (MIGS) and metagenome sequences (MIMS) (Yilmaz et al., 2011).

It is a good data management practice for BRCs to ensure the availability of written documentation. Good documentation ensures data consistency, especially in medium or large BRCs where two or more staff have to jointly work on the same tasks or projects. Documentation may be created internally or referred to the vendor's specifications, international standards or community-derived open guidelines. The data standard and documentation are also useful in data sharing amongst BRCs and amongst research communities. Therefore, generally, open and widely implemented standards such as XML schemas are recommended as they are always accompanied by comprehensive documentation to ensure interoperability. An example of the open biological data standard with excellent documentation is Darwin Core which is a standard data format for biodiversity data that includes a glossary of terms and practical examples (Wieczorek et al., 2012).

File formats for preservation

At a minimum, for the long-term preservation of data, data must be saved in reusable file formats and stored in multiple secure locations in order to be accessible in the future. For example, information of microbial cultures in a small collection or BRC is usually recorded in the spreadsheet and stored in proprietary file formats such as XLS, XLSX, XLSB and XLSM. As such, different versions of Microsoft Excel, proprietary software, may be needed for the data to be read and saved reliably. Such proprietary file formats are readable by proprietary software, but the software can be changed by the company owner at any time. Moreover, proprietary software might be able to run only on a specific proprietary operating system such as Microsoft Windows and MacOS.

Unlike the proprietary file formats, open file formats such as CSV, XML and JPEG 2000 are set out in agreed standards for the purpose of interoperability, so that any proprietary or non-proprietary software can use the open formats freely (The Linux Information, 2007). Therefore, it is recommended that the open, non-proprietary and widely adopted file formats may provide greater confidence of long-term digital preservation. In addition, lossless file formats are more appropriate for data preservation than lossy formats. Lossy formats (e.g., JPEG image file format) entail some loss of data, whilst lossless formats (e.g., TIFF image file format with header metadata and lossless image compression) allow the original data to be perfectly preserved even after compression. Using lossy compression/encoding, the data from the original file will be irreversibly lost, although it results in significantly reduced file size suitable for distribution/access purposes. For best practices, the images of microbial strains and other graphic information are preferred to be preserved in TIFF or other lossless-compressed image formats.

Different data types usually come with different data formats with or without metadata. Some formats store data and basic metadata in the same file and some formats are designed specifically to store data of specific aspects. For example, the sequence data can be presented as a plain-text file in either FASTA, FASTQ or GenBank format. FASTA is the simplest format amongst these, as it contains only the nucleotide or amino acid sequences (each nucleotide or amino acid is represented by a single-letter code) and sequence ID (with a brief description for some samples), making it the most convenient format used for parsing and sequence analysis. FASTQ is similar to FASTA, but it contains additional information about base read quality score. GenBank can store sequence data, annotation data and metadata in the same file. Metadata can help users to easily identify, retrieve and convert the data stored in GenBank format in the future.

The conversion between data formats can inadvertently result in partial information loss. For example, when sequence data are extracted from a FASTQ file and converted into a FASTA file, the base quality score data is irrecoverably discarded. In a real scenario, the data curator and data administrator may maintain data in a single, as complete as possible, non-standard data format stored in the database. The internal documentation is recommended for facilitating good data management practice and data consistency as mentioned before. The standard format of choice can be generated on demand from stored data via user interface or API in order to reduce redundancy, maintenance and update costs and to maximise efficiency.

Even the format of the same data standards can be upgraded to a new version to support a new data schema. Even if the upgrade comes with some degree of backward compatibility, it can bring about technical difficulties or even data loss. Caution is needed when implementing an upgrade of data standards. The backup is highly advisable and, if possible, the upgrade should be performed on a dedicated test database before implementing it on the real production one.

Obsolescence monitoring

With the rapid evolution of digital technologies, a fundamental practice of data preservation is to regularly monitor the obsolescence status of file formats, hardware, software and storage media. The following are potential forms of digital obsolescence.

Obsolescence of formats and media

The physical carrier-based or file-based formats of the data record can become obsolete. For example, current media such as CDs and DVDs are becoming obsolete and no longer commercially available. Data preserved in these media should be migrated to other stable, robust and reliable storage. A variety of storage media for BRC, including hard disk drives and cloud storage, can be chosen based on the size of data, accessibility and budget. It is best practice to put two copies of the files on different media/storage systems. If possible, every 3–5 years, all preserved information should be migrated from an older media to a newer media.

Obsolescence of hardware and software

Most computer hardware has a useable life of 3–5 years. Sometimes, the new computer models will also introduce new peripherals and storage devices that are incompatible with the old computers. Moreover, the replacement of obsolete hardware may need new software in order to operate it. For example, some of the Apple software running on Intel-based Mac may need the updates to be able to run on ARM-based Mac after Apple switches some of its Mac to ARM processors in 2020. Besides the computing hardware, the analytical hardware, along with its required software and analysis result format such as genomic sequencing and chemical analysis, can become obsolete. The changes of hardware and other software that support the functionality of the software can force the old software and file formats into retirement.

Currently, several biological information repository and analysis tools are implemented on the web-based platform. The update of a web browser can lead to incompatibility and failure of the web-based software. Moreover, it is quite common that a website hosting the software or biological resources is inaccessible for intended users. The web-based bioinformatic tools and resources can be discontinued and inaccessible by the changes of URL and API.

Software obsolescence can also occur by the termination of sales/distribution, support and maintenance of the software from its developer or vendor. For example, any form of support for Microsoft Windows XP ended in April 2014 (Microsoft, 2020). To keep the data accessible, digital information from older hardware and software needs to be transferred to the current ones.

Obsolescence monitoring can help the BRC to minimise the risk of digital data becoming inaccessible. The optimal procedure to react to obsolescence can be planned to secure the long-term preservation.

Multi-copy and multi-location storage

Data preservation requires a series of recurrent activities necessary to ensure that the data will be maintained in the *secured repository* with standard metadata and remains *accessible* and *useable* over time. To avoid several risks to the long-term preservation of information, it is essential to make a backup or have two complete copies stored in different locations, so that information recovery is possible when a system failure occurs. However, if two copies are stored in the same geographic location, both copies may not be able to be recovered in the event of a local disaster like a fire, flooding or sabotage.

For best practices, it is recommended to keep three complete copies in locations with three different disaster threats. Cloud services and participation in online networks of digital preservation are optional to further decrease risks from data loss.

Summary

Generally, preservation of microbial strains and facilitation of its uses in biotechnological applications are the key roles of BRCs. Proper microbial preservation methods, efficient stock culture management, periodic quality control assessment and reliable data preservation all contribute to the success of BRCs to effectively serve the need of users in industry, research and medicine. An array of preservation methods pertinent to long-term or short-term storage of microbial strains currently exist and suitable preservation methods will help minimise the risk of decrease in cell viability, change in genetic content and loss of cellular functions. Well-organised management of stock cultures and the rigorous quality control check will ensure that the microbial resource of the best quality is always available for customers and researchers. Lastly, data integrity is crucial for sustainable preservation. Selection of standardised data and metadata format, obsolescence monitoring and data copies not only mitigate the risk of information loss, but they ascertain that the information will always be accessible for retrieval. With these measures, the BRCs can strive to facilitate the utilisation of microorganisms for sustainable biotechnological development.

References

Abadias, M., Teixidó, N., Usall, J., Benabarre, A., VIñas, I., 2001. Viability, efficacy, and storage stability of freeze-dried biocontrol agent *Candida sake* using different protective and rehydration media. J. Food Protect. 64, 856–861.

Achour, M., Mtimet, N., Cornelius, C., Zgouli, S., Mahjoub, A., Thonart, P., Hamdi, M., 2001. Application of the accelerated shelf-life testing method (ASLT) to study the survival rates of freeze-dried *Lactococcus* starter cultures. J. Chem. Technol. Biotechnol. 76, 624–628.

Ahirwar, S., Soni, H., Prajapati, B.P., Kango, N., 2017. Isolation and screening of thermophilic and thermotolerant fungi for production of hemicellulases from heated environments. Mycology 8, 125–134.

ATCC, 2020. Preservation and Recovery of Filamentous Fungi. Available: https://www.atcc.org/~/media/PDFs/Technical%20Bulletins/tb02.ashx. (Accessed 30 September 2020).

Balfour-Cunningham, A., Boxall, N.J., Banning, N., Morris, C., 2017. Preservation of salt-tolerant acidophiles used for chalcopyrite bioleaching: assessment of cryopreservation, liquid-drying and cold storage. Miner. Eng. 106, 91–96.

Ballhausen, M.-B., VAN Veen, J.A., Hundscheid, M.P.J., DE Boer, W., 2015. Methods for baiting and enriching fungus-feeding (mycophagous) rhizosphere bacteria. Front. Microbiol. 6, 1416.

Baron, E.J., 1996. Classification. In: Baron, S. (Ed.), Medical Microbiology, fourth ed. University of Texas Medical Branch at Galveston, Galveston (TX).

Bascomb, S., Manafi, M., 1998. Use of enzyme tests in characterization and identification of aerobic and facultatively anaerobic gram-positive cocci. Clin. Microbiol. Rev. 11, 318–340.

Baur, F.J., Ensminger, L.G., 1977. The association of official analytical chemists (AOAC). J. Am. Oil Chem. Soc. 54, 171–172.

Behjati, S., Tarpey, P.S., 2013. What is next generation sequencing? Arch. Dis. Child. Educ. Pract. Ed. 98, 236–238.

Berney, M., Hammes, F., Bosshard, F., Weilenmann, H.-U., Egli, T., 2007. Assessment and interpretation of bacterial viability by using the LIVE/DEAD BacLight kit in combination with flow cytometry. Appl. Environ. Microbiol. 73, 3283.

Bisen, P.S., Debnath, M., Prasad, G.B.K.S., 2012. Microbes : Concepts and Applications. Wiley-Blackwell, Hoboken, N.J.

Bleidorn, C., 2016. Third generation sequencing: technology and its potential impact on evolutionary biodiversity research. Syst. Biodivers. 14, 1–8.

Bouchet, V., Huot, H., Goldstein, R., 2008. Molecular genetic basis of ribotyping. Clin. Microbiol. Rev. 21, 262–273.

Bozoğlu, T.F., Özilgen, M., Bakir, U., 1987. Survival kinetics of lactic acid starter cultures during and after freeze drying. Enzym. Microb. Technol. 9, 531–537.

Brehm-Stecher, B.F., Johnson, E.A., 2004. Single-cell microbiology: tools, technologies, and applications. Microbiol. Mol. Biol. Rev. 68, 538–559.

Broadbent, J.R., Lin, C., 1999. Effect of heat shock or cold shock treatment on the resistance of *Lactococcus lactis* to freezing and lyophilization. Cryobiology 39, 88–102.

Butinar, L., Sonjak, S.A., Zalar, P., Plemenitas, A., GUNDE-Cimerman, N., 2005. Melanized halophilic fungi are eukaryotic members of microbial communities in hypersaline waters of solar salterns. Bot. Mar. 48.

CABRI, 1998. Laboratory Procedures for Microorganisms: Flow Diagram. Available: http://cabri.org/guidelines/micro-organisms/M300Ap2.html. (Accessed 24 August 2020).

Capriles, C., Mata, S., Middelveen, M., 1989. Preservation of fungi in water (Castellani): 20 years. Mycopathologia 106, 73–79.

Caviedes, L., Lee, T.-S., Gilman, R.H., Sheen, P., Spellman, E., Lee, E.H., Berg, D.E., Montenegro-James, S., 2000. Rapid, efficient detection and drug susceptibility testing of *Mycobacterium tuberculosis* in sputum by microscopic observation of broth cultures. J. Clin. Microbiol. 38, 1203–1208.

Chian, R.-C., 2010. Cryobiology: an overview. In: Chian, R.-C., Quinn, P. (Eds.), Fertility Cryopreservation. Cambridge University Press.

Choi, Y.W., Hyde, K.D., Ho, W.H., 1999. Single spore isolation of fungi. Fungal Divers. 3, 29–38.

Choma, I.M., Grzelak, E.M., 2011. Bioautography detection in thin-layer chromatography. J. Chromatogr. A 1218, 2684–2691.

Clark, A.E., Kaleta, E.J., Arora, A., Wolk, D.M., 2013. Matrix-assisted laser desorption ionization-time of flight mass spectrometry: a fundamental shift in the routine practice of clinical microbiology. Clin. Microbiol. Rev. 26, 547–603.

Clark, W.A., Loegering, W.Q., 1967. Functions and maintenance of a type-culture collection. Annu. Rev. Phytopathol. 5, 319–342.

Crowley, E., Bird, P., Fisher, K., Goetz, K., Boyle, M., Benzinger JR., M.J.,, Juenger, M., Agin, J., Goins, D., Johnson, R.L., 2012. Evaluation of the VITEK 2 Gram positive (GP) microbial identification test card: collaborative study. J. AOAC Int. 95, 1425–1432.

Dahmen, H., 1983. Technique for long-term preservation of phytopathogenic fungi in liquid nitrogen. Phytopathology 73, 241.

Deegan, M., Tanner, S., 2006. Digital Preservation. Facet, London.

Desmond, C., Ross, R.P., O'callaghan, E., Fitzgerald, G., Stanton, C., 2002. Improved survival of *Lactobacillus paracasei* NFBC 338 in spray-dried powders containing gum acacia. J. Appl. Microbiol. 93, 1003–1011.

Dugan, F.M., Tang, J.S., 2000. Culture collections. In: Robinson, R.K., Batt, C.A., Patel, P.D. (Eds.), Encyclopedia of Food Microbiology, vol. I. Academic Press, London.

Dumolin, C., Aerts, M., Verheyde, B., Schellaert, S., Vandamme, T., Van der Jeugt, F., De Canck, E., Cnockaert, M., Wieme, A.D., Cleenwerck, I., Peiren, J., Dawyndt, P., Vandamme, P., Carlier, A., 2019. Introducing SPeDE: high-throughput dereplication and accurate determination of microbial diversity from matrix-assisted laser desorption–ionization time of flight mass spectrometry data. mSystems 4, e00437-19.

Espinel-Ingroff, A., Montero, D., MARTIN-Mazuelos, E., 2004. Long-term preservation of fungal isolates in commercially prepared cryogenic microbank vials. J. Clin. Microbiol. 42, 1257–1259.

Feltham, R.K., Power, A.K., Pell, P.A., Sneath, P.A., 1978. A simple method for storage of bacteria at -76 degrees C. J. Appl. Bacteriol. 44, 313–316.

Fong, Y.K., Anuar, S., Lim, H.P., Tham, F.Y., Sanderson, F.R., 2000. A modified filter paper technique for long-term preservation of some fungal cultures. Mycologist 14, 127–130.

Frisvad, J.C., Thrane, U., 1987. Standardized high-performance liquid chromatography of 182 mycotoxins and other fungal metabolites based on alkylphenone retention indices and UV—VIS spectra (diodearray detection). J. Chromatogr. A 404, 195–214.

Fritze, D., 2008. In: The Proposed Standard MTA of the European Culture Collections' Organization. Microbial Commons Conference, Ghent Belgium.

Ghyselinck, J., van Hoorde, K., Hoste, B., Heylen, K., de Vos, P., 2011. Evaluation of MALDI-TOF MS as a tool for high-throughput dereplication. J. Microbiol. Methods 86, 327–336.

Giampaglia, C.M., de Brito, A.C., Martins, M.C., Ueki, S.Y., Latrilha, F.O., de Oliveira, R.S., Yamauchi, J.U., Da Silva Telles, M.A., 2009. Maintenance of *Mycobacterium tuberculosis* on glass beads. Ann. Clin. Lab. Sci. 39, 51–54.

Goodrich, J.K., DI Rienzi, S.C., Poole, A.C., Koren, O., Walters, W.A., Caporaso, J.G., Knight, R., Ley, R.E., 2014. Conducting a microbiome study. Cell 158, 250–262.

Gorman, R., Adley, C.C., 2004. An evaluation of five preservation techniques and conventional freezing temperatures of -20 degrees C and -85 degrees C for long-term preservation of *Campylobacter jejuni*. Lett. Appl. Microbiol. 38, 306–310.

Griffin, P.C., Khadake, J., Lemay, K.S., Lewis, S.E., Orchard, S., Pask, A., Pope, B., Roessner, U., Russell, K., Seemann, T., Treloar, A., Tyagi, S., Christiansen, J.H., Dayalan, S., Gladman, S., Hangartner, S.B., Hayden, H.L., Ho, W.W.H., Keeble-Gagnère, G., Korhonen, P.K., Neish, P., Prestes, P.R., Richardson, M.F., Watson-Haigh, N.S., Wyres, K.L., Young, N.D., Schneider, M.V., 2018. Best practice data life cycle approaches for the life sciences. F1000Research 6, 1618.

Guermazi, W., Marangoni, R., Sellami-Kammoun, A., Elloumi, J., Drira, Z., Aleya, L., Ayadi, H., Maalej, S., 2009. Microalgal cryo-preservation using dimethyl sulfoxide (Me_2SO) coupled with two freezing protocols: influence on the fatty acid profile. J. Therm. Biol. 35, 175–181.

Gunde-Cimerman, N., Zalar, P., DE Hoog, S., Plemenitaš, A., 2000. Hypersaline waters in salterns – natural ecological niches for halophilic black yeasts. FEMS Microbiol. Ecol. 32, 235–240.

Gupta, B., Reddy, N., Kotasthane, A., 2011. Molecular characterization and mating type analysis of oyster mushroom (*Pleurotus* spp.) using single basidiospores for strain improvement. World J. Microbiol. Biotechnol. 27, 1–9.

Hamaki, T., Suzuki, M., Fudou, R., Jojima, Y., Kajiura, T., Tabuchi, A., Sen, K., Shibai, H., 2005. Isolation of novel bacteria and actinomycetes using soil-extract agar medium. J. Biosci. Bioeng. 99, 485–492.

Hamzah, T.N.T., Lee, S.Y., Hidayat, A., Terhem, R., Faridah-Hanum, I., Mohamed, R., 2018. Diversity and characterization of endophytic fungi isolated from the tropical mangrove species, *Rhizophora mucronata*, and identification of potential antagonists against the soil-borne Fungus, *Fusarium solani*. Front. Microbiol. 9.

Hartsell, S.E., 1953. The preservation of bacterial cultures under paraffin oil. Appl. Microbiol. 1, 36–41.

Hartsell, S.E., 1956. Microbiological process report. Maintenance of cultures under paraffin oil. Appl. Microbiol. 4, 350–355.

Henriksen, S.D., 1978. Chapter I Serotyping of bacteria. In: Bergan, T., Norris, J.R. (Eds.), Methods in Microbiology. Academic Press.

Homolka, L. (Ed.), 2013. Methods of Cryopreservation in Fungi. Springer, New York.

Hubalek, Z., 2003. Protectants used in the cryopreservation of microorganisms. Cryobiology 46, 205–229.

Hwang, S.W., 1966. Long-term preservation of fungus cultures with liquid nitrogen refrigeration. Appl. Microbiol. 14, 784–788.

Iacobellis, N.S., Devay, J.E., 1986. Long-term storage of plant-pathogenic bacteria in sterile distilled water. Appl. Environ. Microbiol. 52, 388–389.

ISO, 2018. ISO 20387 Biotechnology – Biobanking – General Requirements for Biobanking. Available: https://www.iso.org/standard/67888.html. (Accessed 30 September 2020).

Iwase, T., Tajima, A., Sugimoto, S., Okuda, K., Hironaka, I., Kamata, Y., Takada, K., Mizunoe, Y., 2013. A simple assay for measuring catalase activity: a visual approach. Sci. Rep. 3, 3081.

Jackman, J., 2011. The microbe: the basics of structure, morphology, and physiology as they relate to microbial characterization and attribution. Chem. Phys. Sign. Microb. Forensics 13–34.

Janda, J.M., Abbott, S.L., 2007. 16S rRNA gene sequencing for bacterial identification in the diagnostic laboratory: pluses, perils, and pitfalls. J. Clin. Microbiol. 45, 2761–2764.

Janda, J.M., Abbott, S.L., Khashe, S., Kellogg, G.H., Shimada, T., 1996. Further studies on biochemical characteristics and serologic properties of the genus Aeromonas. J. Clin. Microbiol. 34, 1930–1933.

Jiang, C.Y., Dong, L., Zhao, J.K., Hu, X., Shen, C., Qiao, Y., Zhang, X., Wang, Y., Ismagilov, R.F., Liu, S.J., Du, W., 2016. High-throughput single-cell cultivation on microfluidic streak plates. Appl. Environ. Microbiol. 82, 2210–2218.

Joux, F., Lebaron, P., 2000. Use of fluorescent probes to assess physiological functions of bacteria at single-cell level. Microb. Infect. 2, 1523–1535.

Juozaitis, A., Willeke, K., Grinshpun, S.A., Donnelly, J., 1994. Impaction onto a glass slide or agar versus impingement into a liquid for the collection and recovery of airborne microorganisms. Appl. Environ. Microbiol. 60, 861–870.

Kaewkla, O., Franco, C.M., 2013. Rational approaches to improving the isolation of endophytic actinobacteria from Australian native trees. Microb. Ecol. 65, 384–393.

Kato, S., Yamagishi, A., Daimon, S., Kawasaki, K., Tamaki, H., Kitagawa, W., Abe, A., Tanaka, M., Sone, T., Asano, K., Kamagata, Y., 2018. Isolation of previously uncultured slow-growing bacteria by using a simple modification in the preparation of agar media. Appl. Environ. Microbiol. 84, e00807–e00818.

Ke, Z., Su, Y.Y., Lei, C., 2013. An optimized protocol of single spore isolation for fungi. Cryptog. Mycolog. 34, 349–356.

Kim, T.H., Kubica, G.P., 1973. Preservation of mycobacteria: 100 percent viability of suspensions stored at -70°C. Appl. Microbiol. 25, 956–960.

Kimura, T., Horikoshi, K., 1988. Isolation of bacteria which can grow at both high pH and low temperature. Appl. Environ. Microbiol. 54, 1066–1067.

Krumnow, A.A., Sorokulova, I.B., Olsen, E., Globa, L., Barbaree, J.M., Vodyanoy, V.J., 2009. Preservation of bacteria in natural polymers. J. Microbiol. Methods 78, 189–194.

Kulkarni, G.A., Chitte, R.R., 2015. Preservation of thermophilic bacterial spores using filter paper disc techniques. J. Bioprocess. Biotech. 5.

Kumar, S.S., Ghosh, A.R., 2019. Assessment of bacterial viability: a comprehensive review on recent advances and challenges. Microbiology 165, 593–610.

Kurtböke, D.I., 2017. Bioactive actinomycetes: reaching rarity through sound understanding of selective culture and molecular diversity. In: Kurtböke, D.I. (Ed.), Microbial Resources: from functional Existence in Nature to Applications. Elsevier, Academic Press.

Lakkireddy, K., Kües, U., 2017. Bulk isolation of basidiospores from wild mushrooms by electrostatic attraction with low risk of microbial contaminations. Amb. Exp. 7, 28.

Lee, C.A., 2010. Open archival information system (OAIS) reference model. In: Bates, M.J., Maack, M.N. (Eds.), Encyclopedia of Library and Information Sciences, third ed. Taylor & Francis.

Lee, C.H., Steiner, T., Petrof, E.O., Smieja, M., Roscoe, D., Nematallah, A., Weese, J.S., Collins, S., Moayyedi, P., Crowther, M., Ropeleski, M.J., Jayaratne, P., Higgins, D., Li, Y., Rau, N.V., Kim, P.T., 2016. Frozen vs fresh fecal microbiota transplantation and clinical resolution of diarrhea in patients with recurrent *Clostridium difficile* infection: a randomized clinical trial. JAMA 315, 142–149.

Liao, C.H., Shollenberger, L.M., 2003. Survivability and long-term preservation of bacteria in water and in phosphate-buffered saline. Lett. Appl. Microbiol. 37, 45–50.

Lu, J.J., Perng, C.L., Lee, S.Y., Wan, C.C., 2000. Use of PCR with universal primers and restriction endonuclease digestions for detection and identification of common bacterial pathogens in cerebrospinal fluid. J. Clin. Microbiol. 38, 2076–2080.

Malik, K.A., 1990. A simplified liquid-drying method for the preservation of microorganisms sensitive to freezing and freeze-drying. J. Microbiol. Methods 12, 125–132.

Malik, K.A., Hoffmann, P., 1993. Long-term preservation of yeast cultures by liquid-drying. World J. Microbiol. Biotechnol. 9, 372–376.

Mazur, P., 2004. Principles of cryobiology. In: Fuller, B.J., Lan, N.J., Benson, E.E. (Eds.), Life in the Frozen State. CRC Press, Boca Raton, Florida.

Microsoft, 2020. Windows XP End of Support. Available: https://www.microsoft.com/en-us/microsoft-365/windows/end-of-windows-xp-support. (Accessed 27 October 2020).

Ming, L.C., Rahim, R.A., Wan, H.Y., Ariff, A.B., 2009. Formulation of protective agents for improvement of *Lactobacillus salivarius* I 24 survival rate subjected to freeze drying for production of live cells in powderized form. Food Bioprocess Technol. 2, 431.

Mirza, B.S., Rodrigues, J.L.M., 2012. Development of a direct isolation procedure for free-living diazotrophs under controlled hypoxic conditions. Appl. Environ. Microbiol. 78, 5542.

Morkus, P., Zolfaghari, M., Parrello, D., Csordas, M., Malikov, M., Rose, J., Choi, K.B., Filipe, C.D.M., Latulippe, D.R., 2019. Optimization of microorganism preservation conditions for the development of an acute toxicity bioassay for biocides. Chemosphere 221, 45–54.

Mosmann, T., 1983. Rapid colorimetric assay for cellular growth and survival: application to proliferation and cytotoxicity assays. J. Immunol. Methods 65, 55–63.

Nagel, J.G., Kunz, L.J., 1972. Simplified storage and retrieval of stock cultures. Appl. Microbiol. 23, 837–838.

Nagy, K., Abraham, A., Keymer, J.E., Galajda, P., 2018. Application of microfluidics in experimental ecology: the importance of being spatial. Front. Microbiol. 9, 496.

Ňancucheo, I., Rowe, O.F., Hedrich, S., Johnson, D.B., 2016. Solid and liquid media for isolating and cultivating acidophilic and acid-tolerant sulfate-reducing bacteria. FEMS Microbiol. Lett. 363.

Napoli, C., Marcotrigiano, V., Montagna, M.T., 2012. Air sampling procedures to evaluate microbial contamination: a comparison between active and passive methods in operating theatres. BMC Publ. Health 12, 594.

National Information Standards Organization, 2001. Understanding Metadata. NISO Press.

O'hara C, M., 2005. Manual and automated instrumentation for identification of Enterobacteriaceae and other aerobic gram-negative bacilli. Clin. Microbiol. Rev. 18, 147–162.

O'neill, G.L., Ogunsola, F.T., Brazier, J.S., Duerden, B.I., 1996. Modification of a PCR Ribotyping method for application as a routine typing scheme for *Clostridium difficile*. Anaerobe 2, 205–209.

Obara, Y., Yamai, S., Nikkawa, T., Shimoda, Y., Miyamoto, Y., 1981. Preservation and transportation of bacteria by a simple gelatin disk method. J. Clin. Microbiol. 14, 61–66.

OECD, 2001. Biological Resource Centres: Underpinning the Future of Life Sciences and Biotechnology Organization for Economic Cooperation and Development. OECD Publishing, Paris.

OECD, 2007. OECD Best Practice Guidelines for Biological Resource Centres. Available: http://www.oecd.org/sti/emerging-tech/oecdbestpracticeguidelinesfor biologicalresourcecentres.htm. (Accessed 26 October 2020).

Palmfeldt, J., RåDSTRöm, P., Hahn-Hägerdal, B., 2003. Optimisation of initial cell concentration enhances freeze-drying tolerance of *Pseudomonas chlororaphis*. Cryobiology 47, 21–29.

Pernicova, I., Novackova, I., Sedlacek, P., Kourilova, X., Koller, M., Obruca, S., 2020. Application of osmotic challenge for enrichment of microbial consortia in polyhydroxyalkanoates producing thermophilic and thermotolerant bacteria and their subsequent isolation. Int. J. Biol. Macromol. 144, 698–704.

Postgate, J.R., Hunter, J.R., 1961. On the survival of frozen bacteria. Microbiology 26, 367–378.

Prakash, O., Nimonkar, Y., Shouche, Y.S., 2013. Practice and prospects of microbial preservation. FEMS Microbiol. Lett. 339, 1–9.

Rattan, A.K., 2018. Data integrity: history, issues, and remediation of issues. PDA J. Pharm. Sci. Technol. 72, 105–116.

Rupnik, M., Janezic, S., 2016. An update on *Clostridium difficile*; toxinotyping. J. Clin. Microbiol. 54, 13.

Sakurai, K., Kawasaki, H., 2018. Genetic variation during long-term preservation of bacteria in public culture collections. Int. J. Syst. Evol. Microbiol. 68, 1815–1821.

Sansone, S.-A., Rocca-Serra, P., Field, D., Maguire, E., Taylor, C., Hofmann, O., Fang, H., Neumann, S., Tong, W., Amaral-Zettler, L., Begley, K., Booth, T., Bougueleret, L., Burns, G., Chapman, B., Clark, T., Coleman, L.-A., Copeland, J., Das, S., De Daruvar, A., De Matos, P., Dix, I., Edmunds, S., Evelo, C.T., Forster, M.J., Gaudet, P., Gilbert, J., Goble, C., Griffin, J.L., Jacob, D., Kleinjans, J., Harland, L., Haug, K., Hermjakob, H., HO Sui, S.J., Laederach, A., Liang, S., Marshall, S., Mcgrath, A., Merrill, E., Reilly, D., Roux, M., Shamu, C.E., Shang, C.A., Steinbeck, C., Trefethen, A., Williams-Jones, B., Wolstencroft, K., Xenarios, I., Hide, W., 2012. Toward interoperable bioscience data. Nat. Genet. 44, 121–126.

Satoh, Y., Tajima, K., Nakamoto, S., Xuerong, H., Matsushima, T., Ohshima, T., Kawano, S., Erata, T., Dairi, T., Munekata, M., 2011. Isolation of a thermotolerant bacterium producing medium-chain-length polyhydroxyalkanoate. J. Appl. Microbiol. 111, 811–817.

Schoch, C.L., Seifert, K.A., Huhndorf, S., Robert, V., Spouge, J.L., Levesque, C.A., Chen, W., 2012. Nuclear ribosomal internal transcribed spacer (ITS) region as a universal DNA barcode marker for fungi. Proc. Natl. Acad. Sci. U. S. A. 109, 6241.

Schumann, P., 2011. Peptidoglycan structure. In: Rainey, F., Oren, A. (Eds.), Methods in Microbiology. Academic Press.

Schuster, S.C., 2008. Next-generation sequencing transforms today's biology. Nat. Methods 5, 16–18.

Sivathanu, G., Wright, C.P., Zadok, E., 2005. Ensuring Data Integrity in Storage: Techniques and Applications. Association for Computing Machinery, Fairfax, VA, USA, pp. 26–36.

Slabbinck, B., DE Baets, B., Dawyndt, P., DE Vos, P., 2008. Genus-wide *Bacillus* species identification through proper artificial neural network experiments on fatty acid profiles. Antonie Leeuwenhoek 94, 187–198.

Smith, D., 2012. Culture collections. Adv. Appl. Microbiol. 79, 73–118.

SRI-Indrasutdhi, V., Boonyuen, N., Suetrong, S., Chuaseeharoonachai, C., Sivichai, S., Jones, E.B.G., 2010. Wood-inhabiting freshwater fungi from Thailand: *Ascothailandia grenadoidia* gen. et sp. nov., *Canalisporium grenadoidia* sp. nov. with a key to *Canalisporium* species (Sordariomycetes, Ascomycota). Mycoscience 51, 411–420.

Stahl, D.A., Wagner, M., 2006. The knowledge explosion in environmental microbiology offers new opportunities in biotechnology. Curr. Opin. Biotechnol. 17, 227–228.

Stamp, L., 1947. The preservation of bacteria by drying. Microbiology 1, 251–265.

Steel, K.J., 1961. The oxidase reaction as a taxonomic tool. Microbiology 25, 297–306.

Terveer, E.M., van Beurden, Y.H., Goorhuis, A., Seegers, J.F.M.L., Bauer, M.P., van Nood, E., Dijkgraaf, M.G.W., Mulder, C.J.J., Vandenbroucke-Grauls, C.M.J.E., Verspaget, H.W., Keller, J.J., Kuijper, E.J., 2017. How to: establish and run a stool bank. Clin. Microbiol. Infect. 23, 924–930.

The Linux Information Project, 2007. Free File Format Definition. Available: http://www.linfo.org/free_file_format.html. (Accessed 27 October 2020).

Unagul, P., Suetrong, S., Preedanon, S., Klaysuban, A., Gundool, W., Suriyachadkun, C., Sakayaroj, J., 2017. Isolation, fatty acid profiles and cryopreservation of marine thraustochytrids from mangrove habitats in Thailand. Bot. Mar. 60, 363–379.

van Dijk, E.L., Jaszczyszyn, Y., Naquin, D., Thermes, C., 2018. The third revolution in sequencing technology. Trends Genet. 34, 666–681.

van Teeseling, M.C.F., De Pedro, M.A., Cava, F., 2017. Determinants of bacterial morphology: from fundamentals to possibilities for antimicrobial targeting. Front. Microbiol. 8, 1264.

Vaneechoutte, M., Rossau, R., DE Vos, P., Gillis, M., Janssens, D., Paepe, N., DE Rouck, A., Fiers, T., Claeys, G., Kersters, K., 1992. Rapid identification of bacteria of the Comamonadaceae with amplified ribosomal DNA-restriction analysis (ARDRA). FEMS Microbiol. Lett. 72, 227–233.

Villa, L., Torreblanca, A., Otero, L., Carreño, F., Campo, R., López-Escobar, M., Vazquez, F., 2020. Long term storage of fastidious bacteria (*Neisseria* spp. and *Haemophilus* spp.) with swab preservation at -80°C. J. Microbiol. Methods 175, 105969.

Wang, Z., Liu, X., Dacanay, A., Harrison, B.A., Fast, M., Colquhoun, D.J., Lund, V., Brown, L.L., Li, J., Altman, E., 2007. Carbohydrate analysis and serological classification of typical and atypical isolates of *Aeromonas salmonicida*: a rationale for the lipopolysaccharide-based classification of *A. salmonicida*. Fish Shellfish Immunol. 23, 1095–1106.

Wieczorek, J., Bloom, D., Guralnick, R., Blum, S., Döring, M., Giovanni, R., Robertson, T., Vieglais, D., 2012. Darwin Core: an evolving community-developed biodiversity data standard. PLoS One 7.

Wilkinson, M.D., Dumontier, M., Aalbersberg, I.J.J., Appleton, G., Axton, M., Baak, A., Blomberg, N., Boiten, J.-W., DA Silva Santos, L.B., Bourne, P.E., Bouwman, J., Brookes, A.J., Clark, T., Crosas, M., Dillo, I., Dumon, O., Edmunds, S., Evelo, C.T., Finkers, R., Gonzalez-Beltran, A., Gray, A.J.G., Groth, P., Goble, C., Grethe, J.S., Heringa, J., T Hoen, P.A.C., Hooft, R., Kuhn, T., Kok, R., Kok, J., Lusher, S.J., Martone, M.E., Mons, A., Packer, A.L., Persson, B., Rocca-Serra, P., Roos, M., VAN Schaik, R., Sansone, S.-A., Schultes, E., Sengstag, T., Slater, T., Strawn, G., Swertz, M.A., Thompson, M., van der Lei, J., VAN Mulligen, E., Velterop, J., Waagmeester, A., Wittenburg, P., Wolstencroft, K., Zhao, J.,

Mons, B., 2016. The FAIR guiding principles for scientific data management and stewardship. Sci. Data 3, 160018.

Winters, R., Winn, W., 2010. A simple, effective method for bacterial culture storage: a brief technical report. J. Bacteriol. Virol. 40, 99–101.

Woke, B., 2007. How cryoprotectants work. In: Chapman, J. (Ed.), Cryonics. lcor Life Extension Foundation, Arizona.

Woo, P.C.Y., Lau, S.K.P., Teng, J.L.L., Tse, H., Yuen, K.Y., 2008. Then and now: use of 16S rDNA gene sequencing for bacterial identification and discovery of novel bacteria in clinical microbiology laboratories. Clin. Microbiol. Infect. 14, 908–934.

World Health Organization, 2015. Guidance on Good Data and Record Management Practices, WHO Headquarters. Avenue Appia 20, 1211 Geneva 27, Switzerland.

Wu, F., Dekker, C., 2016. Nanofabricated structures and microfluidic devices for bacteria: from techniques to biology. Chem. Soc. Rev. 45, 268–280.

Yang, R.-H., Su, J.-H., Shang, J.-J., Wu, Y.-Y., Li, Y., Bao, D.-P.; Yao, Y.-J., 2018. Evaluation of the ribosomal DNA internal transcribed spacer (ITS), specifically ITS1 and ITS2, for the analysis of fungal diversity by deep sequencing. PLoS One 13, e0206428.

Yilmaz, P., Kottmann, R., Field, D., Knight, R., Cole, J.R., Amaral-Zettler, L., Gilbert, J.A., Karsch-Mizrachi, I., Johnston, A., Cochrane, G., Vaughan, R., Hunter, C., Park, J., Morrison, N., Rocca-Serra, P., Sterk, P., Arumugam, M., Bailey, M., Baumgartner, L., Birren, B.W., Blaser, M.J., Bonazzi, V., Booth, T., Bork, P., Bushman, F.D., Buttigieg, P.L., Chain, P.S.G., Charlson, E., Costello, E.K., Huot-Creasy, H., Dawyndt, P., Desantis, T., Fierer, N., Fuhrman, J.A., Gallery, R.E., Gevers, D., Gibbs, R.A., Gil, I.S., Gonzalez, A., Gordon, J.I., Guralnick, R., Hankeln, W., Highlander, S., Hugenholtz, P., Jansson, J., Kau, A.L., Kelley, S.T., Kennedy, J., Knights, D., Koren, O., Kuczynski, J., Kyrpides, N., Larsen, R., Lauber, C.L., Legg, T., Ley, R.E., Lozupone, C.A., Ludwig, W., Lyons, D., Maguire, E., Methé, B.A., Meyer, F., Muegge, B., Nakielny, S., Nelson, K.E., Nemergut, D., Neufeld, J.D., Newbold, L.K., Oliver, A.E., Pace, N.R., Palanisamy, G., Peplies, J., Petrosino, J., Proctor, L., Pruesse, E., Quast, C., Raes, J., Ratnasingham, S., Ravel, J., Relman, D.A., Assunta-Sansone, S., Schloss, P.D., Schriml, L., Sinha, R., Smith, M.I., Sodergren, E., Spor, A., Stombaugh, J., Tiedje, J.M., Ward, D.V., Weinstock, G.M., Wendel, D., White, O., Whiteley, A., Wilke, A., Wortman, J.R., Yatsunenko, T., GLöckner, F.O., 2011. Minimum information about a marker gene sequence (MIMARKS) and minimum information about any (x) sequence (MIxS) specifications. Nat. Biotechnol. 29, 415–420.

Zaghi Junior, L.L., Berteli, M.B.D., DE Freitas, J.D.S., de Oliveira Filho, O.B.Q., Lopes, A.D., Ruiz, S.P., do Valle, J.S., Linde, G.A., Colauto, N.B., 2020. Five-year cryopreservation at -80°C of edible and medicinal basidiomycetes by wheat grain technique. J. Microbiol. Methods 176, 106030.

Zhang, H., Cai, L., Zhang, F., Ge, C., Yang, F., 2020. Vacuum lyophilization preservation and rejuvenation performance of anammox bacteria. J. Biosci. Bioeng. 129, 519–527.

Zhao, G., Zhang, G., 2005. Effect of protective agents, freezing temperature, rehydration media on viability of malolactic bacteria subjected to freeze-drying. J. Appl. Microbiol. 99, 333–338.

Chapter 9

Contribution of CABI and culture collections to a sustainable future through the utilisation of microbial genetic resources

David Smith, Matthew J. Ryan and Thelma Caine
CABI, Bakeham Lane Egham, Surrey, United Kingdom

Chapter outline

Importance of Microbiology Teaching and Microbial Resource Management for Sustainable Futures
https://doi.org/10.1016/B978-0-12-818272-7.00010-9

Introduction

CABI is an international not-for-profit organisation (https://www.cabi.org/), owned by 50 member countries. It improves people's lives by providing information and applying scientific expertise to solve problems in agriculture and the environment. With a presence in more than 20 of its member countries and regional centres in Brazil, China, Ghana, India, Kenya, Malaysia, Pakistan, Switzerland, Trinidad and Tobago, UK, USA and Zambia, CABI has established an extensive global stakeholder network to enable it to deliver its mission. CABI provides access to agricultural and environmental scientific knowledge and improves global food security by knowledge sharing and providing solutions to combat pests, diseases and invasive species, thereby helping farmers to grow more and lose less produce.

The CABI collection

To underpin these activities, it has built up a living collection of 28,000 strains of fungi and 2000 strains of bacteria over its 100 years of research and development in the field. CABI's work in mycology began in 1920, originating as the Imperial Bureau of Mycology. Its remit at that time was to accumulate information and identify plant pathogenic fungi for countries of the British Empire overseas (Aitchison and Hawksworth, 1993; Blight and Ibbotson, 2011). The main aim of this work, which has continued throughout its entire 100-year history, was to help reduce food losses. The Bureau of Mycology became the Imperial Mycological Institute (IMI), subsequently renamed as the Commonwealth Mycological Institute (CMI), CAB International Mycological Institute and the International Mycological Institute (IMI) (Aitchison and Hawksworth, 1993) before being integrated as one of four scientific institutes in the CABI UK Centre. Laboratory work and field-based research has largely involved isolation, growth and identification of fungi and bacteria and their subsequent study to determine their roles in ecosystems. Initially, many strains were received from soil and diseased crops, particularly from CABI's member countries. Following identification, these were deposited in the living and dried reference collections. Numerous strains of industrial importance have also been deposited since then for study and potential use examples include those described in Box 9.3 below. They also include almost 3000 potential biological control agents and strains used as biopesticides, a particular strength of research teams at CABI. The main genera of these include *Aschersonia*, *Aspergillus*, *Aureobasidium*, *Beauveria*, *Colletotrichum*, *Entomophthora*, *Fusarium*, *Geotrichum*, *Hirsutella*, *Hypocrea*, *Isaria*, *Metarhizium*, *Mortierella*, *Nomurea*, *Paecilomyces*, *Penicillium*, *Phomopsis*, *Trichoderma* and *Verticillium*. The details of the strains in the CABI collection can be found in the online catalogue (https://www.cabi.org/services/microbial-services/culture-collection-microorganism-supply/grc/).

CABI information resources

CABI is renowned for its extensive scientific information resources, key products being the Review of Applied Mycology, first produced in 1922, the Index of Fungi (from 1940), the Distribution Maps of Plant Diseases (from 1942), the Bibliography of Systematic Mycology (from 1943) and the Descriptions of Plant Pathogenic Fungi and Bacteria (from 1964). These publications have provided essential resources for mycologists and plant pathologists worldwide. The first edition of the authoritative Dictionary of Fungi was published in 1943 and has been revised periodically since then, the most recent (10th) edition having been published in 2008.

Research in mycology requires high-quality validated reference strains of fungi. The supply and distribution of fungal cultures by the Bureau was first envisaged by Edwin Butler (Director, 1920–35) in a report on activities (Butler, 1921). Isolates received for identification or collected by the scientists formed the nucleus of the CABI living culture collection (https://www.cabi.org/products-and-services/bioscience-services/microorganism-supply-services/).

CABI networking

Through its membership of 50 countries and extensive scientific networks, CABI has unique insights into global challenges. Its wealth of information and living resources have provided the foundation to tackle and mitigate new and emerging plant diseases, invasive species, climate change, pollution, land use change and other threats to biodiversity. One of CABI's flagship projects, Plantwise, has 115 partners in 33 countries, running >3500 plant diagnostic clinics, reaching 30 million farmers and providing expertise to enable them to increase production and improve their livelihoods (https://www.cabi.org/news-article/cabi-plantwise-collaboration-with-ippc/).

Recently, the importance of understanding plant microbe interactions and soil microbial communities has grown and CABI and its partners are now taking a leading role to meet global needs in the understanding, harnessing and modulating the agricultural microbiome. This could lead to such outcomes as restoring soil health and thus benefit society by improving food security. Microbial interventions and solutions require knowledge of not only those microorganisms that can be grown and preserved axenically but whole microbial communities and how they interact with their hosts and the environment. Current technologies enable access to this hidden resource and the development of strategies to harness and manipulate microbial communities to improve crop yields and facilitate successful interventions such as biocontrol of pests, diseases and invasive species (Ryan et al., 2019).

Furthermore, CABI's 100 years' experience in the legal and ethical use of genetic resources has been a key factor in building workable, effective and legitimate partnerships that bring together a critical mass of multidisciplinary entities to deliver the solutions agriculture and the environment needs. This includes establishing legal frameworks to address the different regulatory systems implemented by countries, e.g., on Access and Benefit Sharing. With regard to utilisation of genetic resources, CABI's policy and best practice procedures are adapted to meet the national requirements of its host and provider countries and it is well placed to reflect the requirements, opinions and guidance of stakeholders globally.

CABI's collection of living microorganisms is managed within the Bioscience department by a team of specialist staff who provide microbial expertise, with particular emphasis on agricultural and biotechnological applications, microbial preservation techniques and molecular biology. The collection adopted the World Federation for Culture Collection (WFCC) guidelines and subsequently the Common Access to Biological Resources and Information (CABRI) guidance and has been accredited to ISO standard since 1983.

CABI's work on cryopreservation of individual strains, and more recently, microbiomes, employs technologies that retain strain functionality, an important factor to enable their successful application. In addition to the organisms themselves, CABI has unique expertise in systematically collecting and providing proprietary access to information and data on agriculture and the environment, including publishing the world's largest applied life science research database, CAB Direct (>13 million records). This expertise has enabled CABI to deliver the UK Department for International Development's R4D portal, and to develop and maintain the open-access Plantwise Knowledge Bank. In partnership with other organisations (e.g., EMBRIC, CORBEL, MIRRI, EBRCN, EMbaRC, MOSAICS, EU Microbiome Support and HOMED), CABI has undertaken many research projects (see Table 9.1).

Through these and other programmes, CABI has supported the establishment of in-country microbial collections. Experience gained through leadership of aspects of the OECD Biological Resource Centre Initiative and co-management of the Global Biological Resource Centre Network (GBRCN) has been invaluable in enabling CABI to contribute to the design and implementation of training programmes and to support other institutions in establishment, management and operation of microbial domain Biological Resource Centres (mBRCs).

There are huge numbers of public and private collections around the world that support such work (Smith, 2012). The majority of microbiology departments in universities, microbial-based research institutions and bioindustry have such collections, and although most microbiologists keep their own collections of research and reference strains, many of these are private and inaccessible. However, there are nearly 800 'public service' culture collections registered with the World Data Centre for Microorganisms (WDCM) which together hold 3,233,009 strains, representing a wide range of microorganisms. The WDCM registered collections supply their strains for research, many, though not all, having online catalogues. Metadata associated with the

TABLE 9.1 Examples of collaborative research projects.

Project	Period	URL	Overview
EMBRIC - **E**uropean **M**arine **B**iological **R**esearch **I**nfrastructure **C**luster	2015–19	http://www.embric.ugent.be/	29 partners came together to accelerate the pace of scientific discovery and innovation from marine Bio-Resources
CORBEL - Coordinated Research Infrastructures Building Enduring Life-science Services	2015–20	https://www.corbel-project.eu/home.html	13 new biological and medical research infrastructures created a platform for harmonised user access to biological and medical technologies, biological samples and data services required by cutting-edge biomedical research
MIRRI - Microbial Resource Research Infrastructure	2012	https://www.mirri.org/	>50 public biorepositories from create a unique pan-European high-performance platform adding value to known and yet unknown microbial biodiversity for the bioeconomy and bioscience
EBRCN - European Biological Resource Centres Network	2004–06	http://www.cabri.org/FAQ/faq_ebrcn.html	Set common policies and maintain high standards of operation, linking BRC catalogues and other data sets through a single common website (CABRI),co-ordinate responses to new initiatives relevant to living resources
EMbaRC - European Consortium of Microbial Resources Centres	2009–13	http://www.embarc.eu/	Key microbial resource centres in Europe collaborated to improve, coordinate and validate microbial resource delivery to European and International researchers from both public and private sectors
MOSAICS - Microorganisms Sustainable use and Access management Integrated Conveyance System	2004–09	https://bccm.belspo.be/projects/mosaics	Building on MOSAICC, to develop a code of conduct as a practical tool for microbiologists to implement the CBD the project identified the necessary features for a system to be implemented coherently

Continued

TABLE 9.1 Examples of collaborative research projects.—cont'd

Project	Period	URL	Overview
EU Microbiome Support	2018-22	https://www.microbiomesupport.eu/	Creates a collaborative EU network platform to encourage innovation, promote knowledge exchange and focus microbiome research to impact on the bioeconomy
UK Crop Microbiome CryoBank (UK-CMCB)	2020–25	https://www.cabi.org/news-article/for-when-the-chips-are-down-preserving-uk-soil-microbial-biodiversity-for-sustainable-agriculture/	Establish a cryopreserved and characterised crop microbiome resource. The focus will be on the microbiomes of major UK crops from a variety of different soil types obtained from across the UK.

collections in the WDCM Culture Collections Information Worldwide (CCINFO) database can provide good background to the strains and their properties and over 130 of these collections have made their strains available via the Global Catalogue of Microorganisms (GCM) which lists over 450,000 of the WDCM collection's strains, representing over 55,000 species (http://gcm.wfcc.info). The collections vary in size and coverage and are often linked more closely through regional or national organisations such as the European Culture Collections' Organisation (ECCO) or the UK Biological Resource Centre Network (UKBRCN). They are also linked in the short term through project consortia or permanent infrastructure, such as the Belgian Coordinated Collections of Microorganisms (BCCM) and the currently establishing Microbial Resources Research Infrastructure (MIRRI). The WDCM Analyzer of Bio-resource Citations states that 79,224 strains from 131 registered collections from 50 countries have been cited in 145,133 papers published in 50,307 journals since January, 1953. These organisms have a multitude of beneficial uses as producers of biologically active derivatives for drug production and in biodegradation, bioremediation, biotransformation and interventions in agriculture for soil and plant health and biological control of pests and diseases (Overmann and Smith, 2017). The focus here is on how mycological resources can address key global issues and how they can be better organised and coordinated to underpin future advancement in science and innovation in healthcare, food security, the environment, education and the bio-based economy. For many years, organisations such as WFCC and ECCO collections have worked together towards common goals. The extent to which collections adapt and change to new technology, advancing science and societal needs is illustrated by the range of development, networking and joint research projects listed in Tables 9.1 and 9.5.

CABI operations

However, the access, movement and use of microorganisms is not always straight forward as a multitude of legislative and regulatory issues impact on such activities. CABI is active in the area of compliance and has worked with culture collection organisations (WFCC, ECCO) and research partners to address shipping, biosecurity, access and benefit sharing (Nagoya Protocol) and Intellectual Property Rights in particular. These have been addressed elsewhere (Kelley and Smith, 1997; Rohde et al., 2013; Ryan et al., 2019; Smith et al., 2017a, 2017b; Smith and Ryan, 2019; Verkley et al., 2019).

Collections must operate to internationally recognised criteria if they wish to provide services to research and collaborate in providing coordinated activities. Many collections have adopted general standards that can be applied to microbiology laboratories such as Good Laboratory Practice and several established by the International Standards Organization (ISO), e.g., ISO 17025, ISO Guide 34 and the ISO 9000 series (Martin et al., 2015). Others have drawn up their own standards, for example, the Association Française de Normalisation has developed the French standard NF S96-900 ‘Quality of biological resource centres (BRCs) – Management system of a BRC and quality of biological resources from human or micro-organism origin’ (AFNOR, 2008). The Brazilian network of collections worked with their accreditation bodies led by IMMETRO - National Institute of Metrology, Quality and Technology to design and publish the standard NIT-Dicla061 Aplicaçõese Requisitos Adicionais Acreditação ABNT NBR ISO/IEC 17025 dos Centros de Recursos Biológicos for the accreditation of BRCs (www.inmetro.gov.br/credenciamento/pdf/nit_dicla_061.pdf). Some simply follow the guidance of culture collection organisations such as the WFCC guidelines (http://www.wfcc.info/guidelines/) or the CABRI guidelines (www.cabri.org). In any event, collections need to ensure they provide high-quality resources for research.

Range of holdings in the CABI collection and their uses

The CABI collection now holds around 28,000 fungi, including yeasts, and over 2000 bacteria. The collection’s first catalogue was produced in 1953 and listed 1418 strains comprising 263 genera and 796 species which included 195 strains that had been transferred from the NCTC. The catalogue provided background on the partnership that made up the UK organisation of culture collections (Anon, 1953). The preface states that whilst the function of the collection was partly ancillary to the taxonomic work of the Commonwealth Mycological Institute, it is also a public collection, distributing living cultures of fungi of interest in plant pathology, industry, taxonomic research and education. Special sets of cultures were supplied for the purpose of testing mould-proofing treatments of commercial products. Manufacturers were invited to consult the institute regarding these tests. The collection also included many strains used for the synthesis of organic substances, and others which were used in physiological assay, soil analysis, etc. In essence, these remain the key objectives of the CABI collection today. At the time, CMI had a booklet in preparation for school and university teachers giving an annotated list of fungi useful in education and which also gave instructions for growing them and preparing them for classes (Dade and Gunnel, 1966). CABI has continued in that tradition, playing a role in transferring knowledge and skills in mycology.

CABI's assigned role to incorporate the UK National Collection of Fungus Cultures initiated a partnership that continues today through the UK Biological Resource Centre Network (UKBRCN - www.ukbrcn.org). A list of other units of the United Kingdom organisation of collections in 1947 is given in Table 9.2; 8 additional collections joined this network (total of 14 collections) now forming the UKBRCN; these are listed in Table 9.3.

In 1953, cultures from CMI were supplied to industry and to trade research associations in the United Kingdom and British Commonwealth at l0 shillings per culture (50 pence) including packaging and post. Donors of cultures were given free exchange (an equal number of strains to that deposited and they were entitled to access their 'gift culture' free) although the maintenance was not guaranteed. Free exchange and access to their own cultures remains the offer to current depositors. Table 9.4 lists some interesting strains that appeared in the first edition of the catalogue that remain available today via the CABI collection online catalogue (https://www.cabi.org/services/microbial-services/culture-collection-microorganism-supply/grc/). Notable is the Fleming strain IMI 15378 which was originally deposited in the NCTC as *Penicillium notatum.* When it arrived at CMI, George Smith, a mycologist at the London School of Hygiene and Tropical Medicine where the NCTC was held originally, noted on the CMI accession form that 'the identification was doubtful' and he was proven correct in that it is now known to be *Penicillium rubens*. Other strains demonstrate CABI's early involvement in biocontrol, e.g., *Beauveria bassiana* isolated from insect pests and used as a natural enemy against them to prevent damage to crops and stored products.

Agricultural sector

CABI holds representatives of pathogens for use in epidemiological studies, biocontrol, soil fertility, plant health and biomolecules in biocides and strains that provide a source of genes for genetic modification of crops. The collection maintains more than 5000 plant pathogens (Ryan et al., 2012a) and holds over 3000 potential biological control agents, including 21 fungal parasites of other

TABLE 9.2 **United Kingdom organisation of collections 1953 to date.**

Collection	Location in 1953	Location in 2020	Link
National Collection of Fungus Cultures	Commonwealth Mycological Institute, Kew	CABI Genetic Resource Collection CABI UK Centre, Egham	https://www.cabi.org/products-and-services/bioscience-services/microorganism-supply-services/
Fungi and yeasts pathogenic to man and animals	London School of Hygiene and Tropical Medicine	Now the National Collection of Pathogenic Fungi (NCPF), PHE	https://www.ukbrcn.org/who-we-are/national-collection-of-pathogenic-fungi-ncpf/
Wood-rotting fungi	Forest Products Research Laboratory	Now absorbed into the CABI collection	https://www.cabi.org/products-and-services/bioscience-services/microorganism-supply-services/
Yeasts	The Brewing Industry Research Foundation	Now the National Collection of Yeast Cultures (NCYC) at the Quadram Institute Bioscience, Norwich	https://www.ncyc.co.uk/
National Collection of Industrial Bacteria	Teddington	Now the NCIMB Ltd., Aberdeen	https://www.ncimb.com/
Bacteria pathogenic to plants	Sub-department of Mycology and Plant Pathology, Botany School, Cambridge	Now the National Collection of Plant Pathogenic Bacteria (NCPPB), Food and Environment Research Agency	https://www.fera.co.uk/ncppb
Bacteria pathogenic to man and animals	Central Public Health Laboratory	Now the National Collection of Type Cultures (NCTC), PHE	https://www.ukbrcn.org/who-we-are/national-collection-of-type-cultures-nctc/
Bacteria of milk and dairy products, bacteria from soil	Agricultural Research Council at Reading as the national Collection of Dairy and Food Organisms	Now absorbed into NCIMB Ltd, Aberdeen	https://www.ncimb.com/

TABLE 9.3 Additional members of the UKBRCN.

Collection	Organisms	Location	Link
Animal and Plant Health Agency (APHA)	Animal and plant microbes	Department of Food and Rural Affairs,	http://apha.defra.gov.uk/apha-scientific/services/culture/index.htm
CCAP is a Biological Resource Centre (BRC)	Cyanobacteria, Algae, Protozoa	Scottish Association for Marine Science campus, Dunstaffnage, near Oban	https://www.ukbrcn.org/who-we-are/culture-collection-of-algae-and-protozoa-ccap/
Centre for Environment, Fisheries and Aquaculture Science (Cefas)	Bacteria	Department of Food and Rural Affairs	https://www.cefas.co.uk/
European Collection of Authenticated Cell Cultures (ECACC)	Cell lines, Stem cells	Public Health England, Porton Down, Salisbury	https://www.ukbrcn.org/who-we-are/european-collection-of-authenticated-cell-cultures-ecacc/
Kew Scientific Collections	Seed bank, Endophytic Microbes	Royal Botanic Gardens Kew	https://www.kew.org/science/collections-and-resources/collections
National Collection of Pathogenic Viruses (NCPV)	Viruses	Public Health England, Porton Down Salisbury	https://www.ukbrcn.org/who-we-are/national-collection-of-pathogenic-viruses-ncpv/
National Institute for Biological Standards and Controls (NIBSC) – UK Stem Cell Bank	Stem Cells	Medicines and Healthcare products Regulatory Agency	https://www.nibsc.org/
Natural History Museum	Molecular Collections	Natural History Museum, London	https://www.ukbrcn.org/who-we-are/natural-history-museum/

TABLE 9.4 Some interesting first edition CABI catalogue entries still available today.

Name	IMI	History
Absidia glauca Hagem (Syn. A. regnieri) (Lucet and Cost.) Lendnprl	15,405	W.H. Wilkins – NCTC (3098), 1929. Mates with IMI 15406
Absidia glauca Hagem (Syn. A. regnieri) (Lucet and Cost.) Lendnprl	15,406	W.H. Wilkins – NCTC (3099), 1929. Mates with IMI 15405
Aspergillus alliaceus Thom and Church	16,125	Church – CBS – CMI, 1937. NRRL 316
Aspergillus amstelodami (Mangin) Thom and Church	17,455	Isolate used tor testing mould-proofing - G. Smith - CMI, 1947, QM 8405
Aspergillus avenaceus G. Smith	16,140	Type; seed peas. England. 1938. G. E. Turfltt - G. Smith (B. 155).
Aspergillus fumigatus Fresen.	16,030	White egret (aspergillosis of lungs) London Zoo, 1920 (NCTC 367) Fide Soltys Nature 154, 550.
Aspergillus fumigatus Fresen.	25,326	Owl (*Athene noctua*) London Zoo, 1922 (NCTC 1373)
Aspergillus luchuensis Inui	15,953	"Awamori" (alcoholic beverage), Japan. Hanzawa - Thom (4291.3) – NCTC (1017) - CMI, 1941 (= NRRL 356)
Aspergillus niger v. Tiegh	31,275	Man (otomycosis). Pijper (12) and Pullinger – NCTC (2389), 1927.
Aspergillus tamarii Kita	16,129	1923 Fermented cocoa, Ghana
Beauveria bassiana (Bals.) Vuill.	12,943	*Scolytus intricata.* E.J. Butler (as B. globulifera (Speg.) Picard – NCTC (11l42), 1921.
Beauveria bassiana (Bals.) Vuill.	44,229	*Diatraea saccharalis*, Venezuela, 1950. R. E.Pontis.
Beauveria bassiana (Bals.) Vuill.	49,823	*Artona catoxantha* larva, Malaya, 1952. A. Johnston.
Curvularia crepini (Westend.) Boedijn (Syn. *Brachysporium crepini* (Westend.) Sacc.)	605	*Ophioglossum vulgatum*, Warbush, Hickling Broad, Norfolk, 1945. E.A. Ellis.
Dematium nigrum Link	17,467	Bainier Collection – B. Church (4719.3). NCTC 1745, 1924.

Continued

TABLE 9.4 Some interesting first edition CABI catalogue entries still available today.—cont'd

Name	IMI	History
Penicillium camemberti Thom	27,831	ExType. French Camembert cheese Storrs, Connecticut, USA, 19904. C. Thom - NCTC (582), 1920. NRRL (877)
Penicillium chrysogenum Thom	26,210	Penicillin production in submerged culture. Raper – NRRL 1951, B.25, NCTC (7088), 1948
Penicillium commune Thom	39,812	ExType. Cheese, Storrs, Conn., USA, 1904. C. Thom (23); NRRL (890) - G. Smith.
Penicillium glabrum (Wehm) Westl.	28,043	A. Jorgerensens Lab. Copenhagen - A.C. Thayson – NCTC (606) (As *Citromyces*) 1920.
Penicillium roqueforti Thom	24,313	Roquefort cheese, 1904. Ex-Type species, Thom (18)-NCTC (588), 1920. (NRRL 849; ATCC 10110)
Penicillium rubens	15,378	Deposited as *Penicillium notatum* Westl. Contaminant of culture, London, 1928. Sir Alexander Fleming's historical isolation. Fleming—Thom (144. 5112.1) as P. rubrum. NRRL (824) - G. Smith. (= ATCC. 9478). Gives high yield of notatin.
Penicillium vinaceum Gilm. and Abbot	29,189	ExType. Soil, Utah. 1949. Gilman and Abbott – Thom (4894.15) – NRRL (739) - G. Smith, 1949. (Raistrick. 10514).
Penicillium waksmani Zal.	39,746	ExType. Zaleski – CBS – Thom (5010. 36), 1928; NRRL (777) - G. Smith.
Memnoniella echinata (Riv.) Galloway	15,377	Decayed water-bag, S. Africa 1947. A.L. James - CMI, 1946. Strain used in South Africa for cellulose decomposition
Memnoniella echinata (Riv.) Galloway	16,201	Cotton fabric, Calcutta, India, 1943. S.W. Basu (73)
Memnoniella echinata (Riv.) Galloway	45,547	Tent canvas, Solomon Islands.QM1c, 1951. Used to test mould proofing.
Rhizopus oryzae Went and Geerl.	21,600	Central Lab. Manchurian Railway. NCTC (1905), 1924.
Rhizopus oryzae Went and Geerl.	21,601	Central Lab., Manchurian Railway. NCTC (1906), 1924.
Rhizopus oryzae Went and Geerl.	21,602	Lab. Manchurian Railway. NCTC (1907) 1924.
Scopulariopsis brevicaulis (Sacc.) Bain.	17,297	Cheese, USA. M.B. Church Washington (2) – NCTC (580), 1920. (= Raistrick Ad 40)

fungi, plus those for use against specific crop pests and diseases including: Acridids; Citrus weevil; Coffee disease; Grasshoppers; Gryllidae; Locusts; Onion white rot; Orthoptera; *Pythium ultimum;* Root borer; Root grub; *Scapanes australis; Sclerotinia sclerotiorum; Sitona discoideus; Teleogryllus* and Termites. Also, of relevance to agriculture are the environmental strains involved in soil fertility, plant symbiosis and natural enemies of invasive species. A major problem currently is that many microorganisms have become resistant to fungicides and antimicrobials. Collections can not only provide candidates for new antimicrobials and biocides but can also provide comparative data on species before and after they develop resistance. This enables a better understanding of how they have changed and therefore potential routes to overcoming resistance (Box 9.1).

CABI's biotechnology solutions for industry

The CABI collection includes environmental strains from unusual ecosystems and a wide range of man-made sources, examples of which are given in Box 9.2. These have the potential for a huge range of uses. Of importance in biotechnology are bioremediation strains that can be utilised for environmental 'clean-up' of substrates such as coal spoil; industrial waste; lignin, paint spill; paper mill slime; polycaprolactone, polyurethane, polyester; polyether, poly vinyl chloride (PVC), saline soil, starch wastes and waste water. In addition to those strains that can provide a 'microbial fix' for polluted sites, the collection includes many potential inoculants that could be used to re-introduce important environmental strains in land reclamation. Strains from unusual environments include those derived from coal spoil material, dung, nematodes,

BOX 9.1 Target resistance genes to kill or inactivate the resistant disease

Resistant strains have developed through mutation and natural selection in response to environmental factors such as pollution, antibiotic use and application of chemical pesticides. The CABI collection holds strains that were isolated from plants and the environment before these factors became problematic. Combating disease through the understanding of gene function is the modern approach. An important target currently is to determine the genetic differences between the pre-resistant and resistant strains. This will provide insights towards development of active compounds to impede gene function which is responsible for resistance. The ancestral strains held in the CABI collection will facilitate this study. DNA technologies are not limited to living organisms. CABI established a dried reference collection of nearly 400,000 strains of over 32,000 species of fungi collected over an 80-year period that is now housed in the *Fungarium* at the Royal Botanical Gardens, Kew. This collection will also contribute towards finding genetic solutions to resistance strains.

BOX 9.2 The source of environmental strains in the CABI collection

Acetic preserve; Aircraft fuel tank; Alkali pulp manufacture; Animal specimen in formalin; Army socks; Art; Artificial leather; Artificial leather cloth; Asbestos cord; Asbestos substitute building; Aviation fuel; Baby powder; Balsa blocks in river; Balsa wood submerged in patuxent river; Batter pudding; BEA vanguard wing tank; Binocular lens; Bitumen treated cardboard; Bromthymol blue indicator solution; Canteen cover; Cannery (walls and ceiling); Canvas; Canvas bag (Officer's); Canvas in mine; Cellophane (buried in soil); Cellophane (weathering treated); Cellophane in sand; Cellophane paper; Cellulose board; Cellulose fibres; Cellulosic material; Charcoal; Cigar tobacco; Cloth; Coal mine spoilage; Concrete wall; Contact lens; Copper electroplating solution; Copper sulphate solution; Cotton; Creosoted wood; Crude oil; Cutting fluid; Dextrin paste; Diesel; Diso naval fuel; Distilled water container; Earthenware plant pot; Electrical equipment; Electrical fuse; Electrical wire sleeving; Explosives; Film; Flannel bag; Formalin tanks (surface); Fuel filters; Gelatine factory effluent; Glycerine; Glyphosate solution; Hardwood; Heat seal foil material; Hydraulic oil; Ice cream carton; Industrial waste; Iron bark in cooling tower; Italian talcum powder; Kerosene fuel filter; Label (museum); Leather; Leather boot; Lignin; Linseed oil-based red primer; Lubricating oil; Manila fibre; Marine diesel oil; Metal working fluid; Military equipment; Mine timber; Mines; Mould proofing; Mattress cover; Mussels in preserving fluid; Nylon; Oil palm trunk; Ointment; Optical instrument; Paint; Paint (emulsion with chromate); Paint (phenolic tung-Zn oxide/chromate); Paint (red lead); Painted beach box; Painted surface metal amplifier case; Painted wood; Paint work; Palm oil; Paper; Paper pulp; Paracetamol tablet; Parachute nylon; Paradise nuts; Paraffin wax; p-digunidino-benzoyl-L-glutamic acid hydrochloride; Pilot plant using kerosene for copper extraction; *Pinus strobus* boards; Plaster and wallpaper; Plastic; Plastic covered nylon sail cover; Polycaprolactone polyurethane; Polyester-based Polyurethane sheet; Polyester polyurethane; Polyester polyurethane foam; Polyether polyurethane; Polypropylene divers hose; Polyurethane; Polyurethane-coated fabric; Polyurethane foam; Polyurethane footwear; Polyurethane products; Polyurethanes used in soil burial experiments; Potassium thiocyanate containing mercuric nitrate; Preservative treated telegraph pole; PVA emulsion; PVA emulsion paint; PVC; PVC-covered wallpaper; Quinine containing medicine; Radio equipment; Radio set; Rubber; Saline soil; Sawdust; Seed -Organomercury treated; Shoe leather; Silk screen of duplicating machine; Starch wastes; Stone; Sugar sack; Synthetic rubber; Talcum powder; Tanning liquor; Tent canvas; Textiles; Vinyl pyridine latex; Vinyl wall covering; Wallpaper; Waste water; Water bag; Wine bottle cork; Wood; Wood pulp and Zinc-based oil paint.

river foam, rock and sewage. Also, amongst the CABI holdings are strains tolerant or resistant for example to acid, alkali, benzimidazole, benomyl fungicide, copper, phenol, mercury and thiabendazole. Such strains are useful when organisms are needed that can grow in harsh polluted environments. The collection also includes many species adapted to high- and low-temperature environments including an extensive collection of Antarctic strains.

Other man-made environments may present organisms with yet unknown benefits (Box 9.3). For example, knowledge of food spoilage organisms and understanding their origins may help to avoid future occurrences and prevent them becoming problematic.

The collection has great potential for characterising strains and finding properties that can be harnessed to provide new products. The limited screening of cultures held at CABI thus far has revealed 52 different enzymes produced by 130 strains. These include amylase, cellulase, lipase and proteolytic enzymes. The primary and secondary metabolites recorded as products from the CABI collection include alcohols, alkaloids, amino acids, peptides and proteins, carbohydrates, hydrocarbons, ketones, lipids and fatty acids, organic acids, pigments, plant hormones and growth regulators, steroids and sterols, toxins, carcinogens, mutagens, teratogens, vitamins, coenzymes and growth factors. Assay strains in the collection include those used in the assay of antibiotics and vitamins; p-Aminobenzoic acid; B317-9-9a Arginine 10; B46004-Arginine 1; Choline; Copper Mulders strain in soils; Inositol; Magnesium in soils; Molybdenum in soils; Phosphorous in soils; Potassium in soils; Proline and Zinc in soils.

The collection of human pathogenic strains is not one of the aims of CABI. However, there are many strains of relevance to the medical field. Not least are the 48 strains that produce 20 antibiotics and the 1000 strains of opportunistic human pathogens (Box 9.4). Of greater significance are the producers of metabolites with medical applications.

BOX 9.3 The source of food spoilage strains held in the CABI collection

Apple; Apricot; Australian apple; Baker's yeast (mouldy); Blackcurrant; Blue peas (dried); Bread; Butter; Cake; Canned blueberries; Canned peaches; Cashew nuts; Cheese; Chocolate; Chocolate covered sponge biscuit; Chocolate sponge biscuit; Chopped Chinese dates; *Citrus sinensis*; *Citrus paradisi;* Coconut (desiccated); Cooked meat; Corn (mouldy field); Corn and rice cake; Corn, barley and sorghum grain; Cottage cheese; Crumpets; Dairy products; Dried fish; Dried prunes; Fermented cocoa; Fermented rice grains; Fruit-based gel (baby food); Fruit sauce (sweet); Gelatine; Grape; Grapes (stored); Groundnuts; Hazelnut; Hops (stored); Irn Bru drink; Jam; Lemon ice tea; Lemon juice; Lime flavour carton drink; Liquorice; Madeira cake; Maize (stored); Malt extract; Margarine; Meat; Milk; Milled rice; Millet alcoholic beverage; Nigerian groundnuts; Orange; Pecans (weevil damaged); Pepper; Pistachio nuts; Plum juice; Rye bread; Salami sausage; Salt fish; Salted anchovy paste; Sausage product; Savoury preserve; Soy products; Stick liquorice; Stored food; Sugar; Table jelly; Tobacco; Tobacco (cured); Tobacco (pipe); Walnut kernel; Wheat bread and Wine.

BOX 9.4 Fungi and the immuno-compromised

There are over 400 fungal species associated with disease in man, yet only 31 species and 9 other genera are considered to belong to hazard group 2 or above by the UK Advisory Committee on Dangerous Pathogens (ACDP, UK). The majority of fungal human pathogens are opportunistic and mainly infect immuno-compromised individuals. It is necessary to determine what makes these fungi different from all the others. Targeted genome sequencing can help reveal the difference between these organisms and their close relatives to help us understand which organisms have the capacity to cause infection and why. It may then be possible to target the genes involved and produce effective drugs. The 1000 opportunistic pathogenic strains and their great numbers of relatives in the CABI collection can make this possible.

There are around 2000 strains in the collection that have been screened at CABI for products, less than 8% of the holdings. Of these, 750 strains have been shown to produce 442 metabolites of potential.

Test strains

CABI holds and provides strains for proficiency testing and mould resistance testing, specifically for the following: Aerosols; Antimicrobial preservative testing; ASTM G21; Fungus resistance test BS2011 PART 2J; Fungus resistance test DEF 133; European standard EN152 (I and II); Fungal toxicity of alcohols and fatty acids; Fungus resistance test MIL 810D (MIL C 62122b); Wall covering adhesive and Wood preservatives. Often the requirement of the standard is that these strains are grown on a specific substrate or growth medium to validate the test. To ensure these strains retain their properties to challenge the materials tested, they are maintained under long-term storage such as freeze-drying or cryopreservation. Some of the recommended strains were isolated in the 1940's and have been maintained in culture collections ever since. The WDCM provides a catalogue produced to enable broader and easier access to the reference strains listed by the ISO Technical Committee 34 SC 9 Joint Working Group 5 and by the Working Party on Culture Media of the International Committee on Food Microbiology and Hygiene (ICFMH-WPCM) in their publication *Handbook of Culture Media for Food and Water Microbiology* (http://refs.wdcm.org/home.htm).

CABI services based on the collection

The Imperial Bureau of Mycology, which later became the Commonwealth Mycological Institute, was established in 1920 to identify fungal pathogens of crops. This service, run by CABI currently through its Bioscience operation, has continued for 100 years. When the UK National Collection of Fungus Cultures was founded in 1947, the Bureau received a grant of £500 (Report, 1947), and soon after, culture supply for research became a key objective for the collection. The UK Government continued to partially fund the collection until 1989, by which time dedicated Biochemistry and Industrial Mycology laboratories had been established with the aim of generating revenue lines to sustain the collection (Aitchison and Hawksworth, 1993). CABI's business model is similar to many other culture collections (Smith et al., 2014) relying on a mix of product, service and research revenue lines. CABI also benefits from support from member countries to help them maintain key strains from research.

Microbial Identification Service

The CABI Microbial Identification Service provides identification of filamentous fungi, yeasts and bacteria for a broad range of industries and for plant pathologists, plant health authorities and researchers worldwide working in agriculture, horticulture, forestry, quarantine, biodiversity assessment and biological control. From 1920, when the service was established, fungal identification was based on examination of morphological features and comparison with taxonomic keys. CABI taxonomists were instrumental in publishing a number of these (Ellis, 1971, 1976; Ellis and Ellis, 1997, 1998; Sutton, 1990). As samples were received and identified from around the world, the living and dried reference collections were established; the Fungarium collection of dried specimens (Herb IMI) in 1945 and the living collections were consolidated in 1947. These provided essential reference material to support taxonomic work. Identification of plant pathogenic and environmental bacteria began in 1959 and the service was further extended in the early 1990s to include yeasts. Initially, identification of bacteria and yeasts was based on phenotypic and biochemical methods, these being superseded in the 1990s by the introduction of fatty acid analysis using gas liquid chromatography (Aitchison and Hawksworth, 1993).

Identification methods based on molecular sequencing were introduced in 2003 and CABI now provides a service accredited to ISO 17025 for the molecular identification of filamentous fungi, yeasts and bacteria. For bacteria, this involves sequencing part of the 16S rRNA gene subunit, and for filamentous fungi and yeasts, sequencing of the internal transcribed spacer region of the rRNA gene is undertaken. Where necessary, additional loci, including beta-tubulin, calmodulin and translation elongation factor 1α, are sequenced to provide definitive identification for fungi. Results from each sample are matched against global databases of sequenced organisms and the data interpreted by specialists to determine the identity. Accurate identifications are essential for pharmaceutical, food and manufacturing industries in environmental monitoring of production areas, sterility testing, validation of reference organisms used in quality control and to establish the identity of microorganisms used in development of novel products. For plant pathologists, plant health authorities and researchers working in agriculture, the service assists in detection of disease organisms, reduction of pre- and post-harvest losses and avoiding the spread of non-indigenous microorganisms in quarantine and international trade.

In recent years, identification resources at CABI have been strengthened by the acquisition of advanced technologies. These include the matrix-assisted laser desorption and ionisation time-of-flight mass spectrometry, Bruker Microflex instrument and Biotyper software that determine the mass of acid-soluble proteins (including ribosomal proteins) in a given sample and then compares the profile of peaks obtained against a database of known samples. In addition, the Next Generation Sequencer Illumina MiSeq provides a powerful detection tool for environmental DNA which enables the detection of tiny amounts of living material in complex sample mixtures. This tool also enables increased resolution of particular organisms of interest through whole genome sequencing of bacteria, full ribosomal RNA gene clusters of fungi, plants and invertebrates, mitochondrial genomes in fungi, plants and invertebrates and chloroplast genomes of plants. Added to these, the MinION (Oxford Nanopore Technologies) provides a high-throughput sequencing tool as an excellent partner for the MiSeq in whole genome studies and field-based work. Its long-read technology is best suited to enabling 'scaffold creation' in genomics studies with the MiSeq. Another technology, the qPCR Eppendorf Realplex Mastercycler, enables reactions to be observed in real time to determine the relative amount of a particular organism or group of organisms from a given DNA sample and can also be used to determine levels of gene expression. When used with RNA starting product, it has the potential to reveal absolute quantification of the target organism or gene. This method allows very sensitive, accurate and robust detection of given microorganisms (e.g., quarantine organisms) in complex samples, particularly when *allied with the use of taxon-specific probes* (*and following the 'gold standard' MIQE guidelines*) (Photograph 1).

As methodologies for fungal identification have changed from morphological to molecular-based approaches and phylogenetic studies have revealed differences in relationships between taxa from those previously described, fungal classification has reached a stage of flux. This has been exacerbated by the implications of the recent move to 'one fungus = one name' (Taylor, 2011; Taylor et al., 2016). This principle was adopted into the International Code of Nomenclature for algae, fungi and plants (Shenzhen Code) by the 19th International Botanical Congress Shenzhen, China, July 2017 (https://www.iapt-taxon.org/nomen/main.php).

Providing the correct scientific name of an organism is of vital importance as the name provides the link to all available information about the organism, an essential factor for assessing the properties of the organism and its potential

PHOTOGRAPH 1 Dr Alan Buddie on sequencer which was introduced in CABI for the identification of fungi and bacteria.

for beneficial use in agriculture, the environment and human health. Identification of fungi using molecular sequencing methods is, however, made difficult by the fact that public sequence databases do not yet cover representatives of all taxa and some sequences in the databases are incorrectly named. In this circumstance, the availability of well-preserved strains (especially type strains) in culture collections is crucial in providing essential reference material to support the identification service.

There are three online repositories designated to serve for the formal registration of new fungal names: Fungal Names, Index Fungorum and MycoBank (Ryan et al., 2019). To underpin taxonomy and the naming of fungi, a core set of data needs to be available, including information on place of publication and deposit of the type specimen. The type can be lodged in a recognised collection as dried a specimen in a fungarium or as a living culture preserved in a metabolically inactive state in a culture collection. MycoBank was launched in 2004 as an online open access repository with the primary aim to register all fungal taxonomic novelties including new names and combinations and provide descriptions and illustrations (Crous et al., 2004; Robert et al., 2013).

Culture supply

As a public services culture collection that publishes its holdings online and through the GCM of the WDCM, CABI gives access to its public collections, supplying organisms to meet specific needs. Indeed, 90% of holdings are strains unique to CABI, with a wide range of properties as described above and collected from 142 countries. These strains can be supplied as active cultures growing on agar plates or slopes, as freeze-dried ampoules for easy storage and growth when needed, or as DNA or culture extracts. They are provided for use under CABI's conditions of supply which incorporates an MTA (Material Transfer Agreement) (https://www.cabi.org/wp-content/uploads/PDFs/Bioscience/genetic-resource-cat-condition-of-sale-web-form-6.pdf). They are provided with information on growth, health and safety and information, and their legal status in regard to the Nagoya Protocol. A supply charge is made to contribute to the costs of maintenance and supply. Special arrangements can be made for bulk supply to support research projects and strains required for specific purposes can be preselected.

Deposit

As a key function of a public service collection, CABI accepts deposits into its open collection that meet its accession criteria and remit. Essentially, the strains accessed are fungi or bacteria of agricultural or environmental interest and species that are not already covered by holdings. CABI will endeavour to accept key strains from published works to ensure the science can be repeated and further work undertaken, and to protect the investment made in the studies. Many scientific journals now require cited strains to be made available in culture collections for future use. Where deposits do not meet the CABI criteria for open deposit, or if the depositor wishes to restrict access, CABI offers a safe deposit service which provides the secure, confidential and long-term deposit of valuable microorganisms. The organisms are preserved by freeze-drying and/or cryopreservation using liquid nitrogen (Photograph 2). The WDCM reports that there are 359 collections around the world offering a storage service. CABI is an International Depository Authority (IDA). The Budapest Treaty on the International Recognition of the Deposit of Microorganisms for the Purposes of Patent Procedure allows deposits of microorganisms at an IDA to be recognised for the purposes of patent procedure. This allows deposit of a microorganism in a single IDA to support a patent in all of the countries that are signatory to the Budapest Treaty, rather than having to deposit separately in each country. There are 47 IDAs listed on the World Intellectual Property Office (WIPO) Website (https://www.wipo.int/budapest/en/idadb/). The WIPO statistics reports that 107,889 strains were deposited in these IDAs over the period 2001–17.

Microbial investigation, detection and testing service

Culture collections have inherent expertise in isolating microorganisms and can offer such services for all types of fungal and bacterial contamination

PHOTOGRAPH 2 Recovering strains from liquid nitrogen storage at CABI.

problems in homes, workplaces and on products. Of the 800 culture collections listed on the WDCM, 321 offer consultancies and many are capable of offering detection and testing services, for example, the Westerdijk Institute (formerly the Centraalbureau voor Schimmelcultures) but few advertise such services.

CABI's involvement in industrial mycology began in the early 1940s when work was undertaken for the UK Ministry of Food to investigate cases of food spoilage. During the 1950s when regulatory standards for industrial products and processes began to be established, CABI took on an increasingly important role in identifying mould growth on products. By 1983, a Biodeterioration and Industrial Services Laboratory had been established, led by Dennis Allsopp, and in 1984 CABI was approved by the National Laboratory Accreditation Scheme as a centre for carrying out microbial testing to British and Foreign standards (Blight and Ibbotson, 2011). This accreditation to the testing standard ISO 17025 has continued annually since then under National Measurement Accreditation Service (NAMAS) from 1985 and under the United Kingdom Accreditation Service (UKAS) from 1995. The current certification covers testing for resistance to mould growth of electrical equipment and components in accordance with BS EN 60068-2-10:2005 + A1:2018, resistance to mould growth of other materials and products to MIL STD 810 H:2019 Method 508.8, sampling of fungi in air and bulk materials from constructed environments and isolation and enumeration of fungi from environmental samples. Currently, CABI is the only UK laboratory accredited by

UKAS for on-site mould contamination testing. CABI has also developed a fungal sampling kit to enable individuals to collect their own samples of indoor mould contamination and send to CABI for analysis.

A further important aspect of CABI's industrial work is fuel testing based on the methodology produced by the Institute of Petroleum to quantify fungal and bacterial contamination of aviation and marine fuels. A partner in this aspect of work is Conidia Bioscience Ltd, based on-site at Egham.

Contracts, research and consultancies

CABI has a multi-disciplinary staff that includes information scientists, applied microbiologists, entomologists and plant pathologists that have experience of most major crops and commodities, agricultural and environmental situations and value chains. With this breadth of knowledge, CABI has been well placed to investigate the causes of agricultural and industrial microbial problems and provide solutions. CABI's history in solving such problems goes back many years. In the Second World War, CABI assisted the Ministry of Supply to 'tropic proof' the army's equipment, documented in a report dated February 1945. The Imperial Mycological Institute at the time was privy to secret information that military equipment, particularly that used in tropical regions, suffered from mould growth and deterioration. The affected items were wide ranging including boots, tents, transport and storage bags, wireless sets, mosquito netting, canvas and parachutes. CABI reported on the fungi involved and advised on mould-proofing to avoid such damage. The document describing this mycological support given by CABI to the British military was discovered recently https://blog.cabi.org/2020/02/26/secrets-of-military-mould-unearthed-after-being-hidden-for-over-75-years/. A further example of CABI's application of knowledge to solve industrial problems came in 1963–64 when Dr J.J. Elphick and Dr Agnes Onions, Curator of the living collection at the time, investigated microbial deterioration of conveyor belts for the National Coal Board (Aitchison and Hawksworth, 1993). Onions also investigated fungal problems on paint inside box girders of the Tinsley Viaduct, climbing in where most would not dare, taking samples, bringing them back to the laboratory, investigating and offering solutions. Amongst the samples, she discovered *Aspergillus fumigatus*, a hazardous pathogen, which as a result, stopped all work on the interior of box girder bridges affected at the time by mould growth throughout the United Kingdom.

Amorphotheca resinae (now known as *Hormoconis resinae*) was first identified from aviation fuel at the institute, during work in which Onions was heavily involved. As a result, she was asked to monitor fuel for the Concorde prototype, visiting the plane during its construction and development.

Amongst a broad range of contract research and consultancies undertaken since the 1990s, CABI has been commissioned to investigate such diverse aspects as resistance of paint to mould growth; fungi as a potential cause of Sudden Infant Death Syndrome; fungal contamination in warehouses; the significance of fungi in water systems and investigation of microbial contamination in melon crops.

Screening the collection

The diversity of fungal strains held in the CABI Culture Collection provides huge potential to screen for bioactive compounds and products of relevance to healthcare, food security and bioindustry. CABI has entered into several partnerships to screen its collections. A partnership with PharmaLinks, SIDR and Strathclyde Scotland (http://www.pharmalinks.org/) screened culture extracts against an array of assays. Royal Holloway, London University (https://www.royalholloway.ac.uk/) has provided chemical analysis of compounds using MALDI-MS, NMR and GCMS. Screening is continuing through a partnership with the Royal Botanic Gardens, Kew (https://www.kew.org/). The CABI collection acquired fungal cultures from the British Antarctic Survey collection which resulted in several projects funded by NERC to explore the holdings. Despite the fact that only a small proportion of the entire collection has been screened so far, these activities have already provided promising results. Evidence of significant action against human diseases has been found. These include tuberculosis (7 strains), African sleeping sickness (9 strains) and Parkinson's disease (1 strain). Other strains have exhibited antimicrobial activities (3 strains) or effects on diabetes (2 strains).

The collaboration with RBG, Kew included the screening of strains of endophytic fungi that were deposited in the CABI collection from the Iwokrama Forest Reserve, Guyana. Endophytic fungi isolated from leaves of 12 tree species yielded 2492 cultures (64 taxa). Isolates were found to exhibit potent activity against insects, fungi and bacteria. This activity had a hit rate of 1 in three, with three novel molecules discovered for potential product development (Cannon and Simmons, 2002; Kelley *et a*l., 2003).

The WDCM metadata tells us that CABI is not alone in providing such services; indeed, the very existence of many collections depends upon securing such revenue lines. Of the 800 plus collections listed, 116 accept patent deposits, 359 offer storage services, 367 are involved in the distribution of cultures, 400 provide identifications, 322 offer training and 321 offer consultancies. There are numerous business models that collections implement for sustainability but generally this includes some form of grant or institutional support (Smith et al., 2014). The bulk of the funding for the CABI living collection has come from a mix of culture supply fees, services, such as identifications and project revenue.

Expanding from single organisms to microbial communities

Microbial collections typically store strains in 'axenic format'; further, only 1% of microorganisms are culturable. Questions to be addressed therefore are how to underpin the burgeoning microbiome research field and how to capture the 99% of unculturable microbes.

One key project CABI is involved to establish a cryopreserved and characterised crop microbiome resource to underpin UK and international crop research.

The UK-CMCB (Crop Microbiome CryoBank) will provide a comprehensive platform to facilitate research towards optimising plant yield in an integrated crop management framework. This provides an infrastructural baseline for other projects and science-based activities from soil health to invasive species management. This UKRI-funded, CABI-led project includes partners from key UK research institutes (Rothamsted, JIC, Scotland's Rural College and the James Hutton Institute) and also aligns with CHAP, a UK Agritech centre. Cryo DSC research will be undertaken in association with the seed bank team at Kew (Wakehurst Place). The project will utilise state-of-the-art cryopreservation technology to conserve the key components of the functional microbiome. This technology has evolved to incorporate nitrogen-free cooling approaches. Importantly, the project will set the framework for approaches in other crops, plants, animal and environmental systems. It will also facilitate broader mechanisms for the conservation of biodiversity. All metagenomes and genomes will be deposited in a freely accessible database resource after sequence annotation, and provide a microbial genome resource for the research community.

CABI contribution to education and training

CABI has being involved in educational training for over 50 years. Activities have included supervision of masters and PhD students, overseeing undergraduate projects, hosting summer students, providing training attachments and running formal training courses for scientists in the United Kingdom and overseas. A teaching laboratory was an essential part of the new building for the culture collection based at Kew in 1975, and from that time, CABI ran courses in the identification of fungi of relevance to plant pathology and agriculture, one of the first being a 2-week course in 1976 for the Agricultural Development and Advisory Service which continued annually for several years. The first International Course on the Taxonomy and Identification of Fungi of Agricultural Importance took place in 1979, a 6-week course attended by students from 10 countries, funded by the British Council and the Food and Agriculture Organisation (see Photograph 3). A training department was established at CABI in the early 1990s and an extensive range of training

PHOTOGRAPH 3 First 6-week International Course, CMI, Kew which was participants and staff from the first Commonwealth Mycological Institute Training Course held at Kew.

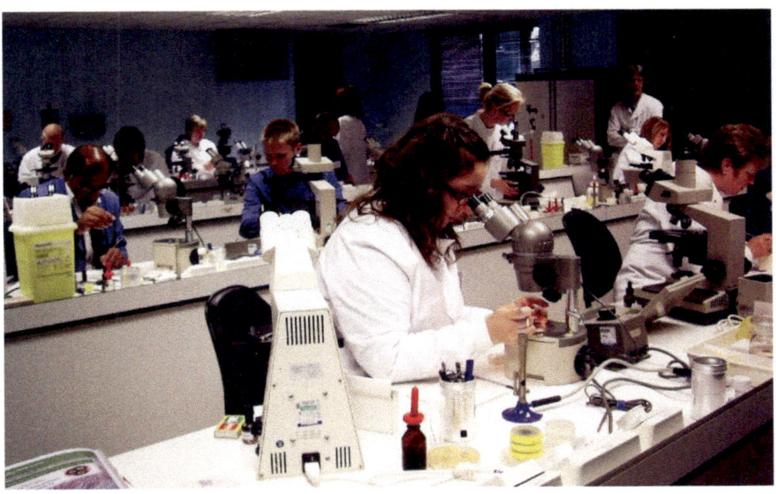

PHOTOGRAPH 4 Identification of Environmental Fungi Training Course CABI 2011 which was participants and trainers at the 2011 Identification of Environmental Fungi Training Course CABI.

courses was introduced. In addition to the 6-week international course, shorter courses covered such topics as preservation, storage and maintenance of microorganisms, culture collection management, mycological techniques, plant pathology and molecular identification methods, all of which continue to be offered by CABI (see Photograph 4). To provide this training, a wide range of specialist course manuals were developed over the years, the first being Dade and Gunnel's Class work with Fungi, price 10 shillings (Dade and Gunnel, 1966). CABI ventured into e-learning in 2010 to produce the *Managing Microbes* course, now installed in CABI Academy. It provides a comprehensive programme of six interactive modules for self-paced learning covering all aspects of collecting, storing and handling microbes and all the associated regulatory guidance. To provide a deeper knowledge of handling and using microorganisms, there is a repository of literature with reference sheets containing live links to international, regional and local regulations. The modules cover all aspects of collecting, developing and managing a microbial culture collection; this includes the following:

- Understanding their characteristics, properties and identification
- Sampling and naming organisms
- Maintaining organisms: growing and storing them, keeping them stable
- Delivering and transporting microorganisms to other laboratories.

The Practice Activity screens at the end of each topic serve as a refresher and enable the participants to apply the knowledge they have learned to working scenarios. On completion of the modules, there is an assessment which allows the participant to test their learning, analyse performance and review the answers given. Throughout the learning experience, the participant can use the Search and Advanced Search features to look for specific information within the core learning screens or e-Library documents. These also serve to facilitate the use of Managing Microbes as a reference source. The course does not replace practical training courses or 'on-the-job' training but serves to provide the theory and direction to towards implementation of best practices. Relevant course modules can be studied in conjunction with CABI laboratory-based training so that the knowledge gained can be implemented in practice. It is always recommended that the e-Library content is read, particularly when the participant's job specifically addresses the topic being presented in the course. Participants in CABI courses are encouraged to seek further assistance where necessary to get direction in implementation of best practice (Smith, 2010; https://assets.publishing.service.gov.uk/media/57a08b0ce5274a31e0000930/Managing-Microbes-Poster-FEMS.pdf).

CABI has also developed the Crop Pest Diagnosis course which is designed to support teachers and trainers in agricultural education institutions and in the workplace. It provides interactive exercises and resources to help users improve their ability to recognise disease symptoms and identify causes. The course guides learners through initial identification of symptoms, then covers causes, both biotic and abiotic and provides a methodology for field-based diagnosis. There are five modules approximately lasting 30 min each (https://www.cabi.org/publishing-products/crop-pest-diagnosis/).

The CABI Academy https://www.cabi.org/products-and-services/academy/ has been established which builds on CABI's scientific and publishing background. Using information from the latest research and innovations, training programmes have been designed for students and professionals. A range of products and services are provided, including a suite of high-quality digital teaching and learning resources. Practical face-to-face training is available as well as programmes comprising self-study, facilitated workshops and digital resources. A benefit of e-learning is that each individual can learn at their own pace. In many cases, this digital training is supplementary to 'hands on' training and learning through experience of real-life situations.

CABI staff have contributed to courses provided by the United Kingdom Federation for Culture Collection on Preservation of Microorganisms and the WFCC training courses. Over the years CABI has worked with universities and provided training on, for example, the University of Kent Fungal Technology MSc Course and the University of Reading and Royal Holloway, University of London biology courses.

CABI has run and contributed to international training courses in many countries. Examples are a 3-week course on Culture Collection Techniques in Kenya in 1984; a 2-week course in Culture Collection Management and Techniques in Assuit, Egypt, in 1987 and a WFCC training course on the Management of Microbial Genetic Resources in Rabat, Morocco, in May 2004. Others have included Australia in 2000, where CABI staff taught on a 2-day pre-conference course on Cryopreservation and Storage of Microorganisms and Cell Cultures and a 2-day post-conference training course in Management and Operation of Culture Collections.

CABI collection contribution to research through publicly funded projects

Much of CABI's work is achieved through partnerships with other organisations and this applies also to its culture collection. Already a partner in the UK culture collections network and strongly embedded in the WFCC from its development in the late 1960s to its establishment in the early 1970s, CABI became a partner in the ECCO in the 1980s. These networking activities became essential to build appropriate consortia to apply for funding from the various research programmes available at the United Kingdom, European and global levels. The fact that the CABI culture collection continues to offer a public service facilitates its access to public funding, without which, participation in most of its European project activities would not have been possible. Table 9.5 lists several of the key projects in which the CABI collection has played a key role. These have brought a critical mass of collections together to provide a more comprehensive and coordinated approach to resource provision for research. Fig. 9.1 demonstrates how a series

TABLE 9.5 CABI's networking: Living Collection related projects post 1986, networking and impact.

Year	Project	Programme	Consortium	Papers	Outputs	Impact
1986–1989	Development of improved techniques for the preservation and maintenance of cultures of fungal strains of biotechnological importance	Biotechnology Action Programme Project (BAP-0028-UK)	CBS; IHEM; CRBIP; MUCL; MNHN;		Recommendations enabled for optimum techniques to adopt for different fungi	New strains made available for use
1986–1994	The Microbial Information Network Europe (MINE)	EC BAP project (BAP-0005-NL)	AUA; CBS; DSMZ; DIMDI; SPP-PS; IHEM; CRBIP; MUCL; CECT/UVEG; CCUG; USP	Aguilar (1990); Allsopp et al. (1989); Hawksworth and Schipper (1989); Gams et al. (1988); Stalpers et al. (1990); Platt (1987a, b), 1988a, b, 1989	Data structure for mBRC data	Modified and adopted for OECD BRC Best Practice and ISO Biobank standard
1996–1999	Common Access to Biological Resources and Information (CABRI)	A demonstration project funded by the European Union ERBBIO4-CT96-0231	ABC; BCCM, CBS, CRBIP, DSMZ, ICLC, NCCB, NCIMB	http://www.cabri.org/guidelines.html	On-line Catalogue and best practices for microbes and data	Modified and adopted for OECD BRC Best Practice and Biobank standard
1999–2003	MOSAICC (Micro-Organisms Sustainable use and Access regulation International Code of Conduct)	Directorate General Research (contract BIO4-CT97-2206). Grant agreement QLG6 2000-30012	IUCN; OECD; RBG, Kew; Novo Nordisk A/S; WFCC; IPM; CCT; INBio; UICC; ARC; BCCM	Davison et al. (1998)	Best practice, standard text, documents for CBD compliance	Provided a basis for EU mBRC compliance with CBD

2003–2006	Elaboration and Diffusion of a Code of Conduct for the Access to and Sustainable Use of Microbial Resources within the Frame work of the Convention on Biological Diversity (MOSAICS)	EU Sixth Framework, Food Quality and Safety Proposal/ Contract no.: 506436	BCRC; SPP-PS; CBS; CNRST; CUP; DSMZ; EFB; FIRDI; JBA; KIEP; MIRCEN; TISTR; ULS; UNU; WFCC	http://bccm.belspo.be/projects/mosaicc	Revised document tested on 4 different continents	Adopted publication by COP
2001–2004	European Biological Resource Centre Network (EBRCN) researching into the co-ordination and harmonisation of collection operation and activities in major centres Europe	DG Research funded project under contract QLRI-CT-2000-00221	ABC; BUESA; CERDIC; DSMZ; SPP-PS; UG; INSERM; CRBIP; IAFB; INRSC: MRC; NBIMCC; NIPH; CBS; SIPH; MUCL; ULB; UL:Latvia;	Smith and Rhode (2005); Smith (2006);	Prepared EU mBRCs for networking; produced best practice for regulatory compliance	Provided input to OECD BRC Task Force leading proposals for Guidance and GBRCN establishment

Continued

TABLE 9.5 CABI's networking: Living Collection related projects post 1986, networking and impact.—cont'd

Year	Project	Programme	Consortium	Papers	Outputs	Impact
2008–2011	General Manager of the GBRCN Demonstration Project through subcontract to DSMZ	Bundesministerium für Bildung und Forschung (BMBF), the German Federal Ministry of Research and Education; Förderkennzeichen: 01GP0901	BELSPO, SPP-PS; BCCM/LMG; BCCM/LMBP; BCCM/IHEM; BCCM/MUCL; CRIA; CBMAI; BCRJ; EMBRAPA; CLIOC; AAFC: CCFC; ACCC; CGMCC; VTT; CRBIP; DSMZ; ICLC; NBRC; WDCM; JKUAT; CBS; UMinho-MUM; UVEG/CECT; BCRC; MUK	Fritze et al. (2010), 2012;	Proof of concept for a Global BRC Network; Delivered the key instruments for establishment	Global funding not available it stimulated regional efforts towards a GBRCN in Africa, Asia, Americas
2009–2012	European Consortium of Microbial Resources Centres (EMbaRC)	EU Seventh Framework Programme Research Infrastructures (INFRA-2008-1.1.2.9: Biological Resources Centres (BRCs) for microorganisms (Grant agreement number: FP7-228310)	INRA; CRBIP; DSMZ; UVEG-CECT; SPP-PS; Uminho-MUM; CBS UGent; MUCL	Smith and Day (2012); Ryan et al. (2012b)	A platform for the MIRRI preparatory phase project	Proposal to ESFRI endorsed to establish the European regional node of the GBRCN

2012–2016	MIRRI – Microbial resources Research Infrastructure	Preparatory phase project funded by the European Commission to develop an ESFRI infrastructure	ABC; DSMZ; CRBIP; SPP-PS; CBS; UGOT; UVEG/CECT; UGENT; INRA; JacobsUni; UMinho-MUM; MUT; CSIC; VKM; IAFB; ICLC	[123–134] Schüngel et al. (2013, 2014); Martin et al. (2015); Stackebrandt et al. (2015); Casaregola et al. (2015); Smith et al. (2013, 2017b); Overmann and Smith (2017); Fritze and Smith (2012); Smith and Fritze (2010); Stackebrandt and Smith (2017); Verkley et al. (2016)	Drafts of documents to establish the MIRRI-ERIC	10 countries are submitting an application to establish the MIRRI ERIC
2015–2019	European Marine Biological Research Infrastructure Cluster (EMBRIC)	HORIZON 2020 - INFRADEV Implementation and operation of cross-cutting services and solutions for clusters of ESFRI; Grant Agreement number: 654008	UPMC; CCMAR; NERC; SAMS; SZN; MSS; MBA; FVB; CNR; USTAN; TUNATECH; Xelect; INRA; UiT; UGENT; CRBIP; HZI; EMBL; UPEM; TAU; UNS; SCALPRO; DSMZ; HCMR; CNRS/IFB; UiB	Brennecke et al. (2018); Pina et al. (2018); Smith et al. (2019)	Workflows across different infrastructures for new products from the sea	New organisms and new compounds for the blue economy

Continued

TABLE 9.5 CABI's networking: Living Collection related projects post 1986, networking and impact.—cont'd

Year	Project	Programme	Consortium	Papers	Outputs	Impact
2015–2019	CORBEL: Coordinated Research Infrastructures Building Enduring Life-science Services project; clustering the Biological and Medical Sciences Research Infrastructures	H2020-INFRADEV-1-2014-1 Funding scheme: Research and Innovation action Proposal number: 654248	EMBL; UMC; ICF; FCRG; UD; BBMRI; IIEAA; EUMC; EATRIS; ECRIN; UL, UK; IRFMN; HHUD; INFRAFRONTIER; HZM; Instruct; CIRMMP; AECSIC; CNRS; SZN; USTAN; FVB; ICSTM; MDCMM; UM; SVU-VUMC; DKZ; DSMZ; JacobsUni; CBS; CSC-TTK; MUW; AZG; UDST		A platform for harmonised user access to biological and medical technologies, biological samples and data services	

For consortium partner acronyms, see list of abbreviations.

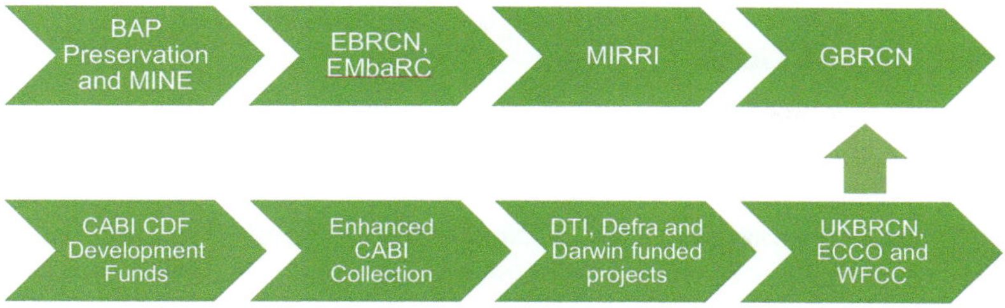

FIGURE 9.1 Demonstrating the linkage between the CABI living collection projects.

of projects have contributed to the establishment of collection infrastructure, the MIRRI being established under the European Strategy for Research Infrastructures (Stackebrandt et al., 2015) with the eventual goal of linking regional initiatives to establish the GBRCN. The GBRCN demonstration project, funded by the German Ministry of Research and Education, brought together 23 collections in 16 countries as proof of concept. By the end of the project, the countries participating numbered 28, with over 100 collections interested in joining such a network (Fritze et al., 2012). The outputs of the project created mechanisms and planning for the implementation of best practices and capacity building, data resources and information systems needed, a legislation compliance and operational framework, the finances and sustainability plan and an overview of its outreach.

CABI's living collection has provided resources and/or preserved strains for many of CABI's research programmes and projects, including the invasive species programme aimed at the prevention, early detection and management of invasive pests. CABI is helping to manage the spread of invasive species, a problem that costs the world almost 5% of global gross domestic product or an estimated 1.4 trillion US$ per year. Millions of the world's most vulnerable people face problems with invasive weeds, insects, plant and animal diseases, which fundamentally threaten the economic growth. Invasive species disproportionately affect vulnerable communities in poor rural areas, especially in developing countries, which depend on natural resources, healthy ecosystems, trade and tourism for their livelihoods (https://www.cabi.org/what-we-do/invasive-species/). CABI holds and maintains strains of many of the fungal species used as in biological control of weeds, for example *Puccinia komarovii* var. *glanduliferae* var. nov; a rust species new to science used in the control of Himalayan Balsam and the leafspot fungus *Mycosphaerella polygoni-cuspidati* used in the control of Japanese Knotweed. The collection also maintains the strain of the key fungus *Metarhizium anisopliae* used in the Mycopesticide Green Muscle, to control locusts.

CABI collection role in delivering the Sustainable Development Goals

The outputs of microbiology research contribute directly and indirectly to several UN Sustainable Development Goals (SDGs). CABI's Medium Term Strategy offers examples of how CABI is contributing to these (https://www.cabi.org/about-cabi/our-mission-strategies/): **SDG 1: No Poverty** – Microbial resources provide indirect opportunities for improved income along value chains, through partnerships with industry to develop new products and processes that will deliver jobs and impact on livelihoods. Amongst CABI's objectives is providing information on good agricultural practices and building capacity to increase productivity and reduce crop losses. CABI also aims to support sustainable economic development in target countries by helping small-holder farmers to increase trade through improved linkages in local and global value chains. **SDG 2: Zero Hunger**. Use of microorganisms as biological control agents to combat agricultural pests and diseases and in products to improve soil health supports this goal by contributing to increased crop productivity. CABI also helps to improve food security, income and livelihoods by developing new crop technologies and management procedures and building local and national capacity to use these effectively. For example, promoting more effective prevention and management of invasive species and co-developing solutions with end users. **SDG 3: Good Health and Well-being** – Microbiology research and the resources generated from it support industry in providing microbial solutions to reduce environmental pollution, and improve well-being for all. The knowledge gained from examining the human microbiome and designing appropriate microbial interventions is contributing to improving human health. The discovery of new antimicrobials to combat disease is an essential area of research today as microorganisms are becoming resistant to the drugs currently used. **SDG 4: Quality Education.** CABI provides a wide range of microorganisms to universities and other training establishments and thus supports education and resources for scientists at all levels across the world. In addition, microbiology research plays an important role in creating knowledge systems, technology and providing knowledge transfer. Use of modern information technology and traditional communication channels ensures key information and learning is made available to all. CABI aims to enhance equitable and inclusive agribusiness growth through activities specifically focused on women and youth through participatory content development and working with local, national and regional partners. Practical tools and services are being developed based on reliable research and data. Through the CABI Academy, CABI is building an e-learning curriculum that helps practitioners and students acquire vital knowledge and skills. **SDG 8: Decent Work and Economic Growth** – this goal links to SDG 1 and the contribution partnerships have with industry that will contribute to sustainable economic growth and full and productive employment.

SDG 9: Industry, Innovation and Infrastructure – Microbial Biological Resource Centres (mBRCs) promote inclusive and sustainable industry and contribute to this goal by providing innovative microbial solutions. **SDG 11: Sustainable cities and communities** – Microorganisms are responsible for extremely important processes including global cycling of nutrients and carbon sequestration. They also underpin processes such composting and waste management, synthesis of biofuels and cleaning up the environment through bioremediation. **SDG 12: Responsible Consumption and Production** – To contribute to this goal, CABI is undertaking activities to promote Integrated Crop Management approaches that minimise the use of agro-chemicals and prevents pollution. CABI also aims to improve the uptake of climate-responsive approaches and technologies by farmers, including climate smart pest management and promoting production of nutritious indigenous crops (particularly fruits, vegetables, grains and pulses) that are better adapted to changing climatic conditions. **SDG 14: Life Below Water** – Conservation of marine microbes and accessing these 'difficult-to-culture organisms' can contribute to our understanding and enable use of the oceans and marine resources for sustainable development and reduce the threat to biodiversity. **SDG 15: Life on Land** – The work of mBRCs and microbiology research in general can promote protection, restoration and the sustainable use of terrestrial ecosystems, management of forests and combating pollution. **SDG 17: Partnerships for the Goals - Strengthen the means of implementation and revitalise the global partnership for sustainable development.** By working with donors, local, national and regional partners, governments, civil society and private sectors, CABI promotes transfer of knowledge and technology between and within and countries, thus building capacity to support agribusiness. The mBRC community can also enhance market access and trade, working through public private partnerships and establishing common platforms.

In summary, it is clear that the organisms and associated information provided by mBRCs can impact at many levels in achieving the SDGs. No single collection can achieve this alone, providing all the more reason why there is a need to coordinate such effort.

Recommendations for future development

- BRCs need to respond to local challenges and network to get value from investment of public funding
- BRCs must operate to international standards to ensure quality and importantly interoperability of service, function and data
- A global strategy is needed to ensure organisms that are critical to scientific discoveries are preserved for future use
- A mechanism to engage the full potential of the microbial resource (mBRC) community in providing resources and solutions to global challenges is required

- BRCs need to develop robust business models to secure their sustainability
- Key outputs come from research and utilisation of culture collection holdings, but output and value would be enhanced enormously through coordinated effort
- A coordinated effort is required to address microbial communities and the microbiome to release the full potential of microorganisms
- Research funders, science publishers, the research community and bio-industry need to work together to make the most of the potential of the hidden microbiology resource and harness the outputs of microbial communities

Discussion, lessons learned and conclusions

To maximise the potential of collections to provide resources to the research community, they require robust and sustainable business plans and the ability to secure funding. This will enable work to be undertaken on characterisation of strains to facilitate their recognition as candidates for research. This chapter describes the types of research programmes that can enhance the scientific value of holdings, and the range of services that can be provided as revenue lines. Securing of projects requires hard work and planning to be competitive. A strategy to address several key issues is needed:

- A sound foundation and a good understanding of the capacity and skills of the research team
- Partnerships with people and organisations with complementary skills
- An understanding of the research programmes and the direction and fundamental principles behind them.
- Making use of the organisations' processes, people, skills and the project writing team
- Working beyond the project itself with a long-term strategy to achieve specific goals

There is not one example of any single collection making huge profits from the work they undertake. Most are not-for-profit and each has different business models (Smith et al., 2014). Access to the new and expanding range of technologies can enhance the revenue potential, but it is not always necessary for every collection to incorporate them. They can be accessed through partnerships or bought in as needed. Inevitably, research is focussed on host institutional strategy and often limited by local policy. For example, CABI's research is directed towards delivering input to enable microbial interventions to be made to improve global agriculture, feed the world's expanding populations and reduce the huge losses in crop production. To serve science and provide solutions to global challenges, coordinated effort of culture collections and improved infrastructure is of vital importance.

Maintaining a living collection project portfolio requires the following:

- Maintaining the public service role of the collection in conservation and supply of strains
- The engagement of colleagues in the culture collection and microbiological society communities.
- Cultivating partnerships with external research teams
- A publications strategy that addresses the long-term goals laid down in the collections strategy and which provides evidence of capacities to address anticipated project calls
- Horizon scanning for current trends, new technologies, current science and opportunities
- Interdisciplinary research team collaboration
- Engaging the appropriate specialist staff for project writing and delivery

Securing project funding is becoming increasingly difficult and more competitive but remains one of the key revenue lines to ensure the future sustainability of the collection. Collections must learn lessons from past activities, become more strategic in their thinking and follow the European trend to establish research infrastructure. A focussed critical mass is more likely to make a difference and attract funding. The background work in spotting opportunities, building consortia and demonstrating capacity are all important in addition to writing a good proposal.

It is evident that BRCs, specifically microbial culture collections, provide resources to underpin research and many new products that utilise microorganisms have resulted from the study of organisms from collections. However, this function can be made much more efficient and effective with better coordination and linkage with funders and the research community. Researchers can access the holdings of collections, but as yet, there is no single platform that provides the information on all strains held. The GCM holds information on only 14% of the available strains. Collection organisations do their best to coordinate access but more needs to be done to facilitate the uptake and use of organisms and their associated data. The greatest opportunity to be able to understand, harness and moderate this hidden resource of microbial communities (99% of the microorganisms in the environment) is by coordinating activities in the microbiome and using the networks of BRCs to achieve this. In addition to the organisms the mBRCs currently hold, there is the potential for coordinating networks of depositors to provide samples to the collections in targeted isolation programmes. This would provide access to new organisms and their potential novel chemistry. A global effort is needed and the establishment of the GBRCN to provide coordinated access and the microbiome research community via appropriate global infrastructures should be undertaken. It has begun in Europe with MIRRI, the Asian BRC Network has been established and the USCCN coordinates activities in the United States. The foundations for great achievements are in place. Researchers, funders, journal editors and collections need to work together to coordinate strategy and introduce mechanisms to accelerate access to microbial communities essential for our survival and the future of our planet.

Abbreviations

AAFC Agriculture and Agri-Food Canada
ABC Advanced Biotechnology Centre
ACCC Agricultural Culture Collection China
AECSIC Agencia Estatal Consejo Superior de Investigaciones Científicas
ARC Agricultural Research Council, Plant Protection Research Institute (Republic of South Africa)
ASTM ASTM International, formerly known as American Society for Testing and Materials
AUA Agriculture, University of Athens
AZG Academisch Ziekenhuis Groningen
BAP Biotechnology Action Programme
BCCM Belgian Co-ordinated Collections of Microorganisms
BCRC Bioresource Collection and Research Center
BCRJ Rio de Janeiro Cell Bank
BEA British European Airways
BELSPO Belgian Federal Science Policy Office
BRC Biological Resource Centre
BS British Standard
BUESA Budapest University of Economic Sciences and Administration
CABI CAB International
CBMAI Coleção Brasileira de Micro-organismos de Ambiente e Indústria
CBS Centraalbureau voor Schimmelcultures now Westerdijk Institute
CCFC Canadian Collection of Fungus Cultures
CCINFO Culture Collections Information Worldwide
CCMAR Centro de Ciencias do Mar do Algarve
CCUG Culture Collection University of Gothenburg
CDF CABI Development Fund
CECT Spanish Type Culture Collection
CERDIC Centre Europeen de Recherche et de Developpement en Information et Communication Scientifiques
CGMCC General Microbiological Culture Collection Center
CIRMMP Consorzio Interuniversitario Risonanze Magnetiche di Metallo Proteine
CLIOC Coleção de Leishmania
CMI Commonwealth Mycological Institute
CNR Consiglio Nazionale delle Ricerche
CNRS/IFB Centre National de la Recherche Scientifique/Institut Français de Bioinformatique
CNRST Centre National de la Recherche Scientifique et Technique
CORBEL Coordinated Research Infrastructures Building Enduring Life-science Services
CRBIP Centre de Ressources Biologiques de l'Institut Pasteur
CRIA Centro de Referência em Informação Ambiental
CSC Center for Science Ltd, Tieteen Tietotekniikan Keskus

CSIC-IMEDEA Mediterranean Institute for Advanced Studies
CUP Catholic University of Peru
Defra Department for Environment, Food and Rural Affairs
DIMDI Deutsche Institut für Medizinische Dokumentation und Information
DKZ Deutsches Krebsforschungszentrum
DSMZ Leibniz Institute DSMZ - German Collection of Microorganisms and Cell Cultures GmbH (Deutsche Sammlung von Mikroorganismen und Zellkulturen)
DTI Department of Trade and Industry was replaced by Department for Innovation, Universities and Skills and Department for Business, Enterprise and Regulatory Reform
EATRIS European Advanced Translational Research Infrastructure
EBRCN European Biological Resource Centre Network
ECCO European Culture Collections' Organisation
ECRIN European Clinical Research Infrastructure Network
EFB European Federation of Biotechnology
EMbaRC European Consortium of Microbial Resources Centres
EMBL European Molecular Biology Laboratory
EMBRAPA Empresa Brasileira de Pesquisa Agropecuária
EMBRIC European Marine Biological Research Infrastructure Cluster
EN European Standards
ESFRI European Strategy Forum for Research Infrastructures
EUMC Erasmus University Medical Centre Rotterdam
FCRG Fundació Centre de Regulació Genómica
FIRDI Food Industry Research and Development Institute
FVP Forschungsverbund Berlin E.V.
GBRCN Global Biological Resource Network
GCM Global Catalogue of Microorganisms
GCMS Gas Chromatography Mass Spectrometry
HCMR Hellenic Centre for Marine Research
HHUD Heinrich-Heine-Universitaet Duesseldorf
HOMED Holistic Management of Emerging forest pests and Diseases
HZI Helmholtz-Zentrum fuer Infektionsforschung GMBH
HZM Helmholtz Zentrum Muenchen Deutsches Forschungszentrum Fuer Gesundheit und Umwelt GmbH
IAFB Institute of Agricultural and Food Biotechnology
ICF Institut de Ciències Fotòniques
ICLC Interlab Cell Line Collection
ICSTM Imperial College of Science, Technology and Medicine
IDA International Depositary Authority
IHEM Institute of Hygiene and Epidemiology, Mycology
IIEAA Foundation of Biomedical Research of the Academy of Athens
IMI International Mycological Institute
INBio Instituto Nacional de Biodiversidad, Department of Bioprospection (Costa Rica)
INRA Institut National de la Recherche Agronomique
INRSC Instituto Nazionale per la Ricerca Sul Cancro
INSERM Institut National de la Santé et de la Recherche Médicale
INSTRUCT Instruct Academic Services Limited
IPM Industrial Platform for Microbiology (Belgium)
IRFMN Istituto di Ricerche Farmacologiche Mario Negri
IUCN International Union for Conservation of Nature
JacobsUni Jacobs University Bremen

JBA Japan Bioindustry Association
JKUAT Jomo Kenyatta University of Agriculture and Technology
KIEP Korean Institute for International Economic Policy
LMBP Laboratory of Molecular Biology and Plasmids now BCCM/GeneCorner Plasmid Collection
LMG Laboratorium voor Microbiologie
MALDI-ToF Matrix-Assisted Laser Desorption/Ionisation
MBA Marine Biological Association of the United Kingdom
mBRC microbial domain Biological Resource Centre
MDCMM Max Delbrueck Centrum Fuer Molekulare Medizin
MINE Microbial Information Network Europe
MIQE Minimum Information for Publication of Quantitative Real-Time PCR Experiments
MIRCEN UNESCO – Microbial Resources Centres Network
MIRRI Microbial Resources Research Infrastructure
MNHN Muséum National d'Histoire Naturelle
MOSAICS Microorganisms Sustainable use and Access management Integrated Conveyance System
MRC Medical Research Council
MS Mass Spectrometry
MSS Marine Scotland Science
MUCL Mycothèque de l'Université catholique de Louvain
MUM Micoteca da Universidade do Minho
MUT Mycotheca Universitatis Taurinensis
MUW Medical University of Vienna
NBIMCC National Bank for Industrial Microorganisms and Cell Cultures
NBRC National Institute of Technology and Evaluation, Biological Resource Center
NCCB Netherlands Culture Collection of Bacteria
NCIMB National Collection of Industrial, Food and Marine Bacteria
NERC Natural Environment Research Council
NGS Next Generation Sequencing
NIPH National Institute of Public Health
NMR Nuclear Magnetic Resonance
OECD Organisation for Economic Co-operation and Development
PVA Polyvinyl alcohol
PVC Polyvinyl chloride
qPCR quantitative Polymerase Chain Reaction
RBG Royal Botanic Gardens
RTCA Radio Technical Commission for Aeronautics
SAMS The Scottish Association for Marine Science
SDG Sustainable Development Goal
SIPH Scientific Institute of Public Health
SPP-PS Service Public Fédéral de Programmation Politique scientifique
SVU-VUMC Stichting Vrije Universiteit, Vrije Universiteit Medical Center
SZN Stazione Zoologica Anton Dohrn
TAU Tel Aviv University
TISTR Thailand Institute of Scientific and Technological Research
UD University of Dundee
UDST Universita Degli Studi di Torino
UGENT University of Gent
UGOT Culture Collection, University of Gothenburg

UiB Universitet i Bergen
UICC Universitas Indonesia Culture Collection, Laboratory of Microbiology (Indonesia)
UiT The Arctic University of Norway
UKAS United Kingdom Accreditation Service
UKBRCN United Kingdom Biological Resource Centre Network
UKFCC United Kingdom Federation for Culture Collections
UL Latvia - University of Latvia
UL University of Liverpool
ULB Universite Libre de Bruxelles
ULS University of Ljubljana - Fac. Natural Sciences and Engineering
UM University of Manchester
UMC University Medical Centre
UNS Université de Nice Sophia Antipolis
UNU United Nations University
UPEM Université Paris Est Marne La Vallée
UPMC Université Pierre et Marie Curie
USCCN United States Culture Collection Network
USMI IRCCS Ospedale Policlinico San Martino
USTAN The University Court of the University of St Andrews
UVEG Universidad Virtual del Estado de Guanajuato
VKM All-Russian Collection of Microorganisms, G.K. Skryabin Institute of Biochemistry and Physiology of Microorganisms, Russian Academy of Sciences
VTT VTT Technical Research Centre of Finland
WDCM World Data Centre for Microorganisms
WFCC World Federation for Culture Collections
WIPO World Intellectual Property Office

Acknowledgements

CABI is an international intergovernmental organisation, and we gratefully acknowledge the core financial support from our member countries (and lead agencies) including the United Kingdom (Department for International Development), China (Chinese Ministry of Agriculture), Australia (Australian Centre for International Agricultural Research), Canada (Agriculture and Agri-Food Canada), Netherlands (Directorate-General for International Cooperation) and Switzerland (Swiss Agency for Development and Cooperation). See https://www.cabi.org/about-cabi/who-we-workwith/key-donors/for full details.

References

Aitchison, E.M., Hawksworth, D.L., 1993. IMI: Retrospect and Prospect. CABI, Wallingford.

Anon, 1953. Catalogue of the Culture Collection of the Commonwealth Mycological Institute. Commonwealth Mycological Institute, Commonwealth Agricultural Bureau, Kew.

AFNOR, 2008. French Standard NF S96-900 "Quality of Biological Resource Centers (BRCs) - Management System of a BRC and Quality of Biological Resources from Human or Microorganism Origin". Association française de normalisation, Paris, France.

Allsopp, D., Hawksworth, D.L., Platt, R., 1989. The CAB international mycological institute culture collection database, microbial culture information service (MiCIS) and microbial information network Europe (MINE). Int. Biodeterior. 25, 169–174.

Aguilar, A. (Ed.), 1990. European Laboratory without Walls: In the Field of MINE, the Microbial Information Network. Commission of the European Communities, Brussels.

Blight, D., Ibbotson, R., 2011. CABI a Century of Scientific Endeavour. CAB Internationa, Wallingford, UK, ISBN 978-1-84593-873-4, p. 171.

Brennecke, P., Ferrante, M.I., Johnston, I.A., Smith, D., 2018. A collaborative European approach to accelerating translational marine science. Journal of Marine Science and Engineering. Special Issue "Methodologies for Outreach in the Marine Sciences" J. Mar. Sci. Eng. 6, 81. https://doi.org/10.3390/jmse6030081.

Butler, E.J., 1921. The imperial Bureau of mycology. Trans. Br. Mycol. Soc. 7, 168–172.

Cannon, P.F., Simmons, C.M., 2002. Diversity and host preference of leaf endophytic fungi in the Iwokrama Forest Reserve, Guyana. Mycologia 94 (2), 210–220.

Casaregola, S., Vasilenko, A., Romano, P., Robert, V., Ozerskaya, S., Kopf, A., Glöckner, F.O., Smith, D., 2015. An Information System for European culture collections: the way forward. SpringerPlus. 2016 5, 772. https://doi.org/10.1186/s40064-016-2450-8. URL: http://www.springerplus.com/content/5/1/772.

Crous, P.W., Gams, W., Stalpers, J.A., Robert, V., Stegehuis, G., 2004. Myco-Bank: an online initiative to launch mycology into the 21st century. Stud. Mycol. 50, 19–22.

Dade, H.A., Gunnel, J., 1966. Class Work with Fungi. Notes for Teachers. Commonwealth Mycological Institute, Commonwealth Agricultural Bureaux, Kew, Surrey, England.

Davison, A., Brebandere, J. de, Smith, D., 1998. Microbes, collections and the MOSAICC approach. Microbiology Australia 19 (1), 36–37.

Ellis, M.B., 1971. Dematiaceous Hyphomycetes. International Mycological Institute.

Ellis, M.B., 1976. More Dematiaceous Hyphomycetes. International Mycological Institute.

Ellis, M.B., Ellis, J.P., 1997. Microfungi on Land Plants an Identification Handbook. The Richmond Publishing Co Ltd, UK.

Ellis, M.B., Ellis, J.P., 1998. Microfungi on Miscellaneous Substrates an Identification Handbook. The Richmond Publishing Co Ltd, UK.

Fritze, D., Smith, D., Martin, D., 2010. Third Intermediate Report on the GBRCN Demonstration Project. GBRCN Secretariat, Germany, ISBN 978-3-00-031458-2.

Fritze, D., Martin, D., Smith, D., 2012. Final Report on the GBRCN Demonstration Project. GBRCN Secretariat, Germany, ISBN 978-3-00-038121-8.

Fritze, D., Smith, D., 2012. Microbial Resource Research Infrastructure (MIRRI) facilitating improved pathways to discovery. In: Paterson, R., Simoes, M.F., Pereira, L., Santos, C.L., Lima, N. (Eds.), Biological Resource Centres: Closing the Gap between Science and Society. Conference presentation (abstract); ECCO XXXI, Micoteca da Universidade do Minho, Braga, Portugal, p. 65 [ISBN: 978-072-97916-5-9].

Gams, W., Hennebert, G.L., Stalpers, J.A., Jansens, D., Schipper, M.A.A., Smith, J., Yarrow, D., Hawksworth, D.L., 1988. Structuring strain data for the storage and retrieval of information on fungi and yeasts in MINE, Microbial Information Network Europe. J. Gen. Microbiol. 134, 1667–1689.

Hawksworth, D.L., Schipper, M.A.A., 1989. Criteria for consideration in the accreditation of culture collections participating in MINE, the Microbial Information Network Europe. MIRCEN Journal 5, 277–281.

Kelley, J., Cannon, P., Simmonds, M., Pingal, R., Simmons, C., 2003. The Iwokrama programme: an approach to the sustainable exploitation of genetic resources. In: Lima, N., Smith, D. (Eds.), Biological Resource Centres and the Use of Microbes, XXII ECCO Meeting Proceedings Book, pp. 185–195.

Kelley, J., Smith, D., 1997. Depositing Micro-organisms as Part of the Patenting Process. European BioPharmaceutical Review. Ballantyne Ross Ltd, London, UK.

Martin, D., Stackebrandt, E., Smith, D., 2015. MIRRI promoting quality management systems for microbiology. EC Microbiology 2.2, 278–287. https://www.ecronicon.com/ecmi/pdf/ECMI-02-000036.pdf.

Overmann, J., Smith, D., 2017. Microbial resource centers contribute to bioprospecting of bacteria and filamentous microfungi. In: Russell Paterson, Nelson (Eds.), Topics in Biodiversity and Conservation, Professor David Hawksworth. Bioprospecting – Successes, Potential and Constraints, vol. 16. Springer, ISBN 978-3-319-47933-0, pp. 51–79 (Print) 978-3-319-47935-4 (Online). http://link.springer.com/book/10.1007/978-3-319-47935-4.

Pina, M., Colas, P., Cancio, I., Audic, A., Bosser, L., Canario, A., Gribbon, P., Johnston, I.A., Kervella, A.E., Kooistra, W.H.C.F., Merciecca, M., Magoulas, A., Nardello, I., Smith, D., Pade, N., Robinson, D., Schoen, A., Schultz, F., Kloareg, B., 2018. The European Marine Biological Research Infrastructure Cluster: an alliance of European research infrastructures to promote the blue bio-economy. In: Rampello, P.H., Trincone, A. (Eds.), Grand Challenges in Marine Biotechnology, Grand Challenges in Biology and Biotechnology. Springer Nature, pp. 405–421. https://doi.org/10.1007/978-3-319-69075-9_10. ISBN 978-3-319-69074-2.

Platt, R., 1987a. Microbial information network Europe (MINE). UKFCC Newsletter 16, 5.

Platt, R., 1987b. The CMICC database and the MINE project at the CAB international mycological institute. Binary 11, 21–23.

Platt, R., 1988a. CMI named to be the UK Node of MINE. USFCC Newsletter 18 (4), 6.

Platt, R., 1988b. Update on microbial information network Europe (MINE). Biotechnol. Inf. News 10, 2.

Platt, R., 1989. CAB international mycological institute (CMI) - the UK node of MINE. Methods Mol. Cell Biol. 1, 43–44.

Report, 1947. Specialist Conference on Culture Collections of Micro-organisms. (S.O. Code No. 47-139.). HMSO, London.

Robert, V., Vu, D., Amor, A.B.H., van de Wiele, N., Brouwer, C., Jabas, B., Szoke, S., Dridi, A., Triki, M., ben Daoud, S., Chouchen, O., Vaas, L., de Cock, A., Stalpers, J.A., Stalpers, D., Verkley, G.J.M., Groenewald, M., dos Santos, F.B., Stegehuis, G., Li, W., Wu, L., Zhang, R., Ma, J., Zhou, M., Gorjón, S.P., Eurwilaichitr, L., Ingsriswang, S., Hansen, K., Schoch, C., Robbertse, B., Irinyi, L., Meyer, W., Cardinali, G., Hawksworth, D.L., Taylor, J.W., Crous, P.W., 2013. MycoBank gearing up for new horizons. IMA Fungus 4 (2), 371–379.

Rohde, C., Smith, D., Martin, D., Fritze, D., Stalpers, J., 2013. Code of conduct on biosecurity for biological resource centres: procedural implementation. Int. J. Syst. Evol. Microbiol. 63, 2374–2382. https://www.microbiologyresearch.org/content/journal/ijsem/10.1099/ijs.0.051961-0.

Ryan, M.J., McCluskey, K., Verkleij, G., Smith, D., 2019. Fungal biological resources to support international development: challenges and opportunities. World J. Microbiol. Biotechnol. 35, 139. https://doi.org/10.1007/s11274-019-2709-7.

Ryan, M.J., Ritchie, B.J., Smith, D., 2012a. Maintenance and storage of fungal plant pathogens. In: Lane, C., Beales, P., Hughes, K. (Eds.), Fungal Plant Pathogens – Principles and Protocols. CAB International, Wallingford, UK, ISBN 978 1 84593 668 6. http://bookshop.cabi.org/?page=2633&pid=2265&site=191.

Ryan, M.J., Verkleij, G., López-Ocaña, L., Munaut, F., Arahal, D.R., Simões, M.F., Pereira, L., Santos, C., Lima, N., Kasulyte-Creasey, D., Declerck, S., Crahay, C., Buddie, A.G., Smith, D.,

2012b. Inter-laboratory evaluation, development and validation of fungal preservation regimes used in different European biological resources centres (BRCs) Conference presentation (abstract). In: Paterson, R., Simoes, M.F., Pereira, L., Santos, C.L., Lima, N. (Eds.), Biological Resource Centres: Closing the Gap between Science and Society. ECCO XXXI, Micoteca da Universidade do Minho, Braga, Portugal, p. 57 [ISBN: 978-072-97916-5-9].

Schüngel, M., Smith, D., Bizet, C., Stackebrandt, E., the MIRRI Consortium, 2014. The role of the European microbial resource research infrastructure project. Enliven: Microb Microbial Tech 1 (1), 001.

Schüngel, M., Stackebrandt, E., Bizet, C., Smith, D., 2013. Mirri - the microbial resource research infrastructure: managing resources for the bio-economy. EMBnet.Journal 19 (1), 5–8. http://journal.embnet.org/index.php/embnetjournal/article/view/706.

Smith, D., 2006. WFCC updates from work programs quality matters work program (ECWP); capacity building. WFCC Newsletter (42). http://www.wfcc.info/pdf/newsletter/Newsletter%20No.42%20(Jul.%202006).pdf.

Smith, D., 2010. Managing Microbes e-learning tool. In: Proceedings of the 12th International Conference on Culture Collections. Florianópolis, Brazil, 26 September – 01 October 2010. CRIA, SBM, WFCC, Brazil, p. 80.

Smith, D., 2012. Culture collections. Adv. Appl. Microbiol. 79, 73–118.

Smith, D., Day, P., 2012. European Consortium of Microbial Resource Centres: Microbial Resources Success Stories. EMbaRC, CABI, UK.

Smith, D., da Silva, M., Jackson, J., Lyal, C., 2017a. Explanation of the Nagoya Protocol on access and benefit sharing and its implication for microbiology. Microbiology 163, 289–296. https://doi.org/10.1099/mic.0.000425. http://mic.microbiologyresearch.org/content/journal/micro/10.1099/mic.0.000425.

Smith, D., Fritze, D., 2010. European Culture Collections - the future is MIRRI. Microbiol. Today 37 (4), 256–258.

Smith, D., Fritze, D., Stackebrandt, E., 2013. Public service collections and biological resource centres of microorganisms. In: Rosenberg, Eugene DeLong, E.F., Stackebrandt, E., Lory, S., Thompson, F. (Eds.), The Prokaryotes: Prokaryotic Biology and Symbiotic Associations, fourth ed., vol. XX, pp. 267–304.

Smith, D., Martin, D., Novossiolova, T., 2017b. Microorganisms: good or evil, MIRRI provides biosecurity awareness. Curr. Microbiol. 74, 299. https://doi.org/10.1007/s00284-016-1181-y.

Smith, D., McCluskey, K., Stackebrandt, E., 2014. Culture Collection funding models and BRC business plans. SpringerPlus 3, 81. http://www.springerplus.com/content/3/1/81.

Smith, D., Rohde, C., 2005. The WFCC on biosecurity. WFCC Newsletter (41). http://www.wfcc.info/pdf/newsletter/Newsletter%20No.41%20(Aug.%202005).pdf.

Smith, D., Ryan, M.J., 2019. International postal, quarantine and safety regulations. Australia publication special issue: Sustainable Use and Preservation of Microbial Resources Microbiology 40 (3). https://doi.org/10.1071/MA19032. http://microbiology.publish.csiro.au/?paper=MA19032.

Smith, D., Buddie, A.G., Goss, R., Overmann, J., Lepleux, C., Brönstrup, M., Kloareg, B., Meiners, T., Brennecke, P., Ianora, A., Bouget, F.-Y., Gribbon, P., Pina, M., 2019. Marine resources for discovery pipelines – an ocean of opportunity for biotechnology? World J. Microbiol. Biotechnol. https://doi.org/10.1007/s11274-019-2685-y.

Stackebrandt, E., Schüngel, M., Martin, D., Smith, D., 2015. The microbial resource research infrastructure MIRRI: strength through coordination. Microbial resource management'. Microorganisms 3 (4), 890–902. https://doi.org/10.3390/microorganisms3040890. http://www.mdpi.com/2076-2607/3/4/890/html.

Stackebrandt, E., Smith, D., 2017. Expanding the 'digital protologue' database (DPD) to 'archives of microbiology': an offer to scientists and science. Arch. Microbiol. https://doi.org/10.1007/s00203-017-1369-y. http://link.springer.com/article/10.1007/s00203-017-1369-y.

Stalpers, J.A., Kracht, M., Jansens, D., De Ley, J., Van der Toorn, J., Smith, J., Claus, D., Hippe, H., 1990. Structuring strain data for storage and retrieval of information on bacteria in MINE, Microbial Information Network Europe. Syst. Appl. Microbiol. 13, 92–103.

Sutton, B.C., 1990. The Coelomycetes. Commonwealth Mycological Institute, Kew.

Taylor, J.W., 2011. One Fungus = One Name: DNA and fungal nomenclature twenty years after PCR. IMA Fungus 2, 113–120. https://doi.org/10.5598/imafungus.2011.02.02.01.

Taylor, J.W., Göker, M., Pitt, J.I., 2016. Choosing one name for pleomorphic fungi: the example of *Aspergillus* versus *Eurotium, Neosartorya* and Emericella. Taxon 65 (3), 593–601. https://doi.org/10.12705/653.10.

Verkley, G., Martin, D., Smith, D., 2016. Microbial Resource Research Infrastructure Best Practice Manual on Access and Benefit Sharing. https://doi.org/10.5281/zenodo.284881 and ABSCH November 2016 published on line MIRRI (May 2016). https://zenodo.org/record/284881. https://absch.cbd.int/api/v2013/documents/F1C80F1C-1EB7-F02A-CEED-E7D523F17079/attachments/MIRRI%20ABS%20Manual_web.pdf http://www.mirri.org/fileadmin/mirri/media/Dokumente/generalDocs/MIRRI_ABS_Manual_web.pdf.

Verkley, G., Perrone, G., Pina, M., Scholz, A., Zuzuarregui, A., Perugini, I., Turchetti, B., Smith, D., Lima, N., 2019. New ECCO tools for Core Material Deposit and Transfer Agreements to promote compliance in procedures for deposit and supply of cultures from a public microbial culture collection. FEMS (Fed. Eur. Microbiol. Soc.) Microbiol. Lett. fnaa044. https://doi.org/10.1093/femsle/fnaa044.

Chapter 10

Understanding the ecological roles of dictyostelid social amoebae for a sustainable future

Phongthana Pasookhush, Chooseel Bunsuwansakul and Pravech Ajawatanawong

Division of Bioinformatics and Data Management for Research, Faculty of Medicine Siriraj Hospital, Mahidol University, Bangkok, Thailand

Chapter outline

Introduction

Dictyostelid social amoebae or **cellular slime moulds** are a group of free-living eukaryotic organisms that live in soil as single-cell amoebae in various ecosystems worldwide (Cavender et al., 2013; Liu et al., 2020; Swanson et al., 1999). They are fascinating organisms because the unicellular amoebae can combine to form a multicellular slug-like organism when they lack food. Then, the slug-like organism can adapt to be a fungal-like fruiting body. The special life cycle of dictyostelids has attracted scientists to study cell biology and the evolution of multicellular organisms.

The first dictyostelid was discovered in 1869 by Oskar Brefeld, a German botanist and mycologist, while he was examined some fungi from horse dung. He named the first dictyostelid as *Dictyostelium mucoroides* because it produced a fruiting body similar to a fungus, *Mucor mucedo* (Kessin, 2001). there

Importance of Microbiology Teaching and Microbial Resource Management for Sustainable Futures
https://doi.org/10.1016/B978-0-12-818272-7.00007-9

are now more than 150 species of dictyostelids that have been described. Formerly, the whole species of dictyostelids was recognised as the cellular slime mould, which belonged to the Kingdom Fungi. That was because they can produce a reproductive structure, called the 'fruiting body', that morphologically resembles the sporangiophore and sporangia of some fungal species (Whittaker, 1969). With the now more advanced knowledge of cell biology, the dictyostelid social amoebae are clearly unrelated to the microbes in the Kingdom Fungi and have been reclassified as a member of the Kingdom Protista (Baldauf et al., 2000), because they have no cell wall when they live as vegetative amoeba cells. The food uptake process also differs from that of fungi. Instead of secreting extracellular enzymes to digest organic matter and reabsorb it back to the cell as general fungi do, they take up their food, bacterial cells, into the cell by a process called phagocytosis, then digest the food inside the cell. Based on the updated version of the Tree of Life (ToL), all slime moulds are clustered together with a diverse group of amoebae called **Amoebozoa** (Baldauf, 2008; Burki et al., 2020). The lifestyles of microbes in this supergroup vary from free-living microbes, such as slime moulds (*Dictyostelium* spp., *Polysphondylium* spp. and *Acytostelium* spp.) to parasitic microbes, such as *Entamoeba* spp. and *Acanthamoeba* spp. (Baldauf et al., 2004; Baxt and Singh, 2008; Khan, 2003).

Slime moulds have unique processes in their life cycle that differentiate them from the other amoebae, which generally live as single-cell organisms. Once food is limited, individual slime moulds in amoeba form can combine to become a slug-like organism and finally produce a fruiting body (Olive, 1970). This unique life cycle has resulted in them being described as social amoebae. Taxonomic studies of slime moulds divide them into two main groups: plasmodial slime moulds and cellular slime moulds. These two groups of slime moulds differ in the aggregation (combination of several amoeba cells) step. Amoeba cells will aggregate to form a gigantic multinucleated cell in plasmodial slime mould, but cellular slime moulds aggregate to form a multicellular slug (Romeralo and Fiz-palacios, 2013).

Because of the amazing life cycle of the cellular slime moulds, scientists use them as a model organism in cell biology studies, such as cell signalling (Ginsburg and Kimmel, 1997; Williams, 2005; Williams et al., 1999), phagocytosis (Benghezal et al., 2006; Cornillon et al., 2000; Schmith et al., 2013), chemotaxis (Jin et al., 2000; Meyer et al., 2011; Wessels et al., 2006), and also cell differentiation (Fukuzawa et al., 2001; Kawata et al., 1997; Thompson and Kay, 2000). Unfortunately, among the approximately 100 species of cellular slime moulds, *Dictyostelium discoideum* is the only well-studied species. The behaviour of the cellular slime moulds, which constantly interacts with soil bacteria and eukaryotic microbes, has attracted the attention of several investigators as a novel source of antibiotics and bioactive compounds (Kikuchi et al., 2006a,b).

Life cycle of dictyostelids

Slime moulds are essential to the soil ecosystem to turn over minerals (nutrients) and control the bacterial population (Romeralo and Fiz-palacios, 2013; Stout, 1973; Swanson et al., 1999). They can be found in the soil surface, which is riche with organic matter and microbes (Raper, 1984). Dictyostelids use pseudopodia to engulf bacteria and occasionally feed on yeast (Cardelli, 2001; Kessin, 2001). Amoeba haploid cells undergo asexual reproduction in nutrient-rich conditions, and can multiply by binary fission (Romeralo et al., 2012). The cell duplicates the genetic material by mitosis cell division and is finally divided into two haploid daughter cells (see the asexual reproduction cycle in Fig. 10.1). This process is repeated until food becomes scarce. The amoeba cell initially accumulates and secretes prestarvation factor (PSF) that activates the cell to produce and secrete a signalling molecule, such as cyclic adenosine-monophosphate (cAMP) (Shaulsky and Kessin, 2007). This chemical promotes individual amoebae to migrate to the centre, where the signalling molecule has the highest concentration. The state of migration of amoeba cells to the centre is called 'streaming'. Several streams combine to create amoeba cells in a step called 'aggregation' (see aggregation in the multicellular development cycle in Fig. 10.1). In this step, cells are packed tightly, and the shape of aggregation develops into a 'mound', which morphologically resembles a small dune. The mound is differentiated into a multicellular slug, which comprises two major types of cells: prespore and prestalk. The prespore cells are dictyostelid cells, which further develop to become spores. The prestalk cells are sacrificed to become the death stalk cells. A hypothesis has

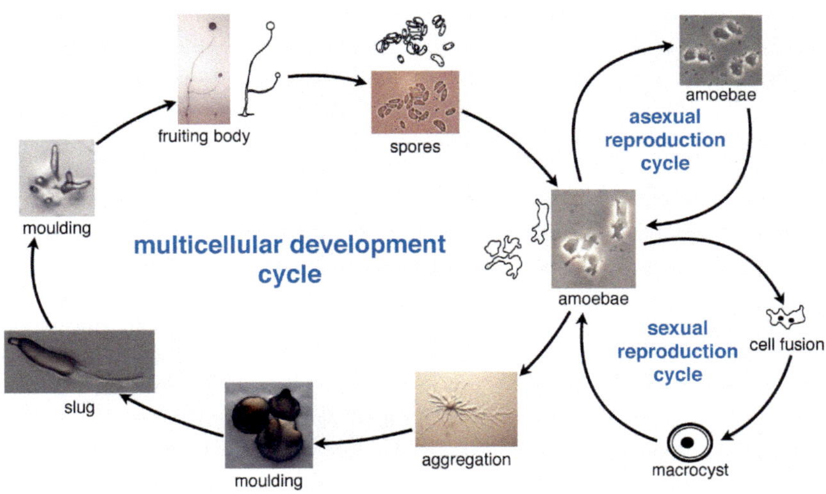

FIGURE 10.1 Life cycle of dictyostelid social amoebae generally comprises three cycles: the asexual cycle, sexual cycle, and multicellular development cycle.

suggested that unhealthy amoeba cells might be programmed to be victims for the development of prestalk cells, whereas healthy cells become prespore cells (Shaulsky and Kessin, 2007).

Slug formation provides several advantages to dictyostelids. One advantage of slug formation is protecting amoebae from animal predators, such as tiny nematodes. That is because the slug develops a mucopolysaccharide sheath to coat the body, which makes the nematode less efficient in attacking the slug (Kessin et al., 1996). Another advantage has promoted the speed of movement to allow dictyostelids to move across soil clods. In laboratory observations, the slug can respond to light by moving in the light source direction (Bonner, 2013; Bonner and Lamont, 2005). This mechanism allows the slug to migrate to the soil surface.

Once the slug moves to the soil surface, the whole slug is differentiated into a fungal-like fruiting body (Romeralo et al., 2012). The slug transforms the whole body from an elliptical shape to a mound-like shape in the laboratory. In this step, the prestalk cells located at the anterior part of the slug turn into a basal structure and attach to the surface, and the prespore cells turn on the top of the basal structure (Shaulsky and Kessin, 2007). Part of the basal structure is projected into the air (or antigravity direction), resembling a Mexican sombrero shape. This step is called 'culmination'. The prestalk cells undergo cell death, and finally, the death-cell stalk or sorophore is formed (Shaulsky and Kessin, 2007; Whittingham and Raper, 1960). The prespore cells develop to be mature spores and move upwards to the top of the stalk. When the spores have matured, numerous spores are released. Dictyostelids use small animals, such as nematodes, snails, bats, and even small birds, as the primary mechanism to disperse their spores (Huss, 1989; Stephenson and Landolt, 1992; Suthers, 1985). This mechanism can escalate spore spreading into remote areas. Finally, a new amoebae will hatch from the spores. Then, the whole life cycle is repeated. However, the overall process of multicellularity development takes place within 24 h after starvation (Bozzaro, 2013).

Sexual reproduction of dictyostelids has been reported in some species (Romeralo et al., 2012). This process is rarely found in nature, and the process of sexual reproduction remains unclear. There have been some efforts to understand the sexual state of cellular slime moulds, such as in *D. mucoroides* and *D. minutum* (Blaskovic and Raper, 1957). However, the best-known species for sexual reproduction is *D. discoideum*. Physical conditions that induce sexual reproduction in dictyostelids are dark, humid and submerged (Nickerson and Raper, 1973). Starvation in those conditions will induce macrocyst formation, which occurs in the sexual life cycles of dictyostelids, instead of the aggregation process. The process begins with the fusion of the two haploid amoeba cells and then undergoes cannibalisation to form a microcyst, while the diploid amoebae cell has recombination and meiosis (Flowers et al., 2010). However, induction of macrocyst formation in the lab remains challenging, and knowledge about its sexual reproduction is also incomplete (Romeralo et al., 2012; Shaulsky and Kessin, 2007).

Systematics and evolution of dictyostelids

Previously, scientists classified dictyostelids based on the morphological characteristics of their fruiting bodies. The anatomy of dictyostelid fruiting bodies comprises two main sections. One is the **sorophore** or stalk, which supports a sorus, and the other is the **sorus**, which is the structure that contains many spore cells. Based on the structures of fruiting bodies, dictyostelids can be classified into three morphotypes, namely the dictyostelid morphotype, polysphondylid morphotype and acytostelid morphotype. Dictyostelids have been classified into three genera (genera *Dictyostelium*, *Polysphondylium* and *Acytostelium*) based on the characteristics of their fruiting bodies (Olive, 1970; Raper, 1984). The fruiting body of dictyostelids in the genus *Dictyostelium* presents a cellular stalk with a terminal sorus. However, sometimes the stalk can be an irregular branch with sori at the tip of the branches (Fig. 10.2A). The fruiting bodies of dictyostelids in the genus *Polysphondylium* present a cellular stalk with a particular branching pattern, called whorl branching (Fig. 10.2B). In some species of *Polysphondylium*, the primary branching can produce a subwhorl branch also. The genus *Acytostelium* is quite different from the previously described genera. They have tiny and delicate fruiting bodies with a non-cellular hollow stalk that typically appears as bushes (Olive, 1970; Raper, 1984) (Fig. 10.2C).

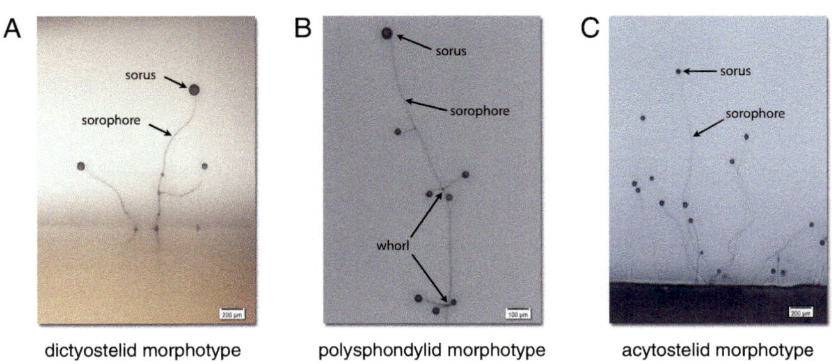

FIGURE 10.2 Characteristics of dictyostelid fruiting bodies in three different morphotypes. Dictyostelids in the genus *Dictyostelium* possess the dictyostelid-morphotype fruiting body composed of a cellular sorophore with unbranched or irregular-branched pattern (A). The genus *Polysphondylium* produces a polysphondylid-morphotype fruiting body containing a cellular sorophore with whorl branching (B). The genus *Acytostelium* produces an acytostelid-morphotype fruiting body composed of acellular sorophores. The fruiting bodies commonly appear as a bush (C).

However, the morphological characteristics of dictyostelids are limited, and several of them are ambiguous and it can be challenging to make a clear decision on their classification at the species level. Therefore, molecular phylogenetics was used for the modern systematics of dictyostelids. The first phylogenetic analysis of dictyostelids was reconstructed from two different markers: the *small subunit of ribosomal rRNA* (*SSU*) gene and the *α-tubulin* gene (Schapp et al., 2006). Although both trees are entirely identical, the *SSU* phylogeny has strong statistical supports in all major clades. However, the molecular phylogeny completely changed the classification based on the morphology of fruiting bodies into four monophyletic clades (Schapp et al., 2006). The tree clustered dictyostelids into four major clades, namely groups 1–4, and the morphological patterns of fruiting bodies were not correlated with the clades on the tree (Schapp et al., 2006). Phylogenetic analysis of the combination of *SSU* and *α-tubulin* gene suggested that 'group 1' could be the basal clade of dictyostelids. However, the deep root of dictyostelids still needed further validation (Fig. 10.3A).

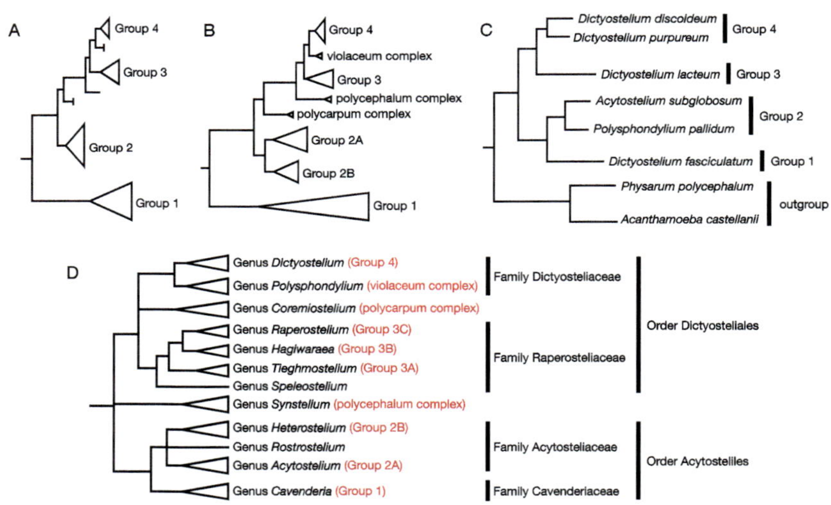

FIGURE 10.3 Simple phylogenies illustrated the history of the systematics and evolution of dictyostelids. Panel A illustrates the first molecular phylogenetic tree that was reconstructed in 2006 with the *SSU* gene (Schaap et al., 2006). The tree classified dictyostelids into four major clades (groups 1–4) plus some unclassified taxa. Moreover, the tree also suggested that the deep root of dictyostelids might be placed between group 1 and the remaining taxa. In 2011, the new phylogeny was reconstructed from a larger dataset (panel B). The phylogeny illustrated eight clades. Five clades (groups 1, 2A, 2B, 3 and 4) are congruent with the previous version of the tree, plus three more clades, namely the polycarpum, polycephalum and violaceum complexes (Romeralo et al., 2011). Panel C illustrates the phylogenetic tree reconstructed from 213 universal protein sequences from eight amoeba genomes suggesting that the early branching of dictyostelids separated them into two subclades: groups 1 + 2 and groups 3 + 4 (Sheikh et al., 2015). The updated version of molecular systematics and evolution of dictyostelids was proposed in 2018 (Sheikh et al., 2018). The new phylogeny classified dictyostelids into 12 genera (panel D).

In 2010, the systematics of dictyostelids was revised again by adding more sequencing samples, which were collected from the 'Global Diversity Survey of Eumycetozoas' project (PBI). This project aimed to investigate new species and study the biodiversity of eumycetozoans, including dictyostelids, worldwide (Romeralo et al., 2011). The new molecular phylogeny was reconstructed from the *SSU rDNA* gene and the internal transcribed spacer (it suggested eight groups of dictyostelids; Fig. 10.3B). Group 2 was separated into two main clades, called 2A and 2B. The other three groups, including groups 1, 3 and 4 remained unchanged. Dictyostelids in group 3 could be potentially divided into three subclades (3A, 3B and 3C). Moreover, the new phylogeny also shows three more complex clades, namely violaceum complex, polycephalum complex and polycarpum complex (Romeralo et al., 2011).

Although the new phylogeny has better supports in all the clades, the root position still needs validation. Moreover, most of the morphological characteristics in the traditional taxonomy are not relevant to the tree (Romeralo and Fiz-palacios, 2013; Schapp et al., 2006; Swanson et al., 2002). A phylogenetic analysis reconstructed from 213 nuclear orthologous genes from the genomic data seems to solve this problem (Sheikh et al., 2015). The multigene phylogeny illustrated the new root position of dictyostelids (Fig. 10.3C). The root separated all dictyostelids into two subclades. One subclade comprises dictyostelids in groups 1 and 2, while the other clade comprises dictyostelids in groups 3 and 4.

Recently, the systematics of dictyostelids was revised in 2018 (Sheikh et al., 2018). This version was updated from eight clades to 12 clades. The current classification classifies dictyostelids into two orders, Dictyosteliales and Acytosteliales, with 12 genera (Fig. 10.3D). However, the genus *Synstelium* is an unclassified order, and could be one of the deep clades on the tree.

The order Acytosteliales comprises two families: Acytosteliaceae (formerly group 2) and Cavenderiaceae (formerly group 1). The family Acytosteliaceae comprises three genera: *Heterostelium*, *Acytostelium* and *Rostrostelium.* The genus *Acytostelium* (formerly group 2A) possesses fruiting bodies with acytostelid morphotype. The genus *Heterostelium* (formerly group 2B) is very diverse in the morphology of sorocarp. The genus *Rostrostelium* comprises a single species, *Rostrostelium ellipticum*. The family Cavenderiaceae comprises a single genus, *Cavenderia*.

Order Dictyosteliales is a relatively large order that comprises two families, Dictyosteliaceae and Raperosteliaceae, and a single genus with an unclassified family, *Coremiostelium*. The family Raperosteliaceae (formerly group 3) consists of four genera: *Speleostelium*, *Tieghemostelium* (formerly group 3A), *Hagiwaraea* (formerly group 3B) and *Raperostelium* (formerly group 3C). The genus *Speleostelium* consists of a single species, *Speleostelium caveatum*. The family Dictyosteliaceae comprises two genera, *Dictyostelium* and *Polysphondylium*.

The genus *Polysphondylium* is correlated to the violaceum complex clade. This genus comprises only four species (*Polysphondylium violaceum*, *P. paragonicum*, *P. fuscans* and *P. laterosorum*). They possess a cellular sorophore with a violet-shade pigment sori. The genus *Dictyostelium* (formerly group 4) is the largest genus of dictyostelids, and has a cellular sorophore. The sorus has no pigment, although some species have purple pigment.

Ecology and diversity of dictyostelids

Dictyostelids are a group of organisms that have adapted to almost all ecosystems on Earth (Bonner, 2013). The species richness of dictyostelids varies by latitude (distance from the equator) and altitude (height above sea level). Dictyostelids reduce their species diversity when isolated from soil samples in higher latitudes or altitudes (Perrigo et al., 2013; Romeralo et al., 2010b; Swanson et al., 1999). However, dictyostelids favour moist soil with a high organic matter (humus) content (Raper, 1984). They can be found around shaded trees in tropical forests and deciduous forests (Swanson et al., 1999). Surprisingly, dictyostelids are also found in extreme environments, such as the desert in Mexico (Cavender et al., 2012) and southwestern Arizona, United States (Cavender and Raper, 1968). However, the abundance of microbes per gram of soil is lower than in tropical and subtropical areas. Moreover, dictyostelids are also found in some lowland tundra regions, such as Alaska (Romeralo et al., 2010a; Stephenson et al., 1997).

The main habitats of the dictyostelids are soil, including sand, clay and submerged sediment. In addition, soil-contaminated substances, such as leaf litter (Seephueak and Petcharat; 2014) and animal dunks (Stephenson and Landolts, 1992), are also habitats for dictyostelids. Interestingly, human boots have been proven to be another dispersal factor in helping dictyostelids spread everywhere (Perrigo et al., 2012). There has been no attempt to survey the dictyostelids in aquatic environments; even other amoebae, such as *Acanthamoeba* spp., can be isolated from lakes or rivers (Khan, 2003; O'Dell, 1979).

Physical properties, such as temperature, pH, UV, soil moisture content and types of bacteria in the soil, affect dictyostelid growth and species richness. Laboratory studies of some species of dictyostelids have indicated that the optimum temperature is around 22°C, and the optimum pH is in the narrow range around pH 6.5−7 (Romeralo et al., 2010b, 2012). Dictyostelid cells in both amoebal form and spore are sensitive to UV light (Romeralo and Lado, 2006); that is why they are mostly found slightly beneath the surface, at about 0.5−1 cm in depth. Moisture in soil is one of the crucial factors in spore germination and the development of the fruiting body (Singh, 1947b). At 19% moisture content, *D. mucoroides* spores can germinate on the eighth day of incubation. However, the spores can germinate faster (on the third day of incubation) if the spores are cultured in a 33% moisture content medium (Singh, 1947b). Dictyostelids can feed on a large variety of bacterial species

(Raper, 1937). However, to cultivate dictyostelids in a laboratory, *Klebsiella aerogenes* or *Escherichia coli* are commonly used as food sources (Raper, 1984).

Following a systematic survey of the species diversity of dictyostelids during 1880–2020 by reviewing 62 publications, we concluded that the dictyostelids were mainly studied in America, Asia, Europe and Australia. There have been very few studies performed in Africa (Fig. 10.4). Considering the genus level of diversity, four countries, including the United States, Mexico, Peru and Argentina, have high genus-level diversity (Fig. 10.4). However, low diversity of dictyostelids has been found in some European and African countries, such as Iceland, the UK, Italy, Morocco, Sierra Leone and Liberia.

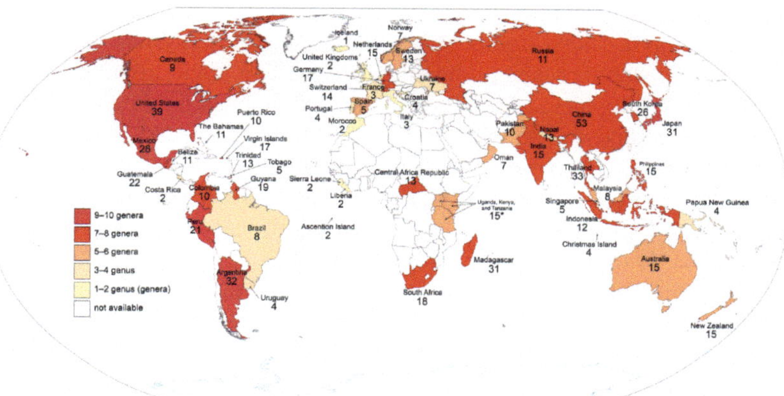

FIGURE 10.4 Global distribution of 62 publications (Agnihothrudu, 1956; Cavender, 1969; Cavender, 1972; Cavender, 1976; Cavender and Hopka, 1986; Cavender and Kawabe, 1989; Cavender and Lakhanpal, 1986; Cavender and Raper, 1968; Cavender and Vadell, 2000; Cavender and Vadell, 2006; Cavender et al., 1995; Cavender et al., 2002; Cavender et al., 2004; Cavender et al., 2010; Cavender et al., 2012; Cavender et al., 2013; Cavender et al., 2016; Cavender et al., 2018; Dagamac and dela Cruz, 2020; Frøyen and Langvad, 1984; Hagiwara, 1990; Hagiwara, 1991; Hagiwara, 1996; Hagiwara, 1998; Landolt and Wong, 1998; Landolt et al., 2006; Landolt et al., 2008a; Landolt et al., 2008b; Li et al., 2020; Liu and Li, 2010a; Liu and Li, 2010b; Liu and Li, 2011; Liu et al., 2020; Liu et al., 2019; Nelson et al., 1967; Perrigo et al., 2012; Perrigo et al., 2013; Perrigo et al., 2020; Rai and Tewari, 1961; Rai and Tewari, 1963; Reddy et al., 2010; Rollins et al., 2010; Romeralo et al., 2009; Romeralo et al., 2010a; Rukseree et al., 2018; Sathe et al., 2010; Seephueak and Petcharat, 2014; Shim et al., 2002; Singh, 1947a; Srirampur et al., 2018); Stephenson et al., 1997; Stephenson et al., 1999; Swanson et al., 1999; Tieghem, 1880; Torres et al., 2011; Traub et al., 1981; Vadell and Cavender, 2007; Vadell et al., 2011; Vadell et al., 2018; Waddell, 1982; Xavier de Lima, 2013; Yulo and dela Cruz, 2012) that surveyed dictyostelids from 1880 to 2020. Each country is highlighted in different colours based on the number of discovered dictyostelid genera per country. The highlighted colours range from red (9–10 genera), orange (7–8 genera), light orange (5–6 genera), and yellow (1–2 genera), while white indicates non-availability of data. The number of dictyostelid species is presented underneath the country name. The asterisk for Uganda, Kenya and Tanzania indicates a single survey that covered all three countries.

There are six genera of dictyostelids, including genus *Cavenderia*, *Coremiostelium*, *Dictyostelium*, *Heterostelium*, *Polysphondylium* and *Raperostelium*, distributed in all the continents. The genus *Acytostelium* is distributed in almost all continents, except Australia and New Zealand. The genus *Hagiwareae* has not been reported from Europe but has been found in the other continents. The genera *Rostrostelium* and *Speleostelium* are found in the United States but have not been found elsewhere.

Dictyostelids as sources of bioactive compounds

Bioactive compounds, also known as natural products or secondary metabolites, are small organic compounds produced by organisms. Bioactive compounds are not produced for growth and development but as defence mechanisms, chemical signalling or alternative metabolic pathways (Pichersky and Gang, 2000). Dictyostelids play roles in the soil ecosystem as microbial population modulators and require cell signalling and differentiation for growth and development. Therefore, dictyostelids potentially carry a number of bioactive compounds in their cells. To date, more than 30 bioactive compounds have been discovered from dictyostelids. These compounds can be divided into three groups according to their chemical configuration, including substituted phenyl compounds, carbohydrate derivatives and other compounds (Barnett and Stallforth, 2018; Kubohara and Kikuchi, 2018).

Substituted phenyl compounds comprise most bioactive compounds discovered in the dictyostelids, such as differentiation-inducing factors (DIFs), 4-methyl-5-pentylbenzene-1,3-diol (MPBD) and monochasiols. An important chemical characteristic of this bioactive compound is that it contains at least one benzene ring in the structure. The biological functions of these compounds mainly involve cell differentiation and cell signalling. Some compounds also exhibited antitumour, immune modulation or antibacterial activities. Among these compounds, DIFs are the most well-studied bioactive compounds in dictyostelids. Differential-inducing factors 1 to 3 (DIF-1 to DIF-3) were isolated from *D. discoideum* as morphogens for stalk-cell differentiation. DIFs are one of the first differentiation inducers identified, which bear the unique structure of chlorinated alkylphenone, representing a new class of biological compound (Morris et al., 1987, 1988). Due to its unique chemical structure and being the first differentiation inducer ever identified, they have been heavily studied in terms of biological functions and potential applications. DIFs play roles in cell differentiation and chemotaxis in *D. discoideum* (Kuwayama and Kubohara, 2009; Morris et al., 1987, 1988). Apart from differentiation-inducing compounds, DIFs also act as a modulator of chemotaxis. DIF-1 reduced chemotaxis, whereas DIF-2 enhanced chemotaxis in cAMP-stimulated *D. discoideum* (Kuwayama and Kubohara, 2009). DIFs also proved to be good candidates for the development of diabetes (Kuwayama and Kubohara, 2009; Morris et al., 1987, 1988), Alzheimer (Myre et al., 2009), anti-*Trypanosoma* (Nakajima-Shimada et al., 2013) and antitumour drugs (Asahi et al., 1995; Kanai et al., 2003; Kubohara, 1997, 1999; Kubohara et al., 1995). DIF-1 has been shown to promote glucose uptake in cell lines (Omata et al., 2007) and reduce blood sugar levels in diabetic mice via intramuscular injection and oral administration (Kawaharada et al., 2016; Kubohara et al., 2008). This has suggested that DIF-1 might be a potential drug for the treatment of diabetes and obesity. DIF-1 was also a promising drug candidate for Alzheimer disease. A study has shown that DIF-1 reduced amyloidogenic processing in CHO−7W cells by decreasing the amyloid-β protein precursor (APP) level (Myre et al., 2009). DIF-3 was a novel lead compound for the development of an anti-*Trypanosoma* drug. The synthetic DIF-3 derivatives exhibited strong anti-*Trypanosoma* activity in vitro and in vivo, with no adverse effects (Nakajima-Shimada et al., 2013). In addition, DIFs have been shown to possess antitumour activity. DIF-1 and DIF-3 can inhibit the proliferation of tumour cells. However, they can promote the cell differentiation of various tumour cell types, such as mouse erythroleukaemia (E8) cells (Asahi et al., 1995), human leukaemia K562 cells (Asahi et al., 1995; Kubohara, 1999), rat pancreatic AR42J cells (Kubohara et al., 1995), human leukaemia HL-60 cells (Kubohara, 1997) and human gastric cancer cells (Kanai et al., 2003).

Numerous phenyl compounds were isolated after the discovery of DIFs. Since these compounds have a similar structure, these compounds were believed to be involved in cell differentiation and to be good candidates for antitumour drug development. MPBD, a polyketide, was identified as an alternative DIF apart from DIF-1 in *D. discoideum* and *D. mucoroides* (Saito et al., 2006; Kikuchi et al., 2006a,b). MPBD was able to induce differentiation in DIF-1-mutant *D. discoideum* strain (Saito et al., 2006) and inhibit tumour cell proliferation (K562 and HL-60 cell) (Kikuchi et al., 2006a,b). Interestingly, MPBD exhibited antimicrobial activity against *E. coli* and *Bacillus subtilis* (Murata et al., 2016). Monochasiols A−H are chlorinated alkylresorcinols with different side-chains isolated from the fruiting body of *D. monochasioides*. Although the biological function is still unknown, monochasiol A inhibited interleukin 2 (IL-2) production in human T lymphocyte cells. Interestingly, monochasiols were produced from enzymes responsible for the production of DIF-1 and MPBD (Kikuchi et al., 2017). This suggests that possible novel types of bioactive compounds in the dictyostelids have not been discovered. Other phenyl compounds include (1) **dictyoquinone**, a prespore cell-inducing factor with alkylbenzoquinone core (Takaya et al., 2014), (2) **dibenzofuran**, and its derivatives (AB0022A and Pf-1−2) (Sawada et al., 2000; Kikuchi et al., 2013), (3) **brefelamide**, an aromatic amide (Kikuchi et al., 2005), (4), **dictyobiphenyls** and **dictyoterphenyls** with aromatic cores (Haruhisa et al., 2012), (5) **dictyomedins**, with 4-aryl dibenzofuran core (Yoshiaki et al., 2001) and (6) ***Polysphondylium tenuissimum*-1−5 (Pt-1−5)** and ***Polysphondylium pseudo-candidum*-1 (Ppc-1)**, prenylated and geranylated aromatic compounds (Suzuki et al., 2015). These compounds, including dictyoquinone, dibenzofuran, dibenzofuran derivatives and dictyomedins, were involved in cell differentiation in dictyostelids (Kikuchi et al., 2013; Sawada et al., 2000; Takaya et al., 2014; Yoshiaki et al., 2001). However, the biological functions of brefelamide, dictyobiphenyls and dictyoterphenyls, and Pt-1−5 and Ppc-1 remain unknown. Various compounds, including dibenzofuran and its derivatives, brefelamide, dictyobiphenyls and dictyoterphenyls, and Pt-1−5 and Ppc-1 were able to inhibit the proliferation of several tumour cells (Haruhisa et al., 2010, 2012; Kikuchi et al., 2013; Zhang et al., 2016). Interestingly, AB0022A, a dibenzofuran derivative, exhibited antimicrobial activity against various Gram-positive bacteria (Sawada et al., 2000). Ppc-1, a prenylated and geranylated aromatic compound, can induce fatty acid secretion in adipocytes, suggesting that the Ppc-1 might be a potential drug candidate for treating obesity (Suzuki et al., 2015). These results demonstrated that substituted phenyl compounds are primarily involved in cell differentiation in distyostelids. In contrast, some compounds exhibit antitumour, antidiabetic and antimicrobial properties, which have been proven to be great candidates for drug development.

Carbohydrate derivatives, consisting of two compounds, dictyoglucosamines and furanodictines, are mainly involved in cell signalling pathways. Dictyoglucosamine A and B are amino sugars with a fatty acid as a side chain isolated from methanol extracts of the fruiting bodies of *D. purpureum* and *D. discoideum*, respectively. Dictyoglucosamine A was first isolated from the fruiting body of *D. purpureum*, whereas dictyoglucosamine B was isolated from *D. discoideum*'s fruiting body. A biological study revealed that dictyoglucosamines could induce a neuronal process of PC-12 cells (Haruhisa et al., 2002). Furanodictine A and B are *N*-acetylglucosamine and *N*-acetylmannosamine derivatives isolated from the fruiting body of *D. discoideum*. Furanodictine A and B were the first 3,6-anhydrosugars bioactive compounds ever isolated from nature. Furanodictines were able to promote neuronal processes in the presence of nerve growth factors, and furanodictine B could induce the neuronal process of PC-12 cells (Kikuchi et al., 2001). Both carbohydrate derivative compounds exhibited neuronal process induction, which suggested the involvement of both compounds in cell signalling in dictyostelids.

The remaining compounds are a group of compounds that do not share the core structure. These other compounds include glorin, dictyopterin, discadenine, dictyopyrones and terpines. Glorin, also called acrasin, was discovered from *P. violaceum* during aggregation. It has been reported that glorin was the chemoattractant during the aggregation stage of *P. violaceum* (Shimomura et al., 1982). It is the second chemoattractant compound in dictyostelids rather than cAMP, which was used in *D. discoideum*. Glorin is also involved in regulating mRNA during the aggregation stage via a quorum sensing-like system (Asghar et al., 2012). Apart from cell signalling, some compounds, including diccadenine, dictyopterin and dictyopyrones, are involved in cell growth and differentiation. Dictyopterin is a natural isomer of L-biopterin and was discovered as a significant pterin in a vegetative cell of *D. discoideum* (Klein et al., 1990). Dictyopterin was required for the vegetative growth of *D. discoideum* since the reduced form has been detected in vegetative cells but not in the spores (Klein et al., 1996). Discadenine is an amino acid derivative isolated from *D. discoideum.* This compound was first reported as a spore germination inhibitor (Aoki et al., 2007), and the chemical structure was discovered, hence the name 'discadenine' (Hiroshi et al., 1976). Dictyopyrones A−D are a-pyronoids with different hydroxyl groups isolated from the fruiting bodies of various *Dictyostelium* spp., including *D. discoideum, D. rhizopozium, D. longosporum, D. magnum* and *D. mucoroides* (Kikuchi et al., 2004; Takaya et al., 2000). Later, more compounds were found in *D. firmibasis* and *D. giganteum,* including dihydrodictyopyrones and dictyobispyrones (Haruhisa et al., 2007; Van Hai et al., 2017). The dictyopyrone derivatives in dictyostelids contain a hydroxyl group at C-4 instead of C-3. This is a piece of evidence that dictyostelids have unique biosynthetic pathways. Therefore, dictyostelids could be an extensive and novel bioactive compound repertoire (Haruhisa et al., 2007; Kikuchi et al., 2004; Takaya et al., 2000; Van Hai et al., 2017). The functional analysis revealed that dictyopyrones A−D promote morphogenesis, whereas dictyopyrones A and B inhibit spore formation and promote stalk formation in *D. discoideum* (Arai et al., 2005; Kikuchi et al., 2004; Takaya et al., 2000). In addition, dictyopyrones A and C were also able to inhibit K562 cells (Kikuchi et al., 2004). Terpenes are a large group of aromatic compounds commonly found in plants that have been recently identified in dictyostelids. Some terpenes are identified in dictyostelids, such as (E, E)-α-farnesene, (E)-nerolidol, (E)-β-farnesene and β-barbatene; however, the biological activity of these compounds remains unknown (Chen et al., 2016). Gene expression analysis showed that terpene synthases (TPSs), the enzyme responsible for terpene synthesis, were up-regulated during multicellular development. This suggested that terpenes play a role in developing dictyostelids (Chen et al., 2016). Chen and others also proposed possible roles of these compounds such as that: (1) terpenes were produced in order to increase spore dispersal accommodated by other organisms such as *D. melanogaster* (fruit fly), (2) terpenes are used as a defence mechanism against nematodes, and (3) terpenes are used as signals during multicellular development (Chen et al., 2016).

Numerous bioactive compounds combined with a large diversity both within and between genera strongly suggest that the dictyostelids are excellent organisms for research and academic purposes. The unique structure of the compound and the evidence of a unique biosynthesis pathway indicates numbers of novel bioactive compounds that have not yet been discovered. Differences in core chemical structure and promising biological activity results are evidence for potential drug candidates such as antitumour, antiobesity and antimicrobial drugs.

Dictyostelium genomes

Since *D. discoideum* is commonly used as a model organism, it was first selected for genome sequencing in the early genomics era. The *D. discoideum* genome project was initiated in 1998, and the high-quality genome sequence of *D. discoideum* was announced in 2005 (Eichinger et al., 2005; Loomis et al., 1995). The genome of *D. discoideum* consists of six chromosomes (ranging from 3.5 to 8.5 Mb) with 34 Mb total size. The nucleotide sequence of *D. discoideum* has an extreme A + T bias (77.57%) distributed throughout the genome (Eichinger et al., 2005). The AT bias in *D. discoideum* is the second highest from the *P. falciparum* genome (Gardner et al., 2002). The AT-rich regions within *D. discoideum* are more common in the coding sequences than in the *P. falciparum* genome and mainly encode the same amino acid (glutamine), which was believed to allow easier DNA separation during DNA replication (Glöckner, 2013). Apart from a high AT content, the genome also harbours numerous transposable elements (TEs), about 10% of the genome sequence, and simple sequence repeats (SSRs), which are more than 11% of the genome (Glockner et al., 2001, 2002). Those elements are distributed non-randomly; for example, non-long terminal repeat (non-TLR) retrotransposons are located next to tRNA genes, while SSRs are mainly located at non-coding regions (Eichinger et al., 2005).

The *D. discoideum* genome has been estimated to have about 12,500 genes from a total of 13,541 predicted proteins. Surprisingly, the genome size is larger than double the size of *P. falciparum* and *Saccharomyces cerevisiae* genomes (Gardner et al., 2002; Olsen, 2005). Similar numbers of predicted genes were also observed in the genomes of *D. purpureum*, *P. pallidum* and *D. fasciculatum* (Heidel et al., 2011; Sucgang et al., 2011). Moreover, the *D. discoideum* genome has some more abundant genes than other genomes, such as the lysozyme-producing genes, polyketide synthases (PKSs) and terpene synthases (TPSs). Interestingly, there are at least 16 lysozyme genes present in the *D. discoideum* genome. This feature is very unusual because each organism's genome generally carries only one or a few types of the lysozyme gene (Callewaert and Michiels, 2010). However, *D. discoideum* carries at least four different types of lysozyme genes, including the viral (V) type, chicken (C) type, *Entamoeba* type and amoeba (Aly) types of lysozyme in its genome (Lamrabet et al., 2020).

Genome studies of *D. discoideum* revealed the potential of a large class of secondary metabolite producers in *Dictyostelium*, including polyketide synthases (PKSs) and terpene synthases (TPSs). PKSs are a family of enzyme complexes responsible for producing polyketides (Cane et al., 1998). PKSs are classified into three types based on the configuration of enzyme modules, including types I, II and III (chalcone synthase-like) PKSs (Austin and Noel, 2003; Staunton and Weissman, 2001). The genome of *D. discoideum* has a total of 45 type I PKS genes (designated as DiPSK1−45 corresponding to the order in the genome). Additionally, according to phylogenetic analysis, all PKS genes in *D. discoideum* formed a distinct monophyletic clade from bacteria and fungi (Zucko et al., 2007). Among those 45 PKS genes, five were predicted to be pseudogenes with truncated domains (DiPSK4, DiPKS11, DiPKS12, DiPKS20 and DiPKS43), and two were hybrid type I−type III PKSs (DiPKS1 and DiPKS37) (Eichinger et al., 2005; Zucko et al., 2007). Enzymatic assays suggested that the final product from DiPKS1 was DIF-1, whereas the final product from DiPKS37 was MPBD (Ghosh et al., 2008). Phylogenetic analysis using the KS and ACP domain revealed a close relationship among PKS genes of *D. discoideum* and *D. purpureum* derived from a common ancestor. This close relationship suggested functional conservation between PKS genes in both species. Interestingly, only four *D. purpureum* PKS genes were orthologous to those in *D. discoideum* despite the close relationship among the PKS genes (Sucgang et al., 2011). A larger number of PKS genes in the *D. discoideum* genome than the other dictyostelids suggested that *Dictyostelium* has a high degree of variation in secondary metabolite reservoirs.

TPSs are enzymes responsible for converting oligoprenyl diphosphates into numerous types of terpenes and alcohols. TPSs had been previously found only in plants, fungi and bacteria, and more recently in dictyostelid genomes (Chen et al., 2011, 2016). The number of TPSs in the dictyostelid genome varies from three (*D. fasciculatum*), nine (*D. discoideum*), to 21 (*P. pallidum*). Phylogenetic analysis suggested that the TPS genes could be classified into four clades by the distribution of dictyostelids. Clade I consisted of TPSs from all four groups, whereas clades III and IV consisted of TPSs from only groups 2 and 4, respectively (Chen et al., 2018). These results suggested common TPSs in the dictyostelids and further group-specific expansion of TPSs (Chen et al., 2018), reflecting variation in secondary metabolites.

After the *D. discoideum* genome announcement in 2005, genome sequencing projects and comparative genome analysis projects of dictyostelids were launched. To date, there are 22 genomes of dictyostelids available in the NCBI database. However, some are unannotated-draft-genome sequences that need to be further curated. Several databases for genome analysis of dictyostelids are available, such as DictyBase, Social Amoebas Comparative Genome Browser (SACGB) and Acytostelium Gene Database. DictyBase (https://dictycr.org), one of the most well-known websites, collects genomic resources of *D. discoideum* AX4, *D. purpureum* AX1, *D. fasciculatum* SH3 and *P. pallidum* PN500. The website provides essential laboratory protocols for dictyostelids, such as methods for culturing and maintaining dictyostelids, recipes for media and buffers, and molecular biology methods for handling dictyostelids. DictyBase also hosts Dicty Stock Center (DSC), a central repository for *D. discoideum* and related species and plasmids and commonly used bacteria for dictyostelid cultures. Researchers can either purchase or deposit their interesting strains from this stock centre. They also provide information for researchers interested in the dictyostelids field, such as news, job offers, the annual conference and a community forum. SACGB (http://sacgb.leibniz-fli.de/cgi/index.pl) provides comparative genome analysis tools for dictyostelid genomes including *D. discoideum* AX4, *D. fasciculatum* SH3, *D. lacteum* and *P. pallidum* PN500. It collects various methods for genome analysis including cluster of orthologous (COG) functional categories, best bidirectional hits for proteins analysis, codon usages and gene synteny analysis. *Acytostelium* Gene Database (http://cosmos.bot.kyoto-u.ac.jp/acytodb//cgi-bin/index.cgi?org=as) collects gene annotations of the *A. subglobosum* genome and some comparative genome analysis tools. This database provides similar tools for comparative genomic analysis but only focuses on the genome sequence of the genus *Acytostelium*.

Conclusions

Dictyostelid social amoebae are a group of eukaryotic microbes that are significant in the soil of all ecosystems. They have had long-term coevolution with soil bacteria. Genome analysis reveals that they host many bioactive compounds, such as antibacterial, anticancer, and enzymes. Unfortunately, the previous research on dictyostelids focused on only a handful of species. However, we have tried to show here that the known species of dictyostelids does not reach half of its natural diversity. We believe that dictyostelids are one of the most significant bioresources in nature, and thus should be explored for their bioactive potentials.

Acknowledgements

The authors thank Ms. Kulawadee Wonganan for her time in cultivating and photographing the Thai dictyostelid isolates. This work was supported by a grant from Mahidol University, Thailand Science Research and Innovation (TSRI), and Thailand Research Fund (TRF).

References

Agnihothrudu, V., 1956. Occurrence of Dictyosteliaceae in the rhizosphere of plants in southern India. Experientia 12 (4), 149–150.

Aoki, Y., Yoshida, M., Kawaide, H., Abe, H., Natsume, M., 2007. Isolation and characterization of a spore germination inhibitor from *Streptomyces* sp. CB-1-1, a phytopathogen causing root tumor of melon. Biosci. Biotechnol. Biochem. 71 (4), 986–992.

Arai, A., Goto, Y., Hasegawa, A., Hosaka, K., Kikuchi, H., Oshima, Y., Tanaka, S., Kubohara, Y., 2005. Dictyopyrones, novel alpha-pyronoids isolated from *Dictyostelium* spp., promote stalk cell differentiation in *Dictyostelium discoideum*. Differentiation 73 (7), 377–384.

Asahi, K., Sakurai, A., Takahashi, N., Kubohara, Y., Okamoto, K., Tanaka, Y., 1995. DIF-1, morphogen of *Dictyostelium discoideum*, induces the erythroid differentiation in murine and human leukemia cells. Biochem. Biophys. Res. Commun. 208 (3), 1036–1039.

Asghar, A., Groth, M., Siol, O., Gaube, F., Enzensperger, C., Glockner, G., Winckler, T., 2012. Developmental gene regulation by an ancient intercellular communication system in social amoebae. Protist 163 (1), 25–37.

Austin, M.B., Noel, J.P., 2003. The chalcone synthase superfamily of type III polyketide synthases. Nat. Prod. Rep. 20 (1), 79–110.

Baldauf, S.L., Bhattacharya, D., Cockrill, J., Hugenholtz, P., Pawlowski, J., Simpson, A., 2004. The tree of life: an overview. In: Cracraft, J., Donohue, M. (Eds.), Assembling the Tree of Life. Oxford university press, Oxford, pp. 66–67.

Baldauf, S.L., Roger, A., Wenk-Siefert, I., Doolittle, W., 2000. A kingdom-level phylogeny of eukaryotes based on combined protein data. Science 290 (5493), 972–977.

Baldauf, S.L., 2008. An overview of the phylogeny and diversity of eukaryotes. J. Systemat. Evol. 46 (3), 263–273.

Barnett, R., Stallforth, P., 2018. Natural products from social amoebae. Chemistry 24 (17), 4202–4214.

Baxt, L.A., Singh, U., 2008. New insights into *Entamoeba histolytica* pathogenesis. Curr. Opin. Infect. Dis. 21 (5), 489–494.

Blaskovics, J., Raper, K., 1957. Encystment stages of *Dictyostelium*. Biol. Bull. 113 (1), 58–88.

Burki, F., Roger, A.J., Brown, M.W., Simpson, A.G.B., 2020. The new tree of eukaryotes. Trends Ecol. Evol. 35 (1), 43–55.

Benghezal, M., Fauvarque, M., Tournebize, R., Froquet, R., Marchetti, A., Bergeret, E., Lardy, B., Klein, G., Sansonetti, P., Charette, S.J., Cosson, P., 2006. Specific host genes required for the killing of *Klebsiella* bacteria by phagocytes. Cell Microbiol. 8 (1), 139–148.

Bonner, J., 2013. The evolution of the cellular slime molds. In: Romeralo, M., Baldauf, S.L., Escalante, R. (Eds.), Dictyostelids. Springer, New York.

Bonner, J., Lamont, D., 2005. Behavior of cellular slime molds in the soil. Mycologia 97 (1), 178–184.

Bozzaro, S., 2013. The model organism *Dictyostelium discoideum*. In: Eichinger, L., Rivero, F. (Eds.), *Dictyostelium discoideum* Protocol, second ed. Springer protocol, New York.

Callewaert, L., Michiels, C.W., 2010. Lysozymes in the animal kingdom. J. Biosci. 35 (1), 127–160.

Cane, D.E., Walsh, C.T., Khosla, C., 1998. Harnessing the biosynthetic code: combinations, permutations, and mutations. Science 282 (5386), 63–68.

Cardelli, J., 2001. Phagocytosis and macropinocytosis in *Dictyostelium*: phosphoinositide-based processes, biochemically distinct. Traffic 2 (5), 311–320.

Cavender, J.C., 1969. The occurrence and distribution of Acrasieae in forest soil I Europe. Am. J. Bot. 56 (9), 989–992.

Cavender, J.C., 1972. Cellular slime molds in forest soils of eastern Canada. Can. J. Bot. 50 (7), 1497–1501.

Cavender, J.C., 1976. Cellular slime molds of Southeast Asia. II. Occurrence and distribution. Am. J. Bot. 63 (1), 71–733.

Cavender, J.C., Cavender-Bares, J., Hohl, H.R., 1995. Ecological distribution of cellular slime molds in forest soils of Germany. Bot. Helv. 105, 199–220.

Cavender, J.C., Hopka, C., 1986. Distribution patterns of Ohio soil dictyostelids in relation to physiography. Mycologia 78 (5), 825–831.

Cavender, J.C., Kawabe, K., 1989. Cellular slime molds of Japan. I. Distribution and biogeographical considerations. Mycologia 81 (5), 683–691.

Cavender, J.C., Lakhanpal, T.N., 1986. Distribution of dictyostelid cellular slime molds in forest soils of India. Mycologia 78 (1), 56–65.

Cavender, J.C., Landolt, J.C., Ndiritu, G.G., Stephenson, S.L., 2010. Dictyostelid cellular slime moulds from Africa. Mycosphere 1 (2), 147–152.

Cavender, J.C., Landolt, J.C., Romeralo, M., Perrigo, A., Vadell, E.M., Stephenson, S.L., 2016. New species of Polysphondylium from Madagascar. Mycologia 108 (1), 80–109.

Cavender, J.C., Landolt, J.C., Stephenson, S.L., Cavender, N., Vadell, E., 2004. Comparison of cellular slime mold populations on three Caribbean Islands, including a description of the new species *Polysphondylium equisetoides*. Syst. Geogr. Plants 74 (2), 243–250.

Cavender, J.C., Landolt, J.C., Suthers, H., Stephenson, S., 2012. Dictyostelid cellular slime moulds of Mexico. Mycosphere 3 (3), 336–351.

Cavender, J., Raper, K., 1968. The occurrence and distribution of Acrasieae in forests of subtropical and tropical America. Am. J. Bot. 55 (4), 504.

Cavender, J.C., Stephenson, S.L., Landolt, J.C., Vadell, E.M., 2002. Dictyostelid cellular slime moulds in the forests of New Zealand. N. Z. J. Bot. 40 (2), 235–264.

Cavender, J.C., Vadell, E.M., 2000. The genus *Acytostelium*. Mycologia 92 (5), 992–1008.

Cavender, J.C., Vadell, E.M., 2006. Cellular slime molds of Ohio. Ohio Biol. Surv. Bull. New Ser. XVI (1), 28–29.

Cavender, J.C., Vadell, E., Landolt, J.C., Stephenson, S.L., Baldauf, S.L., Roux, J., Cavender, N., Winsett, K.E., 2018. New dictyostelid cellular slime molds from South Africa. Phytotaxa 383 (3), 233–251.

Cavender, J.C., Vadell, E.M., Landolt, J.C., Winsett, K.E., Stephenson, S.L., Rollins, A.W., Romeralo, M., 2013. New small dictyostelids from seasonal rainforests of Central America. Mycologia 105, 610–635.

Chen, F., Tholl, D., Bohlmann, J., Pichersky, E., 2011. The family of terpene synthases in plants: a mid-size family of genes for specialized metabolism that is highly diversified throughout the kingdom. Plant J. 66 (1), 212–229.

Chen, X., Kollner, T.G., Jia, Q., Norris, A., Santhanam, B., Rabe, P., Dickschat, J.S., Shaulsky, G., Gershenzon, J., Chen, F., 2016. Terpene synthase genes in eukaryotes beyond plants and fungi: occurrence in social amoebae. Proc. Natl. Acad. Sci. U. S. A. 113 (43), 12132–12137.

Chen, X., Kollner, T.G., Shaulsky, G., Jia, Q., Dickschat, J.S., Gershenzon, J., Chen, F., 2018. Diversity and functional evolution of terpene synthases in dictyostelid social amoebae. Sci. Rep. 8 (1), 14361.

Cornillon, S., Pech, E., Benghezal, M., Ravanel, K., Gaynor, E., Letourneur, F., Brückert, F., Cosson, P., 2000. Phg1p is a nine-transmembrane protein superfamily member involved in *Dictyostelium* adhesion and phagocytosis. J. Biol. Chem. 275 (44), 34287–34292.

Dagamac, N.H.A., dela Cruz, T.E.E., 2020. The Philippine slime molds after Dogma's 1975 list—how far have we been? Philipp. J. Syst. Biol. 13 (2), 58–65.

Eichinger, L., Pachebat, J.A., Glockner, G., Rajandream, M.A., Sucgang, R., Berriman, M., Song, J., Olsen, R., Szafranski, K., Xu, Q., Tunggal, B., Kummerfeld, S., Madera, M., Konfortov, B.A., Rivero, F., Bankier, A.T., Lehmann, R., Hamlin, N., Davies, R., Gaudet, P., Fey, P., Pilcher, K., Chen, G., Saunders, D., Sodergren, E., Davis, P., Kerhornou, A., Nie, X., Hall, N., Anjard, C., Hemphill, L., Bason, N., Farbrother, P., Desany, B., Just, E., Morio, T., Rost, R., Churcher, C., Cooper, J., Haydock, S., van Driessche, N., Cronin, A., Goodhead, I., Muzny, D., Mourier, T., Pain, A., Lu, M., Harper, D., Lindsay, R., Hauser, H., James, K., Quiles, M., Madan Babu, M., Saito, T., Buchrieser, C., Wardroper, A., Felder, M.,

Thangavelu, M., Johnson, D., Knights, A., Loulseged, H., Mungall, K., Oliver, K., Price, C., Quail, M.A., Urushihara, H., Hernandez, J., Rabbinowitsch, E., Steffen, D., Sanders, M., Ma, J., Kohara, Y., Sharp, S., Simmonds, M., Spiegler, S., Tivey, A., Sugano, S., White, B., Walker, D., Woodward, J., Winckler, T., Tanaka, Y., Shaulsky, G., Schleicher, M., Weinstock, G., Rosenthal, A., Cox, E.C., Chisholm, R.L., Gibbs, R., Loomis, W.F., Platzer, M., Kay, R.R., Williams, J., Dear, P.H., Noegel, A.A., Barrell, B., Kuspa, A., 2005. The genome of the social amoeba *Dictyostelium discoideum*. Nature 435 (7038), 43–57.

Flowers, J.M., Li, S.I., Stathos, A., Saxer, G., Ostrowski, E.A., Queller, D.C., Strassmann, J.E., Purugganan, M.D., 2010. Variation, sex, and social cooperation: molecular population genetics of the social amoeba *Dictyostelium discoideum*. PLoS Genet. 6 (7), e1001013.

Frøyen, O.J., Langvad, F., 1984. Occurrence and distribution of dictyostelid cellular slime molds in Norway. Nord. J. Bot. 4 (6), 817–822.

Fukuzawa, M., Araki, T., Adrian, I., Williams, J., 2001. Tyrosine phosphorylation-independent nuclear translocation of a *Dictyostelium* STAT in response to DIF signaling. Mol. Cell 7 (4), 779–788.

Gardner, M.J., Hall, N., Fung, E., White, O., Berriman, M., Hyman, R.W., Carlton, J.M., Pain, A., Nelson, K.E., Bowman, S., Paulsen, I.T., James, K., Eisen, J.A., Rutherford, K., Salzberg, S.L., Craig, A., Kyes, S., Chan, M.S., Nene, V., Shallom, S.J., Suh, B., Peterson, J., Angiuoli, S., Pertea, M., Allen, J., Selengut, J., Haft, D., Mather, M.W., Vaidya, A.B., Martin, D.M., Fairlamb, A.H., Fraunholz, M.J., Roos, D.S., Ralph, S.A., McFadden, G.I., Cummings, L.M., Subramanian, G.M., Mungall, C., Venter, J.C., Carucci, D.J., Hoffman, S.L., Newbold, C., Davis, R.W., Fraser, C.M., Barrell, B., 2002. Genome sequence of the human malaria parasite *Plasmodium falciparum*. Nature 419 (6906), 498–511.

Ghosh, R., Chhabra, A., Phatale, P.A., Samrat, S.K., Sharma, J., Gosain, A., Mohanty, D., Saran, S., Gokhale, R.S., 2008. Dissecting the functional role of polyketide synthases in *Dictyostelium discoideum*: biosynthesis of the differentiation regulating factor 4-methyl-5-pentylbenzene-1,3-diol. J. Biol. Chem. 283 (17), 11348–11354.

Ginsburg, G., Kimmel, A., 1997. Autonomous and nonautonomous regulation of axis formation by antagonistic signaling via 7-span cAMP receptors and GSK3 in *Dictyostelium*. Genes Dev. 11 (16), 2112–2123.

Glöckner, G., 2013. Genome analysis of social amoebae. In: Romeralo, M., Baldauf, S., Escalante, R. (Eds.), Dictyostelids: Evolution, Genomics and Cell Biology. Springer Berlin Heidelberg, Berlin. Heidelberg, pp. 35–47.

Glockner, G., Eichinger, L., Szafranski, K., Pachebat, J.A., Bankier, A.T., Dear, P.H., Lehmann, R., Baumgart, C., Parra, G., Abril, J.F., Guigo, R., Kumpf, K., Tunggal, B., Cox, E., Quail, M.A., Platzer, M., Rosenthal, A., Noegel, A.A., Dictyostelium Genome Sequencing, C., 2002. Sequence and analysis of chromosome 2 of *Dictyostelium discoideum*. Nature 418 (6893), 79–85.

Glockner, G., Szafranski, K., Winckler, T., Dingermann, T., Quail, M.A., Cox, E., Eichinger, L., Noegel, A.A., Rosenthal, A., 2001. The complex repeats of *Dictyostelium discoideum*. Genome Res. 11 (4), 585–594.

Hagiwara, H., 1990. Enumeration of the Dictyostelid Cellular Slime Molds Obtained from Mt. Phulchoki in the Kathmandu Valley, Nepal, vol. 2, pp. 23–32.

Hagiwara, H., 1991. A new species and some new records of dictyostelid cellular slime molds from Oman. Bull. Natl. Mus. Nat. Sci. Tokyo Ser. B 17, 109–121.

Hagiwara, H., 1996. Dictyostelids in Japan X. Two new species of *Dictyostelium, D. pseudobrefeldianum* and *D. robustum*. Bull. Natl. Mus. Nat. Sci. Tokyo Ser. B 22, 47–54.

Hagiwara, H., 1998. Dictyostelids in Japan XI. *Dictyostelium giganteum* Singh. Bull. Natl. Mus. Nat. Sci. Tokyo Ser. B 24, 81–84.

Haruhisa, K., Jun, K., Yoshinori, S., Jun-ichi, S., Shigeyoshi, H., Norimichi, N., Yoshiteru, O., 2002. The isolation and synthesis of two novel N-acetyl glucosamine derivatives from *Dictyostelium* cellular slime molds which exhibit neurite outgrowth activity. Tetrahedron Lett. 43 (8), 1477–1480.

Haruhisa, K., Koji, N., Yuzuru, K., Naomi, G., Kohei, H., Yasuo, M., Yoshiteru, O., 2007. Dihydrodictyopyrones A and C: new members of dictyopyrone family isolated from *Dictyostelium* cellular slime molds. Tetrahedron Lett. 48 (33), 5905–5909.

Haruhisa, K., Shinya, I., Koji, N., Yuzuru, K., Yoshiteru, O., 2010. Novel prenylated and geranylated aromatic compounds isolated from *Polysphondylium* cellular slime molds. Tetrahedron 66 (32), 6000–6007.

Haruhisa, K., Yusuke, M., Yasuhiro, K., Yuzuru, K., Yoshiteru, O., 2012. Isolation, synthesis, and biological activity of biphenyl and m-terphenyl-type compounds from *Dictyostelium* cellular slime molds. Tetrahedron 68 (43), 8884–8889.

Heidel, A.J., Lawal, H.M., Felder, M., Schilde, C., Helps, N.R., Tunggal, B., Rivero, F., John, U., Schleicher, M., Eichinger, L., Platzer, M., Noegel, A.A., Schaap, P., Glockner, G., 2011. Phylogeny-wide analysis of social amoeba genomes highlights ancient origins for complex intercellular communication. Genome Res. 21 (11), 1882–1891.

Hiroshi, A., Masaaki, U., Yoshimasa, T., Hazame, S., 1976. Structure of discadenine, a spore germination inhibitor from the cellular slime mold, *Dictyostelium discoideum*. Tetrahedron Lett. 17 (42), 3807–3810.

Huss, M., 1989. Dispersal of cellular slime molds by two soil invertebrates. Mycologia 81 (5), 677.

Jin, T., Zhang, N., Long, Y., Parent, C., Devreotes, P., 2000. Localization of the G Protein βγ complex in living cells during chemotaxis. Science 287 (5455), 1034–1036.

Kanai, M., Konda, Y., Nakajima, T., Izumi, Y., Kanda, N., Nanakin, A., Kubohara, Y., Chiba, T., 2003. Differentiation-inducing factor-1 (DIF-1) inhibits STAT3 activity involved in gastric cancer cell proliferation via MEK-ERK-dependent pathway. Oncogene 22 (4), 548–554.

Kawaharada, R., Nakamura, A., Takahashi, K., Kikuchi, H., Oshima, Y., Kubohara, Y., 2016. Oral administration of *Dictyostelium* differentiation-inducing factor 1 lowers blood glucose levels in streptozotocin-induced diabetic rats. Life Sci. 155, 5662.

Kawata, T., Shevchenko, A., Fukuzawa, M., Jermyn, K., Totty, N., Zhukovskaya, N., Sterling, A.E., Mann, M., Williams, J.G., 1997. SH2 signaling in a lower eukaryote: a STAT protein that regulates stalk cell differentiation in *Dictyostelium*. Cell 89 (6), 909–916.

Kessin, R.H., 2001. *Dictyostelium*: Evolution, Cell Biology, and the Development of Multicellularity, first ed. Cambridge University Press, Cambridge (Chapter2).

Kessin, R., Gundersen, G., Zaydfudim, V., Grimson, M., 1996. How cellular slime molds evade nematodes. Proc. Natl. Acad. Sci. U. S. A. 93 (10), 4857–4861.

Khan, N.A., 2003. Pathogenesis of *Acanthamoeba* infections. Microb. Pathog. 34 (6), 277–285.

Kikuchi, H., Ito, I., Takahashi, K., Ishigaki, H., Iizumi, K., Kubohara, Y., Oshima, Y., 2017. Isolation, synthesis, and biological activity of chlorinated alkylresorcinols from *Dictyostelium* cellular slime molds. J. Nat. Prod. 80 (10), 2716–2722.

Kikuchi, H., Kubohara, Y., Nguyen, V.H., Katou, Y., Oshima, Y., 2013. Novel chlorinated dibenzofurans isolated from the cellular slime mold, *Polysphondylium filamentosum*, and their biological activities. Bioorg. Med. Chem. 21 (15), 4628–4633.

Kikuchi, H., Oshima, Y., Ichimura, A., Gokan, N., Hasegawa, A., Hosaka, K., Kubohara, Y., 2006b. Anti-leukemic activities of *Dictyostelium* secondary metabolites: a novel aromatic metabolite, 4-methyl-5-n-pentylbenzene-1,3-diol, isolated from *Dictyostelium mucoroides* suppresses cell growth in human leukemia K562 and HL-60 cells. Life Sci. 80 (2), 160–165.

Kikuchi, H., Saito, Y., Komiya, J., Takaya, Y., Honma, S., Nakahata, N., Ito, A., Oshima, Y., 2001. Furanodictine A and B: amino sugar analogues produced by cellular slime mold *Dictyostelium discoideum* showing neuronal differentiation activity. J. Org. Chem. 66 (21), 6982–6987.

Kikuchi, H., Saito, Y., Sekiya, J., Okano, Y., Saito, M., Nakahata, N., Kubohara, Y., Oshima, Y., 2005. Isolation and synthesis of a new aromatic compound, brefelamide, from *Dictyostelium* cellular slime molds and its inhibitory effect on the proliferation of astrocytoma cells. J. Org. Chem. 70 (22), 8854–8858.

Kikuchi, H., Saito, Y., Sekiya, J., Okano, Y., Saito, M., Nakahata, N., Kubohara, Y., Oshima, Y., 2006a. Isolation and synthesis of a new aromatic compound, brefelamide, from *Dictyostelium* cellular slime molds and its inhibitory effect on the proliferation of astrocytoma cells. ChemInform 37 (11), 8854–8858.

Kikuchi, H., Sasaki, K., Sekiya, J., Maeda, Y., Amagai, A., Kubohara, Y., Oshima, Y., 2004. Structural requirements of dictyopyrones isolated from *Dictyostelium* spp. in the regulation of *Dictyostelium* development and in anti-leukemic activity. Bioorg. Med. Chem. 12 (12), 3203–3214.

Klein, R., Tatischeff, I., Cotter, D., 1996. Dictyopterin appears at amoebae emergence during spore germination in *Dictyostelium.* C R Acad. Sci. III 319 (4), 289–294.

Klein, R., Thiery, R., Tatischeff, I., 1990. Dictyopterin, 6-(D-threo-1,2-dihydroxypropyl)-pterin, a new natural isomer of L-biopterin. Isolation from vegetative cells of *Dictyostelium discoideum* and identification. Eur. J. Biochem. 187 (3), 665–669.

Kubohara, Y., 1997. DIF-1, putative morphogen of *D. discoideum*, suppresses cell growth and promotes retinoic acid-induced cell differentiation in HL-60. Biochem. Biophys. Res. Commun. 236 (2), 418–422.

Kubohara, Y., 1999. Effects of differentiation-inducing factors of *Dictyostelium discoideum* on human leukemia K562 cells: DIF-3 is the most potent anti-leukemic agent. Eur. J. Pharmacol. 381 (1), 57–62.

Kubohara, Y., Kikuchi, H., 2018. *Dictyostelium*: an important source of structural and functional diversity in drug discovery. Cells 8 (1).

Kubohara, Y., Kikuchi, H., Oshima, Y., 2008. Exploitation of the derivatives of *Dictyostelium* differentiation-inducing factor-1, which promote glucose consumption in mammalian cells. Life Sci. 83 (1718), 608612.

Kubohara, Y., Saito, Y., Tatemoto, K., 1995. Differentiation-inducing factor of *D. discoideum* raises intracellular calcium concentration and suppresses cell growth in rat pancreatic AR42J cells. FEBS Lett. 359 (2–3), 119–122.

Kuwayama, H., Kubohara, Y., 2009. Differentiation-inducing factor-1 and -2 function also as modulators for *Dictyostelium* chemotaxis. PLoS One 4 (8), 6658.

Lamrabet, O., Jauslin, T., Lima, W.C., Leippe, M., Cosson, P., 2020. The multifarious lysozyme arsenal of *Dictyostelium discoideum.* Dev. Comp. Immunol. 107, 103645.

Landolt, J.C., Wong, G.J., 1998. Dictyostelid cellular slime molds from Hawaii. Pac. Sci. 52 (1), 98–103.

Landolt, J.C., Shadwick, J.D., Stephenson, S.L., 2008a. First records of dictyostelids and protostelids from Ascension Island. Sydowia 60, 235–245.

Landolt, J.C., Stephenson, S.L., Cavender, J.C., 2006. Distribution and ecology of dictyostelid cellular slime molds in Great Smoky Mountains National Park. Mycologia 98 (4), 541–549.

Landolt, J.C., Cavender, J.C., Stephenson, S.L., Vadell, E.M., 2008b. New species of dictyostelid cellular slime moulds from Australia. Aust. Syst. Bot. 21 (1), 50–66.

Li, W., Wei, Y., Zou, Y., Liu, P., Li, Z., Gontcharov, A.A., Stephenson, S.L., Wang, Q., Zhang, S., Li, Y., 2020. Dictyostelid cellular slime molds from the Russian Far East. Protist 171 (5), 125756.

Liu, P., Li, Y., 2010a. Dictyostelids from Ukraine 1: two new records of *Dictyostelium*. Mycotaxon 111, 275–278.

Liu, P., Li, Y., 2010b. Dictyostelids from Ukraine 2: two new records of *Dictyostelium*. Mycotaxon 112, 367–370.

Liu, P., Li, Y., 2011. New species and new records of dictyostelids from Ukraine. Mycologia 103 (3), 641–645.

Liu, P., Zhang, S., Zou, Y., Li, Z., Stephenson, S.L., Li, Y., 2020. Distribution and ecology of dictyostelids in China. Fungal Biol. Rev. 34, 170–177.

Liu, P., Zou, Y., Li, W., Li, Y., Li, X., Che, S., Stephenson, S.L., 2019. Dictyostelid cellular slime molds from Christmas Island, Indian Ocean. mSphere 4 (2) e00133–19.

Loomis, W.F., Welker, D., Hughes, J., Maghakian, D., Kuspa, A., 1995. Integrated maps of the chromosomes in *Dictyostelium discoideum*. Genetics 141 (1), 147–157.

Meyer, I., Kuhnert, O., Gräf, R., 2011. Functional analyses of lissencephaly-related proteins in *Dictyostelium*. Semin. Cell Dev. Biol. 22 (1), 89–96.

Morris, H.R., Taylor, G.W., Masento, M.S., Jermyn, K.A., Kay, R.R., 1987. Chemical structure of the morphogen differentiation inducing factor from *Dictyostelium discoideum*. Nature 328 (6133), 811–814.

Morris, H.R., Masento, M.S., Taylor, G.W., Jermyn, K.A., Kay, R.R., 1988. Structure elucidation of two differentiation inducing factors (DIF-2 and DIF-3) from the cellular slime mould *Dictyostelium discoideum*. Biochem. J. 249 (3), 903–906.

Murata, C., Ogura, T., Narita, S., Kondo, A.P., Iwasaki, N., Saito, T., Usuki, T., 2016. Synthesis and SAR of 4-methyl-5-pentylbenzene-1,3-diol (MPBD), produced by *Dictyostelium discoideum*. Bioorg. Med. Chem. Lett. 26 (5), 1428–1433.

Myre, M.A., Washicosky, K., Moir, R.D., Tesco, G., Tanzi, R.E., Wasco, W., 2009. Reduced amyloidogenic processing of the amyloid beta-protein precursor by the small-molecule differentiation inducing factor-1. Cell. Signal. 21 (4), 567–576.

Nakajima-Shimada, J., Hatabu, T., Hosoi, Y., Onizuka, Y., Kikuchi, H., Oshima, Y., Kubohara, Y., 2013. Derivatives of *Dictyostelium discoideum* differentiation-inducing factor-3 suppress the activities of *Trypanosoma cruzi in vitro* and *in vivo*. Biochem. Pharmacol. 85 (11), 1603–1610.

Nelson, N., Olive, L.S., Stoianovitch, C., 1967. A new species of *Dictyostelium* from Hawaii. Am. J. Bot. 54 (3), 354–358.

Nickerson, A., Raper, K., 1973. Macrocysts in the life cycle of the Dictyosteliaceae. I. formation of the macrocysts. Am. J. Bot. 60 (2), 190–197.

O'Dell, W., 1979. Isolation, enumeration and identification of amebae from a Nebraska lake. J. Protozool. 26 (2), 265–269.

Olive, L., 1970. The Mycetozoa: a revised classification. Bot. Rev. 36 (1), 59–89.

Olsen, R.M., 2005. How many protein encoding genes does *Dictyostelium discoideum* have? In: Loomis, W.F., Kuspa, A. (Eds.), *Dictyostelium Genomics*. Wymondham: Horizon Bioscience, pp. 265–278.

Omata, W., Shibata, H., Nagasawa, M., Kojima, I., Kikuchi, H., Oshima, Y., Hosaka, K., Kubohara, Y., 2007. *Dictyostelium* differentiation-inducing factor-1 induces glucose transporter 1 translocation and promotes glucose uptake in mammalian cells. FEBS J. 274 (13), 3392–3404.

Perrigo, A.L., Baldauf, S.L., Romeralo, M., 2013. Diversity of dictyostelid social amoebae in high latitude habitats of Northern Sweden. Fungal Divers. 58 (1), 185–198.

Perrigo, A.L., Romeralo, M., Baldauf, S.L., 2012. What's on your boots: an investigation into the role we play in protist dispersal. J. Biogeogr. 39 (5), 998–1003.

Perrigo, A.L., Vadell, E.M., Cavender, J.C., Landolt, J.C., Liu, P., Stephenson, S.L., 2020. Additional new species suggest high dictyostelid diversity on Madagascar. Mycologia 112 (5), 1026–1042.

Pichersky, E., Gang, D.R., 2000. Genetics and biochemistry of secondary metabolites in plants: an evolutionary perspective. Trends Plant Sci. 5 (10), 439–445.

Rai, J.N., Tewari, J.P., 1961. Studies in cellular slime moulds from Indian soils. Proc. Natl. Acad. Sci. India B 53 (1), 1–9.

Rai, J.N., Tewari, J.P., 1963. Studies in cellular slime moulds from Indian soils. III. On the occurrence of two strains of *Dictyostelium mucoroides*-complex, conforming to the species *Dictyostelium sphaerocephalum* (Oud.) Saccardo and March. Proc. Natl. Acad. Sci. India B 58, 263–266.

Raper, K., 1937. Growth and development of *Dictyostelium discoideum* with different bacterial associates. J. Agric. Res. 55 (4), 289–316.

Raper, K., 1984. The Dictyostelids. Princeton University Press, Princeton.

Reddy, A.K., Balne, P.K., Garg, P., Sangwan, V.S., Das, M., Krishna, P.V., Bagga, B., Vemuganti, G.K., 2010. *Dictyostelium polycephalum* infection of human cornea. Emerg. Infect. Dis. 16 (10), 1644.

Rollins, A.W., Landolt, J.C., Stephenson, S.L., 2010. Dictyostelid cellular slime molds associated with grasslands of the central and western United States. Mycologia 102 (5), 996–1003.

Romeralo, M., Baldauf, S.L., Cavender, J.C., 2009. A new species of cellular slime mold from southern Portugal based on morphology, ITS and SSU sequences. Mycologia 101 (2), 269–274.

Romeralo, M., Cavender, J., Landolt, J., Stephenson, S., Baldauf, S.L., 2011. An expanded phylogeny of social amoebas (Dictyostelia) shows increasing diversity and new morphological patterns. BMC Evol. Biol. 11 (1), 84.

Romeralo, M., Escalante, R., Baldauf, S., 2012. Evolution and diversity of Dictyostelid social amoebae. Protist 163 (3), 327–343.

Romeralo, M., Fiz-palacios, O., 2013. Evolution of dictyostelid social amoebas inferred from the use of molecular tools. In: Romeralo, M., Baldauf, S.L., Escalante, R. (Eds.), Dictyostelids. Springer, New York.

Romeralo, M., Lado, C., 2006. Dictyostelids from mediterranean forests of the south of Europe. Mycol. Prog. 5 (4), 231–241.

Romeralo, M., Landolt, J., Cavender, J., Laursen, G., Baldauf, S.L., 2010a. Two species of dictyostelid cellular slime molds from Alaska. Mycologia 102 (3), 588–595.

Romeralo, M., Moya-Laraño, J., Lado, C., 2010b. Social Amoebae: environmental factors influencing their distribution and diversity across South-Western Europe. Microb. Ecol. 61 (1), 154–165.

Rukseree, K., Wonganun, K., Palittapongarnpim, P., Ajawatanawong, P., 2018. Phylogenetic diversity of 18S rDNA sequences of dictyostelids from Amnat Charoen Province, Thailand. Mycosphere 9 (2), 202–214.

Saito, T., Taylor, G.W., Yang, J.C., Neuhaus, D., Stetsenko, D., Kato, A., Kay, R.R., 2006. Identification of new differentiation inducing factors from *Dictyostelium discoideum*. Biochim. Biophys. Acta 1760 (5), 754–761.

Sathe, S., Kaushik, S., Lalremruata, A., Aggarwal, R.K., Cavender, J.C., Nanjundiah, V., 2010. Genetic heterogeneity in wild isolates of cellular slime mold social groups. Microb. Ecol. 60 (1), 137–148.

Sawada, T., Aono, M., Asakawa, S., Ito, A., Awano, K., 2000. Structure determination and total synthesis of a novel antibacterial substance, AB0022A, produced by a cellular slime mold. J. Antibiot. 53 (9), 959–966.

Schaap, P., Winckler, T., Nelson, M., Alvarez-Curto, E., Elgie, B., Hagiwara, H., Cavender, J., Milano-Curto, A., Rozen, D.E., Dingermann, T., Mutzel, R., Baldauf, S.L., 2006. Molecular phylogeny and evolution of morphology in the social amoebas. Science 314 (5799), 661–663.

Schmith, A., Groth, M., Ratka, J., Gatz, S., Spaller, T., Siol, O., Glöckner, G., Winckler, T., 2013. Conserved gene regulatory function of the carboxy-terminal domain of dictyostelid C-module-binding factor. Eukaryot. Cell 12 (3), 460–468.

Seephueak, P., Petcharat, V., 2014. The biodiversity of dictyostelid cellular slime molds in rubber tree leaf litter in Southern Thailand. Mycosphere 5 (6), 805–813.

Shaulsky, G., Kessin, R., 2007. The cold war of the social amoebae. Curr. Biol. 17 (16), R684–R692.

Sheikh, S., Gloeckner, G., Kuwayama, H., Schaap, P., Urushihara, H., Baldauf, S.L., 2015. Root of dictyostelia based on 213 universal proteins. Mol. Phylogenet. Evol. 92, 53–62.

Sheikh, S., Thulin, M., Cavender, J.C., Escalante, R., Kawakami, S.-I., Lado, C., Landolt, J.C., Nanjundiah, V., Queller, D.C., Strassmann, J.E., Spiegel, F.W., Stephenson, S.L., Vadell, E.M., Baldauf, S.L., 2018. A new classification of the dictyostelids. Protist 169 (1), 1–28.

Shim, K.C., Kil, J.H., Chang, N.K., 2002. Occurrence and distribution of cellular slime Molds in South Korea. Korean J. Ecol. 25 (1), 51–58.

Shimomura, O., Suthers, H.L., Bonner, J.T., 1982. Chemical identity of the acrasin of the cellular slime mold *Polysphondylium violaceum*. Proc. Natl. Acad. Sci. U. S. A. 79 (23), 7376–7379.

Singh, B.N., 1947a. Studies on Soil Acrasieae: 1. Distribution of species of Dictyostelium in soils of Great Britain and the effect of bacteria on their development. Microbiology 1 (1), 11–21.

Singh, B.N., 1947b. Studies on Soil Acrasieae: the active life of species of *Dictyostelium* in soil and the influence there on of soil moisture and bacterial food. J. Gen. Microbiol. 1 (3), 361–367.

Srirampur, A., Reddy, A.K., Siva, S., Kalwad, A., 2018. February. Keratitis due to an unusual pathogenic social amoeba, *Dictyostelium polycephalum*. In: Open Forum Infectious Diseases, vol. 5(2). Oxford University Press, US, p. ofy009.

Staunton, J., Weissman, K.J., 2001. Polyketide biosynthesis: a millennium review. Nat. Prod. Rep. 18 (4), 380–416.

Stephenson, S., Landolt, J.C., 1992. Vertebrates as vectors of cellular slime moulds in temperate forests. Mycol. Res. 96 (8), 670–672.

Stephenson, S., Landolt, J.C., Laursen, G., 1997. Dictyostelid cellular slime molds from Western Alaska, U.S.A., and the Russian Far East. Arct. Alp. Res. 29 (2), 222–225.

Stephenson, S.L., Landolt, J.C., Moore, D.L., 1999. Protostelids, dictyostelids, and myxomycetes in the litter microhabitat of the Luquillo experimental forest, Puerto Rico. Mycol. Res. 103 (2), 209–214.

Stout, J., 1973. The relationship between protozoan populations and biological activity in soils. Am. Zool. 13 (1), 193–201.

Sucgang, R., Kuo, A., Tian, X., Salerno, W., Parikh, A., Feasley, C.L., Dalin, E., Tu, H., Huang, E., Barry, K., Lindquist, E., Shapiro, H., Bruce, D., Schmutz, J., Salamov, A., Fey, P., Gaudet, P., Anjard, C., Babu, M.M., Basu, S., Bushmanova, Y., van der Wel, H., Katoh-Kurasawa, M., Dinh, C., Coutinho, P.M., Saito, T., Elias, M., Schaap, P., Kay, R.R., Henrissat, B.,

Eichinger, L., Rivero, F., Putnam, N.H., West, C.M., Loomis, W.F., Chisholm, R.L., Shaulsky, G., Strassmann, J.E., Queller, D.C., Kuspa, A., Grigoriev, I.V., 2011. Comparative genomics of the social amoebae *Dictyostelium discoideum* and *Dictyostelium purpureum.* Genome Biol. 12 (2), 20.

Suthers, H., 1985. Ground-feeding migratory songbirds as cellular slime mold distribution vectors. Oecologia 65 (4), 526–530.

Suzuki, T., Kikuchi, H., Ogura, M., Homma, M.K., Oshima, Y., Homma, Y., 2015. Weight loss by Ppc-1, a novel small molecule mitochondrial uncoupler derived from slime mold. PLoS One 10 (2), 0117088.

Swanson, A.R., Vadell, E.M., Cavender, J.C., 1999. Global distribution of forest soil dictyostelids. J. Biogeogr. 26 (I), 133–148.

Swanson, A.R., Spiegel, F.W., Cavender, J.C., 2002. Taxonomy, slime molds, and the questions we ask. Mycologia 94 (6), 968–979. PMID:21156570.

Takaya, Y., Hotta, R., Fujiwara, K., Otani, R., Uchiyama, Y., Sakakibara, M., Fukuda, E., Niwa, M., Inouye, K., Oohata, A.A., 2014. Alkylbenzoquinone involved in development of cellular slime molds. Org. Lett. 16 (14), 3660–3663.

Takaya, Y., Kikuchi, H., Terui, Y., Komiya, J., Furukawa, K.I., Seya, K., Motomura, S., Ito, A., Oshima, Y., 2000. Novel acyl alpha-pyronoids, dictyopyrone A, B, and C, from *Dictyostelium* cellular slime molds. J. Org. Chem. 65 (4), 985–989.

Thompson, C., Kay, R., 2000. Cell-fate choice in *Dictyostelium*: intrinsic biases modulate sensitivity to DIF signaling. Dev. Biol. 227 (1), 56–64.

Tieghem, P.E.L. van, 1880. Sur Quelques Myxomycètes A Plasmode Agrégé. Bull. Soc. Bot. France 27, 317–322 (in French). https://www.biodiversitylibrary.org/item/8650#page/3/mode/1up. (Accessed 27 May 2021).

Torres, N.H.A., Yap, J., Ching, M., Yulo, P.R.J., 2011. Occurrence of cellular slime molds (Dictyostelids) In Subic Bay Natural Forest Reserve, Zambales, Philippines. Philipp. J. Syst. Biol. 99–104.

Traub, F., Hohl, H.R., Cavender, J.C., 1981. Cellular slime molds of Switzerland. II. Distribution in forest soils. Am. J. Bot. 68 (2), 172–182.

Van Hai, H., Haruhisa, K., Hikaru, S., Kyoichi, I., Yuzuru, K., Yoshiteru, O., 2017. Production of novel bispyrone metabolites in the cellular slime mold *Dictyostelium giganteum* induced by zinc(II) ion. Tetrahedron 73 (5), 583–588.

Vadell, E.M., Cavender, J.C., 2007. Dictyostelids living in the soils of the Atlantic Forest, Iguazú region, Misiones, Argentina: description of new species. Mycologia 99 (1), 112–124.

Vadell, E.M., Cavender, J.C., Landolt, J.C., Perrigo, A.L., Liu, P., Stephenson, S.L., 2018. Five new species of dictyostelid social amoebae (Amoebozoa) from Thailand. BMC Evol. Biol. 18 (1), 198.

Vadell, E.M., Cavender, J.C., Romeralo, M., Edwards, S.M., Stephenson, S.L., Baldauf, S.L., 2011. New species of dictyostelids from Patagonia and Tierra del Fuego, Argentina. Mycologia 103 (1), 101–117.

Waddell, D.R., 1982. A predatory slime mould. Nature 298, 464–466.

Wessels, D., Srikantha, T., Yi, S., Kuhl, S., Aravind, L., Soll, D., 2006. The Shwachman-Bodian-Diamond syndrome gene encodes an RNA-binding protein that localizes to the pseudopod of *Dictyostelium* amoebae during chemotaxis. J. Cell Sci. 119 (2), 370–379.

Whittaker, R., 1969. New concepts of kingdoms of organisms. Science 163 (3863), 150–160.

Whittingham, W., Raper, K., 1960. Non-viability of stalk cells in *Dictyostelium.* Proc. Natl. Acad. Sci. U. S. A. 46 (5), 642–649.

Williams, R., 2005. Pharmacogenetics in model systems: Defining a common mechanism of action for mood stabilisers. Prog. Neuro-Psychopharmacol. Biol. Psychiatry 29 (6), 1029–1037.

Williams, R., Eames, M., Ryves, W., Viggars, J., Harwood, A., 1999. Loss of a prolyl oligopeptidase confers resistance to lithium by elevation of inositol (1,4,5) trisphosphate. EMBO J. 18 (10), 2734–2745.

Xavier de Lima, V., 2013. First records and human activity on Brazilian Atlantic Forest Dictyostelids. Mycosphere 4 (3), 584–589.

Yoshiaki, T., Haruhisa, K., Yuichi, T., Jun, K., Yasuo, M., Akira, I., Yoshiteru, O., 2001. Novel aromatic substances, dictyomedin A and B, from *Dictyostelium* cellular slime molds and their inhibitory effects on *Dictyostelium* development. Tetrahedron Lett. 42 (1), 61–63.

Yulo, P.R.J., dela Cruz, T.E.E., 2012. Diversity and distribution of cellular slime molds (dictyostelids) in Lubang Island, Occidental Mindoro, Philippines. J. Mycol. 3 (3), 188–194.

Zhang, J., Yamada, O., Kida, S., Matsushita, Y., Murase, S., Hattori, T., Kubohara, Y., Kikuchi, H., Oshima, Y., 2016. Identification of brefelamide as a novel inhibitor of osteopontin that suppresses invasion of A549 lung cancer cell. Oncol. Rep. 36 (4), 2357–2364.

Zucko, J., Skunca, N., Curk, T., Zupan, B., Long, P.F., Cullum, J., Kessin, R.H., Hranueli, D., 2007. Polyketide synthase genes and the natural products potential of *Dictyostelium discoideum*. Bioinformatics 23 (19), 2543–2549.

Chapter 11

Yeast in a megadiverse country for sustainable futures

Enrique Javier Carvajal Barriga, Patricia Portero-Barahona and Alexandra Narváez-Trujillo

Pontificia Universidad Católica del Ecuador, Centro Neotropical para la Investigación de la Biomasa, Colección de Levaduras Quito-Católica (CLQCA), Quito, Ecuador

Chapter outline

Introduction

Ecuador is acknowledged as one of the megadiverse countries globally (Dupérré and Tapia, 2020), not only because of its latitude but also because of its large number of microhabitats created by the orography of the Andes

Importance of Microbiology Teaching and Microbial Resource Management for Sustainable Futures
https://doi.org/10.1016/B978-0-12-818272-7.00001-8

mountain chain that runs north to south of its territory. Ecuador is also one of the under-explored regions in the world in terms of microbial biodiversity. The current worldwide yeast species count has surpassed 2000 (The Yeasts, n.d.), of which many have metabolic features that might be useful in industry. Meanwhile, sampling, collecting and conserving yeast species in Ecuador has not been extensive and primarily has been performed by the group belonging to the Neotropical Center for the Biomass Research (CNIB) at the Pontificia Universidad Católica del Ecuador. The Catholic University Yeast Collection in Quito (CLQCA) was founded in 2006 to preserve yeast biodiversity by collecting from selected substrates found in different ecosystems (CLQCA, n.d.). The data obtained from yeast isolates has positively contributed to explain yeast distribution, abundance and adaptation of these microorganisms in a biodiverse country with four natural regions (Amazonia, Andes, Pacific Coast and the Galapagos archipelago).

Additionally, and of the same importance, the CLQCA has emphasised discovering and studying the metabolic capabilities of Ecuadorian environmental yeast isolates for biotechnological applications such as probiotics, decontamination of pollutants, ancient yeast reanimation methods (Microbial Archaeology), screening of yeast metabolites for the industry, development and innovation of fermentation systems, second-generation ethanol research and development and ancient fermented beverages recreation. Collaboration with yeast researchers' international networks has been of high importance for the progress of yeast projects, including the description of novel yeast species, organisation of symposia and courses, organisation of expeditions to pristine Ecuadorian natural regions, amongst other collaborative outcomes.

This chapter aims to provide an overall view of the most relevant results of the projects carried out at the CLQCA. Also deals with the relevance of *ex situ* preservation of yeasts collected in natural environments, and the biotechnological potential of Ecuadorian yeast species. Finally, the chapter demonstrates how a culture collection can find a niche as a supporter of the local economy by developing new technologies and services for the industry building up sustainability for the future (Smith, 2003).

The significance of microbial collections in a megadiverse country like Ecuador

The CLQCA holds a fraction of Ecuador's mainland and insular territories' biodiversity. The exploration of natural regions allowed our researchers to understand the underlying facts that could explain yeast distribution, abundance and adaptation.

For instance, it has been established that about one-third of the yeast species found in Ecuador's mainland coincides with those species that populate different environments in the Galapagos Islands. Similar studies on plant diversity in the Galapagos archipelago reveal that about 30% of the species inhabiting the islands originated on Ecuador's mainland, and their occurrence is explained by the ocean currents (Humboldt marine current) that carry floating species of plants to the archipelago. Most probably, that was the way yeasts were transported. Plants also carry other organisms such as insects; thus, the dispersion through insect species most likely had its role during the early times of the island's colonisation.

It got clearer that if we check the list of yeast species identified in our collections, there are some species that are quite generalist that they can colonise almost every substrate and region of the country, whilst there are other yeast species that are extremely rare that no more than one isolate was identified amongst thousands of other yeasts. This fact led us to understand that yeast species and populations' fragility can be determined as a function of different parameters: natural regions and the occurrence in different substrates. From the data collected by our researchers, we discovered that about 2/3 of the Ecuadorian yeast species tend to be highly specialised, meaning that they are adapted to a somewhat reduced set of substrates and few natural regions. These data clearly show that culture collections and *ex situ* preservation are crucial to studying the potential applications of yeast in industry and the state of its abundance, as well as the threats these microscopic organisms may face with the perturbation of their natural habitats. Culture collections provide authentic strains for research purposes and are repositories with valuable information to preserve yeasts and habitats where they inhabit. The CLQCA constitutes the repository of Ecuador's yeast biodiversity.

Threats and opportunities of microbial collections in developing countries

Developing countries are generally located in tropical regions of the planet. These zones are characterised by their high biodiversity (Brown, 2014). Nonetheless, the most significant culture collections are in developed countries due to science and technology's financial support and, most importantly, the scientific tradition and availability of trained human talents.

The costs of maintenance in culture collection are high. In the context of scarcity of funds for research in developing countries, establishing a culture collection depends on national financial sources and less on private funds (World Federation for Culture Collections, n.d.). The cost of maintenance per isolate in European Culture collections is about 108−146 € (131−177 US$) (Smith et al., 2014).

At the CLQCA, the average annual cost of maintenance per isolate goes from 80 to 85 US$. The difference with European collections is due to the personnel wages in Ecuador. Still, it is challenging to sustain a microbial collection with state or university funds in developing countries. Frequently, the research grants are channelled to applied science projects. The culture collections as scientific facilities need to find their way to obtain funds.

One strategy is proposing services or biotechnology research that can yield outcomes potentially interesting for the industry. For instance, culture collections may offer services to fermentation industries like brewing, wine or distilling (Daniel and Prasad, 2010). Nowadays, the CLQCA is locally recognised as the yeast culture providers and strain deposit for microbreweries in Ecuador.

Another source of funds is by offering training for microbreweries and other fermentation start-ups, including teaching the basic protocols for scaling up yeast starters, detecting contaminants, equipment sanitation practices and guidance for basic in situ microbiology laboratories implementation, contaminants detection and identification, amongst others.

Finally, the design of other research and development projects that may end as spinoffs has also been regarded as a potential source of funding for the yeast collection. An example of that is the reanimation of a dormant yeast found at the Americas' still-existing oldest brewery at the San Francisco Convent in Quito, Ecuador (Fig. 11.1). The reanimation and identification of a unique *Saccharomyces cerevisiae* indigenous strain was a milestone of Microbial

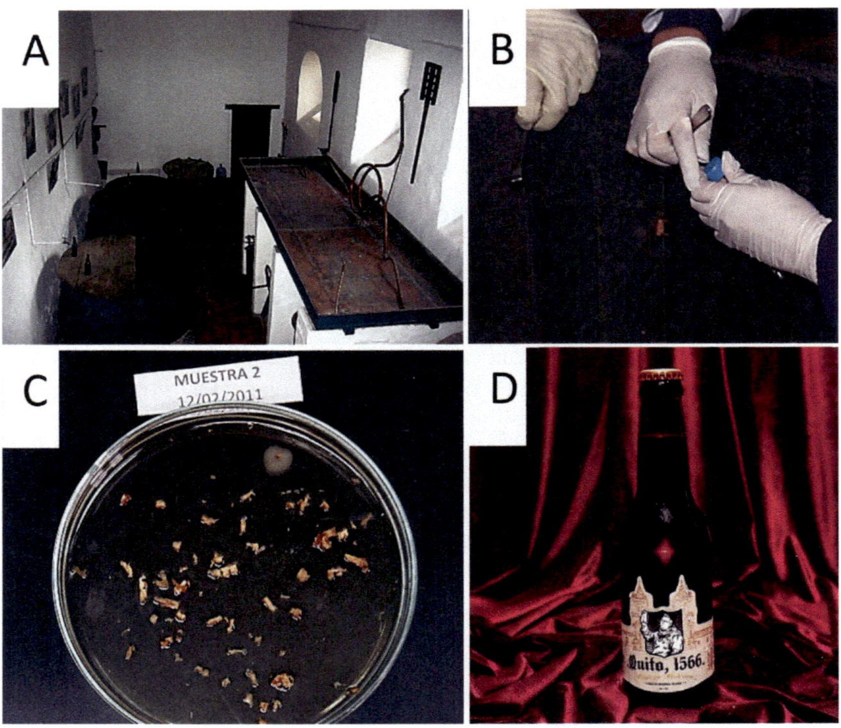

FIGURE 11.1 Recreation of the oldest brew of the Americas through the reanimation of dormant yeasts found in the wooden fermentation barrels from the XVI Century in San Francisco Monastery in Quito, Ecuador. From up left to down right: A. Brewery and fermentation barrels; B. Sampling the barrels; C. Yeast colony growth; D. Recreation of the beer.

Archaeology that brought acknowledgment to our biotechnological approach as a yeast collection.

The recreation of the beer produced more than 450 years by the Franciscan priests is a business with high potential. Nowadays, an external spinoff produces the beer and offers royalties for research at the CLQCA. The model is still in construction. Nevertheless, the beer has been produced and commercialised under the brand '*Quito, 1566*', with success in Ecuador and the United States of America (Dispatch, n.d.).

In developing countries like Ecuador, managing a culture collection can be challenging. On the other side, it is an excellent opportunity to develop new scientific knowledge by contributing to the world's science by discovering and describing novel yeast species. A culture collection is an essential player in helping the local economy by supporting and promoting start-ups dedicated to fermented products and other related activities. Nonetheless, all those activities need coordination and the establishment of clear boundaries to keep the culture collection's central mission intact.

As pointed out, the primary thread to culture collections is sustainability, including financing the specialised personnel and the access to technology, equipment and technical support. Technical problems with freezing may substantially diminish the viability of the isolates. Back-up ultra-freezers are needed for such an emergency. Refrigeration problems became the most common thread to microbial collections in developing countries.

In an intent to preserve culture collections with low-cost methods, the CLQCA has innovated by developing nanoencapsulation of yeast isolates. The concept is relatively simple but needs the proper materials and expertise. It consists of protecting yeast cells with nanomaterials that prevent them from being damaged by environmental factors in the long term (Geng et al., 2019). We synthesised and crafted a nanocellulose special composite (patent pending) that works as a universal binder (Li et al., 2019). The encapsulating substrate is enriched with vital nutrients to facilitate a prompt and safe yeast reanimation at due time.

The composite of yeast, binder and nutrients is subjected to a spherification process to produce porous nanobeads overnight dried in a sterile laminar flow hood and stored in sterile plastic micro vials until the next reanimation. In Fig. 11.2, the microspheres containing yeast cultures are shown at different levels of amplification in three separated SEM images. This way, most yeast isolates can be naturally dried and metabolically inactivated with high protection for at least 3 years. We believe this is a potential solution for culture collections in the developing world, with limited refrigeration capacity and low financing support.

Case studies of sustainable technologies for biotechnology applications

15 years of work on Ecuadorian yeast has produced a collection of 3230 yeast isolates corresponding to 200 different species, recovered from different

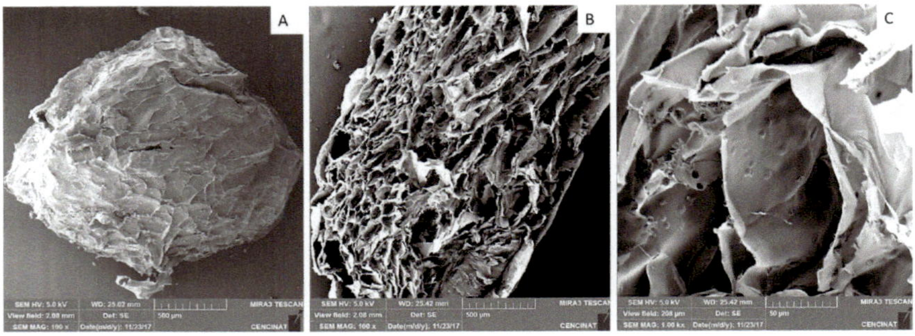

FIGURE 11.2 (A) microsphere of nanocellulose (scale bar 500 μm), (B) transversal section of nanocellulose microsphere (scale bar 500 μm), (C) nano encapsulated yeast cells (scale bar 50 μm).

geographical regions, including the Amazon rainforest, the Galápagos Islands and different bioregions on the Ecuadorian mainland. By applying molecular taxonomy methods, the CLQCA has contributed interesting and novel yeast species such as *Candida carvajalis* (James et al., 2009), *Saturnispora quitensis* (James et al., 2011), *Candida ecuadorensis* (James et al., 2013), *Kazachstania yasuniensis* (James et al., 2015), *Cyberlindnera galapagoensis* (Guamán-Burneo et al., 2015), *Kodamaea transpacifica* (Freitas et al., 2013), *Candida theae* (Chang et al., 2012), *Saccharomycopsis fodiens* (Lachance et al., 2012) and *Wickerhamomyces arborarius* (James et al., 2014). Some of these species have demonstrated strong potential in different biotechnological applications through various projects that have been carried out by the research staff at CLQCA and PUCE.

Candida carvajalis, the very first microorganism described in Ecuador

The CLQCA was established with the aim to catalogue, characterise and compare indigenous yeast species found in the different ecological habitats of the four (climatic) regions of Ecuador (James et al., 2009). Sampling early in the project was extensive and led to identifying a new species, *Candida carvajalis* collected in the Amazon basin. The yeast isolates catalogued as CLQCA-20-11 and CLQCA-20-14 were collected nearby abandoned oil wells in the central Amazonia of Ecuador. The collector was searching debris and rotten vegetal matter aiming to obtain xylose fermenting yeasts. Instead, these two isolates representing a novel yeast species were found. Later, they were fully characterised at molecular and physiological level for their further description.

The discovery and description of the novel species *Candida carvajalis* represented a milestone in discovering yeast biodiversity in this megadiverse country. No other microscopic unicellular organisms had been described before 2009. Culture collections of microscopic fungi, yeast and bacteria have since been established at PUCE and other academic institutions in Ecuador. However, the discovery of other novel microorganism species (Kolařík et al., 2017) is essential to support the sustainable use of Ecuadorian biodiversity.

Candida carvajalis is also an example of the adaptation of a yeast species to the conditions in its environment, an exhibition of a microbes plasticity. The two strains that were identified as this novel species were isolated from rotten leaf debris and the sap of a tree growing nearby to oil-polluted areas next to two abandoned oil wells in Ecuador's Amazon Basin (James et al., 2009). The oil industry has been the main economic resource in Ecuador since the 1960s; this has meant an extensive intervention in the Ecuadorian Amazon Rainforest where the oil reserves are located and unfortunately causing extensive environmental impacts (San Sebastián and Karin Hurtig, 2004). Nevertheless, the microorganisms that adapt to these conditions may present unique physiological capacities to use the hydrocarbons as a carbon source (Marín et al., 2018) regarding endophytic fungi which were isolated from flora in an area contaminated with crude oil. Therefore, even though *Candida carvajalis* has not been evaluated for hydrocarbon degradation, it is possible that its physiological adaptation to the environmental conditions in which it was found includes the breakdown of crude oil hydrocarbons, as has been reported (Türker, 2014).

Three yeasts with high potential for water decontamination

Chromium VI is a polluting metal widely used in different industries such as wood preservation, plastics and pigments production, amongst others. It is a common pollutant in water (Singh et al., 2021). The International Agency for Research on Cancer has classified this element in a list of human carcinogens (Pellerin and Booker, 2000).

A wide range of technologies have been employed to absorb Cr VI from water bodies, including phytoremediation (Ram et al., 2019), membrane filtration, solvent extraction, precipitation, capture with biopolymers (Farooqi et al., 2021) and microbial remediation by utilising various microorganisms, including yeast, bacteria and fungus (Dhal et al., 2013).

Yeasts by themselves possess electrostatic and hydrophobic forces on their surface. This characteristic is given mainly by proteins and mannan oligosaccharides on the cell wall (Li et al., 2020). The electrostatic charges and hydrophobic attractions allow the yeast cells to recognise each other and stick, forming large yeast cell clusters or flocs (El-Kirat-Chatel et al., 2015). Such characteristic has been utilised to adsorb heavy metals in bioremediation (Ojima et al., 2019).

During a set of experiments utilising yeasts for decontamination of Cr VI polluted waters, we intended to measure Ecuadorian yeast species efficiency versus the *S. cerevisiae* control. The yeast species employed were *Kazachstania yasuniensis*, *Kodamaea transpacifica* and *Saturnispora quitensis*. After pre-treatment of the yeast flocs by utilising a cation detergent, the yeast cells were tested as adsorbent micro particles. The results were exciting due to the high efficiency exhibited by *Kodamaea transpacifica* due to a combination of different factors such as surface area, individual cell size and surface charge. Moreover, the performance exhibited by *Saturnispora quitensis* was only slightly lower. They could remove 476.19 and 416.67 mg of Cr (VI)/g of yeast, respectively (Campaña-Pérez et al., 2019).

This work paved the road for non-conventional yeast species to be employed as microbial particles to decontaminate water. The advantage attained by this technology is the easiness and high efficiency of decontamination of polluted water. The three Ecuadorian yeast species' adsorption capacity ranged between 85.8% and 80.7%, whilst *S. cerevisiae* in the same set of experiments showed an efficiency of 75.8% (Campaña-Pérez et al., 2019). The results encourage developing cleaner and sustainable technologies for environmental remediation by utilising nonconventional yeasts from the Ecuadorian biodiversity.

Two Ecuadorian yeast species with probiotic potential in monogastric animals

Yeasts are considered as suitable microorganisms for being used as probiotics in animal production (Jacques and Casaregola, 2008). The principal advantage of using yeast instead of bacteria as probiotics is that yeasts are bigger cells, enhancing the intestine protection against pathogenic organisms (Czerucka et al., 2007). Adherence to microvilli in the intestine lumen is a key factor for probiotic microorganisms to induce the potentially pathogenic microbial flora in the host animal's mucosae. Yeast cell walls are completely covered of sugar moieties, especially mannose arrangements, where pathogenic bacteria get adhered through the interaction between mannose on the yeast surface and bacterial mannose-sensitive fimbriae (Teng et al., 2005). Ecuadorian yeast species were tested for its probiotic potential at verifying the tolerance to temperature, acid pH and bile acids.

This research project's objective was to evaluate the probiotic potential of nine recently described yeast species collected in Ecuador. The results were very promising for two species, namely *Kazachstania yasuniensis* (CLQCA-20-374) (James et al., 2015) and *Saturnispora quitensis* (CLQCA-10-114) (James et al., 2011), that tolerate 37°C, and an acid pH 2.5 for a period of 24 h. Also, both yeasts survived in the presence of bile salts. Both yeast isolates showed high bacterial adherence and agglutination capacity in vitro.

Kazachstania yasuniensis (CLQCA-20-374) was able to aggregate and adhere to four bacteria species: *Salmonella typhimurium*, *Staphylococcus epidermidis*, *Pseudomonas aeruginosa* and *Escherichia coli*. As for

Saturnispora quitensis, it aggregated and adhered to the cell walls of *Salmonella typhimurium, Listeria monocytogenes, Staphylococcus epidermidis* and *Escherichia coli.* In contrast, *Saccharomyces cerevisiae* var. boulardii, which represents the current standard for probiotic yeast, could only aggregate and adhere to *Salmonella typhimurium, Staphylococcus epidermidis* and *Escherichia coli* (Torres et al., 2018).

Antarctic yeast with potential for the biofuels industry

Yeasts have extensively been studied and utilised in the biofuels industry (Petrovič, 2015). Not only because of their natural fermenting capacity for ethanol fuel production but also for its versatility to be engineered. *Saccharomyces cerevisiae* is the most utilised species by far due to several advantages: high ethanol tolerance, wide variety of domesticated strains commercially available, well adaptation to industrial processes, high temperature and pH stress tolerance and reproducibility of processes, amongst others (Lacerda et al., 2020). A wide variety of these yeast strains is available in microbial collections worldwide. At the CLQCA, we were able to collect and characterise at least 120 different wild-type strains of *S. cerevisiae* found in substrates like archaeological fermentation pots, natural environments and chicha fermentation vessels (Piló et al., 2018).

On the other hand, the biodiesel industry registers fewer cases where yeasts have any role in the processes (Spagnuolo et al., 2019). Biodiesel production depends on the transesterification of fatty acids (acyl glycerides), where a methyl group is transferred to the carboxyl-terminal of fatty acids (Supraja et al., 2020). Such chemical reaction needs intensive energy use, and the utilisation of chemicals which effluents need to be treated before discarded. One alternative to chemical processes is the biochemical transesterification by utilising lipase enzymes (Encinar et al., 2019). The resultant methyl esters do not produce contaminants and the biological reaction can be carried out with lower energy consumption.

In 2012, the CLQCA was part of the 16th Ecuadorian expedition to Antarctica. The aim was to search for yeast species growing in extremophile environments, such as glaciers. 9 different yeast species were found out of a 120 yeast isolates collected in the Antarctic Peninsula (Fig. 11.3).

Extremophile yeasts can produce lipase enzymes that may replace the chemical esterification of fatty acids in biodiesel production. A screening of 120 isolates belonging to nine different Antarctic yeast isolates was performed to achieve this aim. Amongst other yeasts, the strain belongs to the species *Geomyces pullulans* and is catalogued as CLQCA-ANT-073 showed promising lipase/esterase activity (Demera et al., 2019).

The enzyme isolated presented high activity at hydrolysing 0.27 U/mL. Its maximum activity was registered at pH 8.0 and 40°C. Moreover, a $K_M = 3.7 \times 10^{-4}$ M was obtained regarding the kinetic parameters, a value

FIGURE 11.3 Collecting Antarctic yeasts at the Ecuadorian Pedro Vicente Maldonado scientific station in the Antarctic Peninsula.

located in the range of industrial lipases (Demera et al., 2019). Nevertheless, there is still a margin of improvement by adjusting the culture medium conditions can be improved to be optimised in order to attain higher production rates or the lipase-esterase enzyme. The enzyme showed kinetics adjusted to the Michaelis–Menten model but exhibits a remarkable variation as the so-called interfacial activation. This work demonstrates the high relevance of culture collections as biotechnological platforms for renewable energy projects.

From the Galapagos Islands to the pharmacy

Xylitol is a five-carbon sugar-alcohol widely used as sweetener and ingredient in toothpaste and chewing gum (Burt, 2006; Riley et al., 2015). Its antibacterial properties contribute to keeping teeth free of cavities and prevents ear infection, inflammation and osteoporosis (Xu et al., 2019). Its sweetening capacity is also beneficial for diabetes patients (Wölnerhanssen et al., 2020).

Industrial production of xylitol is achieved by hydrogenating xylose's C3 in a nickel-catalysed process, which is a costly and energy-demanding process (Ur-Rehman et al., 2015). The alternative biochemical process is by the catalysis of xylitol oxidase enzymes that diverse microorganisms possess, including yeast.

In 2009, an expedition to the Galapagos Islands yielded a subcollection of highly interesting yeast isolates. Two isolates belonging to the species *Candida tropicalis* CLQCA-24SC-125 and a novel species accommodated as *Cyberlindnera galapagoensis* were noticeable for its high xylitol production, yield and productivity (27.1 g/L xylitol, $Y^{xyl}{}_{p/s} = 0.67$ g/g, $Q_p = .38$ g/L and 24 g/L, $Y^{xyl}{}_{p/s} = 0.64$ g/g, $Q_p = .33$ g/L/ h) on sugarcane bagasse hydrolysate (Guamán-Burneo et al., 2015).

Xylitol represents an attractive product in lignocellulosic ethanol biorefineries where pentosans are extracted during biomass pre-treatment (Dasgupta et al., 2019). At the CLQCA, we have connected the culture collection project with studies on biorefinery and pre-treatments of local agricultural and industrial residues such as sugarcane bagasse (Portero-Barahona et al., 2019); also, we developed semi-continuous fermentation systems that facilitate the saccharification and simultaneous fermentation in the second generation ethanol model. The same system (pending patent) would be utilised to produce xylitol and other fermentation products for future sustainable industrial production (Portero Barahona et al., 2020).

Novel yeast strains for fermented beverages and foods

During the past decade, the markets experienced a flourishment of new and diverse crafted fermented beverages. The craft brewery sector experienced significant growth that nowadays about 17,000 craft breweries are operating worldwide (Baiano, 2020). Consumers are keen to experience new flavours and styles of fermented beverages. Yeast strains can impart different flavours and aromas to the product. The opportunity of selecting novel yeast strains with local or regional own characteristics is an engaging activity for craft beer producers (Elzinga et al., 2015).

In Ecuador, chicha (a fermented indigenous beverage made with corn or cassava) has been prepared since pre-Incan times. According to our research, we have collected more than 100 *Saccharomyces cerevisiae* strains from different substrates, including fruits, traditional beverages, flowers, insects and archaeological pieces. The strains were grouped according to their molecular profiles obtained through RFLP-mtDNA analysis (Rodríguez et al., 2011).

At least 15 different yeast RFLP profiles were tested with standard beer wort. Experts evaluated the resulting beer. Remarkably, most of the yeast isolates produced pleasant and some unique organoleptic profiles, competitive with commercial counterparts. Small companies and start-ups dedicated to biotechnology services may exploit the yeast's biodiversity in support of the craft beer revolution that has flourished since 1980 in the United States of America and expanded to the rest of the globe (Pokrivcak et al., 2019).

New yeast strains bring new opportunities to the beer industry by exploring beer styles with low carbohydrate and low alcohol contents (Troilo et al., 2020). Another search of brewers and consumers is focussed on specific flavours appreciated in specialty styles such as phenolics, fruitness and sourness, amongst and the ever-evolving preferences of the markets (Holt et al., 2019).

Raising the dead

Microbial Archaeology refers to the rescue and reanimation of ancient and metabolically inactive microorganisms found in natural or human substrates, mostly, but not limited to, archaeological contexts. Why resuscitate ancient microorganisms? The answer is the genetics they hold. As it is known, during dormancy, microorganisms do not perform cell division. Dormancy refers to the ability a yeast has to extend its lifespan by being still; in consequence, the genetics they had at the moment of dormancy is preserved (as in a time capsule) without significant modifications (Maire et al., 2020).

Bringing back to life microorganisms like yeasts offers new opportunities for scientific studies in terms of evolution, phylogeny, omics approaches and biochemical understanding of life mechanisms, including the aging and death of the cells (Eleutherio et al., 2018). Resuscitated microorganisms can also give new data about ancient cultures, origins, customs, diseases, nourishment and migrations, amongst others. Undoubtedly, microorganisms are somehow the microscopic witnesses of ancient cultures' history because their occurrence in archaeological pieces (such as fermentation vessels) may help shed light on the underlying aspects of daily ancient lives (Brüssow, 2020).

By 2006, the CLQCA was involved in a series of innovative projects searching for new contributions that may impact yeast science. That is how Microbial Archaeology began. At first, the yeast researchers collaborated with Ecuadorian archaeologists who, by that time, were digging ancient indigenous tombs in locations nearby Quito, the capital city of Ecuador. The indigenous cultures from the pre-Incan period used to bury corpses in tombs 15–20 m deep along with food and corn beer in vessels designed for their afterlife trip back to mother earth.

The porous clay material represents an ideal substrate where microorganisms find shelter and may become dormant (Aouizerat et al., 2019). That was how ancient chicha makers (corn beer) used their fermenters. Batch after batch of fresh corn broth added to the fermentation recipients propelled the colonisation and establishment of microorganismal communities that finally after adaptation settled as to form part of the fermenter's internal flora (Tyakht et al., 2021). Thus, fermentation vessels represent a suitable substrate for microbial reanimation and further studies of microbial communities and strains (Piló et al., 2018).

The resuscitation of microorganisms needed special care and conditions. The CLQCA is, to the best of our knowledge, the only microbial collection that developed a complete method of restoration of cell walls and membranes; controlled hydration of dry cells and metabolic activation of those ancient microorganisms, which at least preserved the structure and integrity that allowed them to restart the cell's replication machinery. The resuscitation process takes more time than the conventional refreshing of microorganisms performed routinely in microbiology labs.

The mechanism for the resuscitation of ancient microorganisms is still very poorly understood. The cell processes underlying the metabolic activation involve large amounts of different macromolecules, both structural and functional, such as enzymes, second messengers, growth factors, cofactors, polysaccharides (Orlean, 2012) and all the conditions to allow a delicate and physically compromised cell to reconstruct itself with the help of a special culture medium tailored for this aim. The results obtained were very satisfactory (Fig. 11.4). Up to date, dozens of yeast species have been resuscitated from the ancient fermentation vessels, amongst them, a new and medically interesting yeast species is described, namely *Candida theae* (Chang et al., 2012), and also other yeasts such as local wild-type *Saccharomyces cerevisiae* that fermented indigenous beverages (Piló et al., 2018).

The yeast resuscitation method and ancient yeast applications

If we imagine a 15 centuries-old yeast cell hidden in a microscopic shelter surrounded by hundreds of other similar cells that fell asleep in a tight arrangement embedded in cell wall polysaccharides that protected the cell's integrity and the cluster's structure, that is a *Persister Cell* (Gibney et al., 2020).

The architecture of yeast cell walls is highly dynamic. When the environmental conditions get aggressive for yeasts due to the accumulation of toxic metabolites such as ethanol or nutrient scarcity, the synthesis of mannoproteins is activated in the cell (Orlean, 2012). These mannoproteins are cell wall components synthesised and glycosylated within the Golgi apparatus and the endoplasmic reticulum. The anchoring of the peptide part within the cell wall's inner layers permits the sugar arrangements to emerge from the cell wall surface. These thin and intricate polysaccharide networks exert electrostatic and hydrophobic attractions with other similar sugar arrangements. Thus, cell-to-cell recognition occurs, and the final consequence of such a tight union is the formation of cell clusters – a very primitive but efficient way to protect the species when the environment gets aggressive. When environmental conditions improve, the yeasts can be reactivated for further reproduction.

The cell clusters apparently suffer superficial damages as time passes. Understandably, the cluster's peripheral cells may sacrifice whilst keeping the inner core protected. Nonetheless, this natural protection may have its limits. The degree of deterioration of the core (persister cells) should be understood

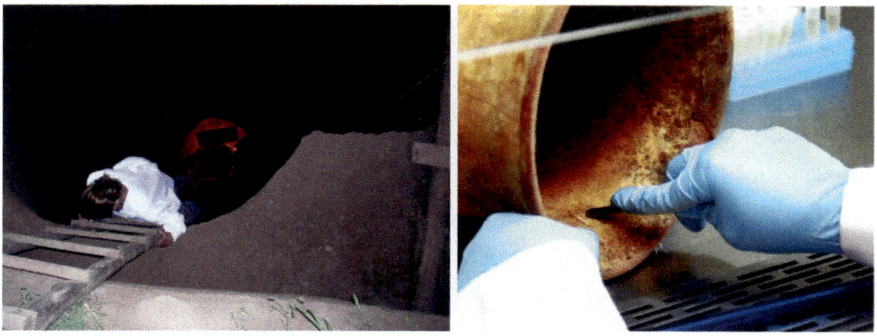

FIGURE 11.4 Raising the dead. Left: Collecting ancient fermentation jars from Quitu-Caras burials (200–680 BCE). Right: Applying the yeasts reanimation method developed at the CLQCA.

on a case-by-case basis. Not all the samples taken from ancient fermentation vessels yield yeast resuscitated. It is a microscopic treasure hunting search. For instance, in a survey carried out in Quito, from 2006 to 2010, from 40 different chicha vessels sampled, only 23 yeast isolates were collected. Some of the vessels yielded various isolates, whilst most others did not yield any yeast.

The culture medium for the resuscitation of yeast was designed practically by adding different elements that can accomplish structural restoration functions, controlled hydration and metabolic activation.

Restoration of yeast cell compartment structures

Compartmentalisation in cells is fundamental for the inter- and intracellular organisation. Cell membranes and cell walls are barriers and the structures holding the cell contents. Moreover, cell membranes are mosaics containing diverse proteins involved in the vast diversity of cell functions. These structures regulate osmotic and transport phenomena through different mechanisms, where embedded proteins have a crucial role. The integrity, flexibility and functionality of cell walls and cell membranes are crucial for an adequate resuscitation of yeasts.

The restoration and fluidisation of cell walls is the first step. The method includes the use of nanocellulose fibrils as universal binders. The nanoencapsulation of ancient cells provides a thin layer that repairs superficial damages and keeps the cells' integrity for further steps.

Additionally, glycerol is added for preserving cells at freezing temperatures because it protects the cells from freeze stress and acts as an enzymes stabiliser (Izawa et al., 2004). The restoration treatment overnight prepares the cells for the next steps. During this time, the cells are gently protected.

Controlled cell's hydration

The interior of the cell presents a hydrogel state. In the cell's interior, thin water layers surround cytoskeleton nanostructures and other functional macromolecules and organelles. This makes the cytosol a gel in equilibrium when the cell is active. In order to hydrate the cell's interior, slow and controlled hydration via water diffusion in a gel medium is performed. The samples are gently placed in wells made on a 0.3%–0.5% agar medium on the surface. This assures the diffusion of water into the previously restored cells.

Metabolic activation

For this aim, the method needs a preculture of yeasts previously subjected to high stress. Briefly, an active culture of *S. cerevisiae* is grown for 3 days in a medium containing 10% ethanol. The stressed yeast cells produce different molecules that are protective for the yeasts. They also contain all the signalling and transduction machinery, the growth factors, ATP, GTP and a whole battery of enzymes, including kinases needed to reanimate ancient yeast cells. This step of the method is key to resuscitation success. Cell biomass is centrifuged and washed several times with buffer at physiological pH. A proportion of 1:10 cell biomass:buffer is submitted to cell disruption by using an ultrasound probe for three to 5 min. The cells need to be cooled at 4°C with ice during the disruption process. A final centrifugation step is needed to separate the cell contents from cell debris. Finally, the cell content is sterilised by filtration with 0,45 μm syringe filters and freshly added to the wells with the agar samples.

After one to 4 weeks, the culture medium wells may present cell growth. It is still unknown why some yeasts take that much time to reanimate, but once the yeast population in a colony is visible, the rest of the process fits the traditional widely known microbiology methods. The resuscitation process has evolved in time, making it more efficient because of the increased isolates recovered from ancient fermentation vessels. It is a remarkable achievement of the CLQCA researchers contributing to recovering lost strains and even novel yeast species found in archaeological rests.

Culture collection management

As a consequence of Ecuador's rising interest in microbiological research, several new culture collections were strengthened or founded in the last 10 years. Moreover, the Environmental Ministry considered the microbial biodiversity to be protected and monitored through the issuance of licenses and permissions. These early steps conducted to the regularisation of microbial collections: quality control practices, operation manuals, tracking and assuring the authenticity of the microbial isolates are a few achievements gotten the last years.

Moreover, another essential step is creating public databases for international exposure of Ecuadorian microbial biodiversity. The databases are the centre of the microbial collection. The databases are ever-increasing and updating platforms containing every single isolate's vital information.

In an effort to optimise the culture collections of the Pontificia Universidad Católica del Ecuador, a few years ago, the CLQCA led the project SICOM-PUCE for its acronym in Spanish to Microbial Collections System at the Pontificia Universidad Catolica del Ecuador.

It has been a conjunct effort of curators and technical personnel to organise the different microbial collections existing under a harmonised and uniform system. The management of such collections can be adapted to a general scheme described on a case-by-case basis. The university microbial collections do not consider for-profit services but can share MTA strains with local and international microbial collections.

Microorganisms culture collections system at PUCE

The dispersion of resources in uncoordinated small culture collections leads to duplication of efforts, inefficient use of financial sources, equipment and personnel. Centralisation of microbial collections is the first step towards the optimisation of efforts, which is crucial to find sustainability in culture collections.

A culture collection's priority is to warrantee the authenticity of its isolates, no matter if it is a not-for-profit organisation or a services company. A culture collection is a repository that is obliged to positively and effectively respond to research or industry requirements. Thus, the personnel's trustworthiness is critical for the product's quality. A practical tracking system is also fundamental for ensuring the final products' quality.

Microorganisms themselves are very demanding and somehow hard to preserve. The equipment needed for its accurate identification and adequate preservation is costly. Most of the culture collections around the world are not economically sustainable if only considering the commercialisation of strains. That is why the culture collections are also involved in research projects that attract industry or governmental institutions' funds.

By 2008, Ecuador's Secretariat of Higher Education, Science and Technology (SENESCYT) launched a call for applied science projects in different areas of high interest in Ecuador's Development Plan. That is how the CLQCA proposed an ambitious project on biorefinery whose acronym was RESETA, or Sustainable Sources for Ethanol. One of this project's objectives was exploring and collecting yeasts in Ecuador's mainland and the Galápagos Islands. We visited every single of the 24 different provinces of the country, where we sampled yeast from a set of 10 different substrates that were predefined: flowers, insects, rotten wood, decaying vegetal matter, excrement, exude, fungi, leaves and sugar cane mills.

In such kind of project, a large amount of data is generated. The priceless information obtained must be accurate and trustworthy since it constitutes the basis for further studies that focus on conservation, ecology, sustainability and biotechnological applications. That is how we designed a roadmap that includes detailed instructions for the scientific and technical personnel involved in the yeast projects, from preparing the adequate equipment for field trips to bringing back living material to the labs, characterising pure cultures and feeding the online database. Yeasts are microscopic organisms themselves, but they need enormous efforts to preserve them. The roadmap scheme is in a flow chart that would help organise other microbial collections, especially in developing countries with still unexplored regions globally. In Fig. 11.5, there is the full track of a sample starting up from the criteria employed for the collection to the identification and accommodation in the phylogenetic tree.

Full description of the database and its importance for science

Data are priceless and fragile. Keeping them cured and safe is as important as preserving microbial isolates' authenticity in a culture collection. Nevertheless, the quality of data and its safe storage and organisation are highly dependent on the personnel's training and professional capabilities.

A microbial collection's curator must keep one eye on the general context and the other on the detail. Administrative, scientific, research and development, legal and regulatory and educational flanks are frequently the curator's tasks. The operational aspects such as quality control, culture management, collection of microorganisms, equipment maintenance and acquisitions of materials and database feeding can be assigned to specialised personnel within

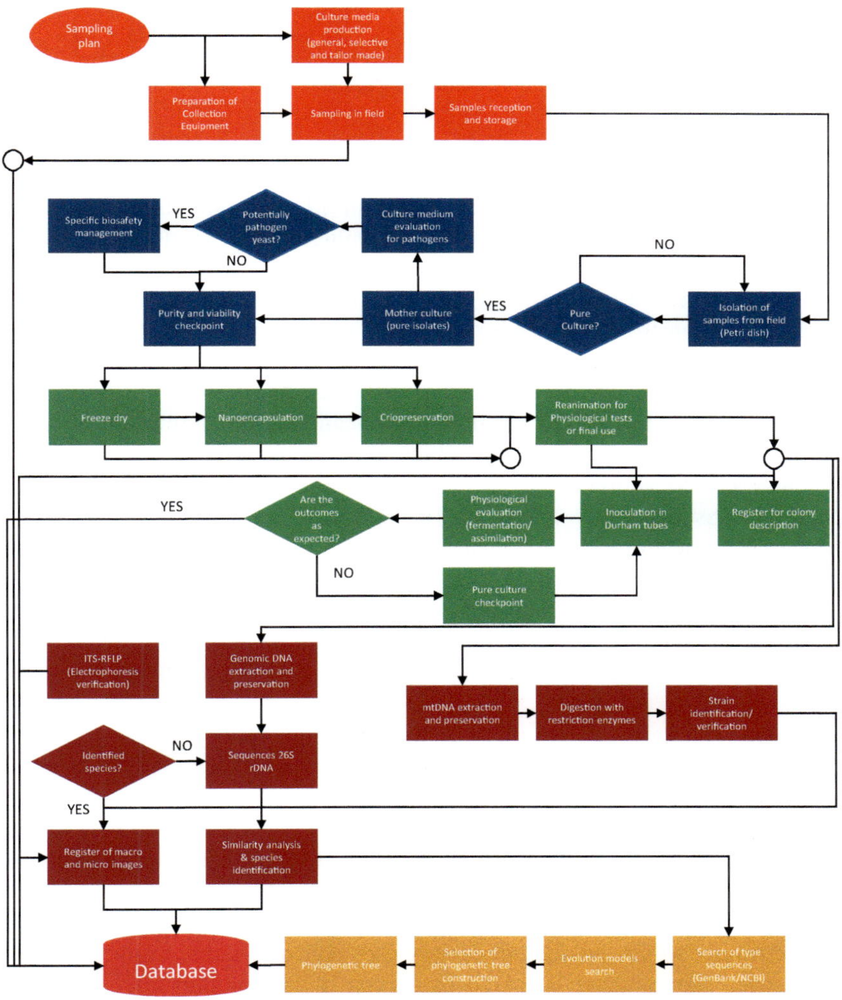

FIGURE 11.5 Flow diagram that shows the track from the yeast collection criteria to the construction of phylogenetic trees at the CLQCA.

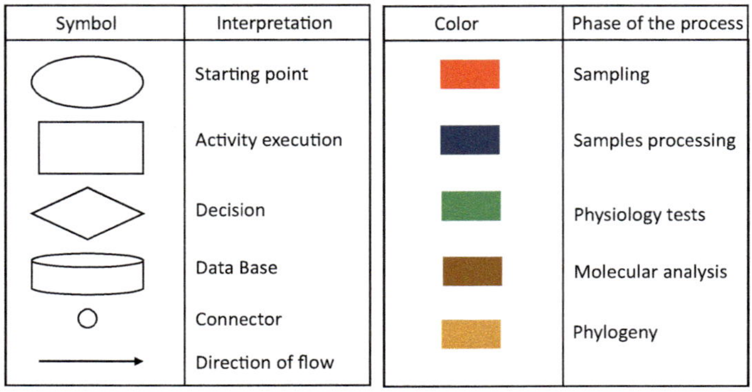

FIGURE 11.5 cont'd

the microbial collection. A multidisciplinary team is part of the success of a microbial collection.

Despite the roles that personnel display in different microbial collections, the database is its core. No collection has value without vital data of its accessions. At the CLQCA, we designed a database adjusted to the local reality and possibilities. We constructed all of it from scratch, but a database is an ever-lasting construct that needs continuous improvement, updating, increase of accessions, new data registration and adaptation to the global needs.

At the CLQCA database, we keep as much data as possible. The crude field data includes date of collection, province, locality, GPS coordinates, altitude, collector's name, substrate from which the sample was collected, culture medium and collection technique, project's name, number of samples taken from the substrate and observations. The field notes are taken in forms directly in the field. Cleanliness and organisation during this data gathering are of the most significant importance.

Other data are taken at the lab, such as the date of reception of the samples, macroscopic and microscopic images, molecular fingerprinting, DNA electrophoresis gels imaging and physiology tests (assimilation, tolerance, fermentation).

Fingerprinting may save much time and resources for species identification. Nevertheless, some of the yeasts that could not be identified through our internal fingerprinting database have to be sequenced for further identification or description.

The crude data are then selected and organised in electronic forms and datasheets and then uploaded to the online database for public access (www.cnib.com.ec). The majority of our data keeps saved and backgrounded in our hardware, whilst the most frequently used data are readily available on the web page. The database is a kind of deposit of information funnelled from different stages of the process: from the collection to the identification and classification.

Once an isolate is completely characterised, it acquires a code that in our yeast collection is in the format: CLQCA (00) (000), where CLQCA represents the acronym in Spanish of 'The Catholic University Yeast Collection in Quito' (web site address??) http://www.cnib.com.ec/index.php/es/clqca; the first figures between parentheses represent the number of Ecuador's province where the isolated was collected and, the second figures between parentheses represent the number of the isolates collected from that province.

Finally, the yeast isolates are preserved by triplicate at −80°C. We also preserve every isolate's DNA in separate fridges and boxes in parallel.

Final remarks

Herein we have mentioned some case studies related to one yeast culture collection with a biotechnology approach. Sustainability in most living microbe collections is endangered due to the scarcity of funds granted to support microbial diversity studies, in contrast to those microbial collections with a medical interest (McCluskey, 2017).

Moreover, during the pandemic caused by SARS-CoV-2, the CLQCA was actively working on the development of nanoencapsulation methodologies that can fight the virus. These technologies were studied since 2013 to preserve yeast cultures at room temperature. It can now respond against the COVID-19, which is airborne. The project where CLQCA contributes its experience and knowledge is carried at the National Institutes of Health in Maryland, USA. We hope to bring new and fresh ideas to microbiologists and epidemiologists through our previous experience in preserving yeasts by nanoencapsulation.

This chapter summarises the most noticeable yeast species described in Ecuador by local and international researchers and their significance in creating a more sustainable future for humanity. The fundamental science studies involving the description and characterisation of yeast through the 'Omics' are appropriate ways of training future researchers in microbiology, bioengineering, nanotechnology, biotechnology, chemistry, ecology and all those scientists that can identify the greatness of the small.

The sustainability of microbial collections is critical to its survival. Many small collections are orphaned due to the lack of funds to continue operations. Not only effort, but also creativity and innovation are needed to support microbial collections. We mentioned some examples of innovation, entrepreneurship and services, but it is also crucial to find the administrative and political support from institutions that rule the management of living organisms' repositories. Sustainability is also linked to fair trade with companies that may be interested in intellectual property transfer and negotiation. Finally, it is of great importance to create the awareness of the role of culture collections in society by effective communication of achievements and impacts. One excellent example is the response against COVID-19 that has been driven by significant culture collections around the world that provided all the information about the coronavirus family, so the scientists understood in a relatively early phase of the pandemic what the origin of the SARS-CoV-2 virus was.

References

Aouizerat, T., Gutman, I., Paz, Y., Maeir, A.M., Gadot, Y., Gelman, D., Szitenberg, A., Drori, E., Pinkus, A., Schoemann, M., Kaplan, R., Ben-Gedalya, T., Coppenhagen-Glazer, S., Reich, E., Saragovi, A., Lipschits, O., Klutstein, M., Hazan, R., 2019. Isolation and characterization of live yeast cells from ancient vessels as a tool in BioArchaeology. mBio 10 (2). https://doi.org/10.1128/mBio.00388-19 e00388-19,/mbio/10/2/mBio.00388-19.atom.

Baiano, A., 2020. Craft beer: an overview. Compr. Rev. Food Sci. Food Saf. https://doi.org/10.1111/1541-4337.12693.

Brown, J.H., 2014. Why are there so many species in the tropics? J. Biogeogr. 41 (1), 8–22. https://doi.org/10.1111/jbi.12228.

Brüssow, H., 2020. Bioarchaeology: a profitable dialogue between microbiology and archaeology. Microb. Biotechnol. 13 (2), 406–409. https://doi.org/10.1111/1751-7915.13527.

Burt, B.A., 2006. The use of sorbitol- and xylitol-sweetened chewing gum in caries control. J. Am. Dent. Assoc. 137 (2), 190–196. https://doi.org/10.14219/jada.archive.2006.0144.

Campaña-Pérez, J.F., Portero Barahona, P., Martín-Ramos, P., Carvajal Barriga, E.J., 2019. Ecuadorian yeast species as microbial particles for Cr(VI) biosorption. Environ. Sci. Pollut. Control Ser. 26 (27), 28162–28172. https://doi.org/10.1007/s11356019-06035-8.

Chang, C.-F., Lin, Y.-C., Chen, S.-F., Carvaja Barriga, E.J., Barahona, P.P., James, S.A., Bond, C.J., Roberts, I.N., Lee, C.-F., 2012. Candida theae sp. Nov., a new anamorphic beverage-associated member of the Lodderomyces clade. Int. J. Food Microbiol. 153 (1–2), 10–14. https://doi.org/10.1016/j.ijfoodmicro.2011.09.012.

CLQCA, (n.d.). Retrieved January 30, 2021, from http://www.cnib.com.ec/index.php/es/clqca.

Czerucka, D., Piche, T., Rampal, P., 2007. Review article: yeast as probiotics – *Saccharomyces boulardii*. Aliment. Pharmacol. Ther. 26 (6), 767–778. https://doi.org/10.1111/j.1365-2036.2007.03442.x.

Daniel, H.-M., Prasad, G.S., 2010. The role of culture collections as an interface between providers and users: the example of yeasts. Res. Microbiol. 161 (6), 488–496. https://doi.org/10.1016/j.resmic.2010.01.011.

Dasgupta, D., Junghare, V., Nautiyal, A.K., Jana, A., Hazra, S., Ghosh, D., 2019. Xylitol production from lignocellulosic pentosans: a rational strain engineering approach toward a multiproduct biorefinery. J. Agric. Food Chem. 67 (4), 1173–1186. https://doi.org/10.1021/acs.jafc.8b05509.

Demera, L.L., Barahona, P.P., Barriga, E.J.C., 2019. Production, extraction and characterization of lipases from the antarctic yeast *Guehomyces pullulans*. Am. J. Biochem. Biotechnol. 15 (2), 75–82. https://doi.org/10.3844/ajbbsp.2019.75.82.

Dhal, B., Thatoi, H.N., Das, N.N., Pandey, B.D., 2013. Chemical and microbial remediation of hexavalent chromium from contaminated soil and mining/metallurgical solid waste: a review. J. Hazard Mater. 250 (251), 272–291. https://doi.org/10.1016/j.jhazmat.2013.01.048.

Dispatch, C. M./F. T. C, (n.d.). Jackie O's, OU Create a Beer Rich with History. The Columbus Dispatch. Retrieved January 28, 2021, from. https://www.dispatch.com/business/20191112/jackie-orsquos-ou-create-beer-rich-withhistory.

Dupérré, N., Tapia, E., 2020. Megadiverse ecuador: a review of Mysmenopsis (Araneae, Mysmenidae) of ecuador, with the description of twenty-one new kleptoparasitic spider species. Zootaxa 4761 (1), 1–81. https://doi.org/10.11646/zootaxa.4761.1.1.

Eleutherio, E., Brasil, A. de A., França, M.B., de Almeida, D.S.G., Rona, G.B., Magalhães, R.S.S., 2018. Oxidative stress and aging: learning from yeast lessons. Fungal Biol. 122 (6), 514–525. https://doi.org/10.1016/j.funbio.2017.12.003.

El-Kirat-Chatel, S., Beaussart, A., Vincent, S.P., Flos, M.A., Hols, P., Lipke, P.N., Dufrêne, Y.F., 2015. Forces in yeast flocculation. Nanoscale 7 (5), 1760–1767. https://doi.org/10.1039/c4nr06315e.

Elzinga, K., Tremblay, C., Tremblay, V., 2015. Craft beer in the United States: history, numbers, and geography. J. Wine Econ. 10, 1–33. https://doi.org/10.1017/jwe.2015.22.

Encinar, J.M., González, J.F., Sánchez, N., Nogales-Delgado, S., 2019. Sunflower oil transesterification with methanol using immobilized lipase enzymes. Bioproc. Biosyst. Eng. 42 (1), 157–166. https://doi.org/10.1007/s00449-018-2023-z.

Farooqi, Z.H., Akram, M.W., Begum, R., Wu, W., Irfan, A., 2021. Inorganic nanoparticles for reduction of hexavalent chromium: physicochemical aspects. J. Hazard Mater. 402, 123535. https://doi.org/10.1016/j.jhazmat.2020.123535.

Freitas, L.F.D., Barriga, E.J.C., Barahona, P.P., Lachance, M.-A., Rosa, C.A., 2013. *Kodamaea transpacifica* f.a., sp. nov., a yeast species isolated from ephemeral flowers and insects in the Galápagos Islands and Malaysia: further evidence for ancient human transpacific contacts. Int. J. Syst. Evol. Microbiol. 63 (Pt_11), 4324–4329. https://doi.org/10.1099/ijs.0.052282-0.

Geng, W., Jiang, N., Qing, G.-Y., Liu, X., Wang, L., Busscher, H.J., Tian, G., Sun, T., Wang, L.-Y., Montelongo, Y., Janiak, C., Zhang, G., Yang, X.-Y., Su, B.-L., 2019. Click reaction for reversible encapsulation of single yeast cells. ACS Nano 13 (12), 14459–14467. https://doi.org/10.1021/acsnano.9b08108.

Gibney, P.A., Chen, A., Schieler, A., Chen, J.C., Xu, Y., Hendrickson, D.G., McIsaac, R.S., Rabinowitz, J.D., Botstein, D., 2020. A tps1Δ persister-like state in *Saccharomyces cerevisiae* is regulated by MKT1. PLoS One 15 (5), e0233779. https://doi.org/10.1371/journal.pone.0233779.

Guamán-Burneo, M.C., Dussán, K.J., Cadete, R.M., Cheab, M.A.M., Portero, P., CarvajalBarriga, E.J., da Silva, S.S., Rosa, C.A., 2015. Xylitol production by yeasts isolated from rotting wood in the Galápagos Islands, Ecuador, and description of *Cyberlindnera galapagoensis* f.a., sp. Nov. Antonie Van Leeuwenhoek 108 (4), 919–931. https://doi.org/10.1007/s10482-015-0546-8.

Holt, S., Miks, M.H., de Carvalho, B.T., Foulquié-Moreno, M.R., Thevelein, J.M., 2019. The molecular biology of fruity and floral aromas in beer and other alcoholic beverages. FEMS Microbiol. Rev. 43 (3), 193–222. https://doi.org/10.1093/femsre/fuy041.

Izawa, S., Sato, M., Yokoigawa, K., Inoue, Y., 2004. Intracellular glycerol influences resistance to freeze stress in *Saccharomyces cerevisiae*: analysis of a quadruple mutant in glycerol dehydrogenase genes and glycerol-enriched cells. Appl. Microbiol. Biotechnol. 66 (1), 108–114. https://doi.org/10.1007/s00253-004-1624-4.

Jacques, N., Casaregola, S., 2008. Safety assessment of dairy microorganisms: the hemiascomycetous yeasts. Int. J. Food Microbiol. 126 (3), 321–326. https://doi.org/10.1016/j.ijfoodmicro.2007.08.020.

James, S.A., Barriga, E.J.C., Barahona, P.P., Harrington, T.C., Lee, C.-F., Bond, C.J., Roberts, I.N., 2014. *Wickerhamomyces arborarius* f.a., sp. Nov., an ascomycetous yeast species found in arboreal habitats on three different continents. Int. J. Syst. Evol. Microbiol. 64 (Pt_3), 1057–1061. https://doi.org/10.1099/ijs.0.059162-0.

James, S.A., Cadet, G.M., Carvajal Barriga, E.J., Barahona, P.P., Cross, K., Bond, C.J., Roberts, I.N., 2011. *Saturnispora quitensis* sp. Nov., a yeast species isolated from the Maquipucuna cloud forest reserve in Ecuador. Int. J. Syst. Evol. Microbiol. 61 (12), 3072–3076. https://doi.org/10.1099/ijs.0.030759-0.

James, S.A., Carvajal Barriga, E.J., Barahona, P.P., Cross, K., Bond, C.J., Roberts, I.N., 2013. *Candida ecuadorensis* sp. Nov., an ascomycetous yeast species found in two separate regions of Ecuador. Int. J. Syst. Evol. Microbiol. 63 (Pt_1), 393–397. https://doi.org/10.1099/ijs.0.047555-0.

James, S.A., Carvajal Barriga, E.J., Bond, C.J., Cross, K., NÃÃ±ez, N.C., Portero, P.B., Roberts, I.N., 2009. *Candida carvajalis* sp. Nov., an ascomycetous yeast species from the Ecuadorian Amazon jungle. FEMS Yeast Res. 9 (5), 784–788. https://doi.org/10.1111/j.1567-1364.2009.00518.x.

James, S.A., Carvajal Barriga, E.J., Portero Barahona, P., Nueno-Palop, C., Cross, K., Bond, C.J., Roberts, I.N., 2015. *Kazachstania yasuniensis* sp. Nov., an ascomycetous yeast species found in mainland Ecuador and on the Galápagos. Int. J. Syst. Evol. Microbiol. 65 (Pt_4), 1304–1309. https://doi.org/10.1099/ijs.0.000102.

Kolařík, M., Spakowicz, D.J., Gazis, R., Shaw, J., Kubátová, A., Nováková, A., Chudíčková, M., Forcina, G.C., Kang, K.W., Kelnarová, I., Skaltsas, D., Portero, C.E., Strobel, S.A., Narváez-Trujillo, A., 2017. Biatriospora (Ascomycota: pleosporales) is an ecologically diverse genus including facultative marine fungi and endophytes with biotechnological potential. Plant Syst. Evol. 303 (1), 35–50. https://doi.org/10.1007/s00606-016-1350-2.

Lacerda, M.P., Oh, E.J., Eckert, C., 2020. The model system *Saccharomyces cerevisiae* versus emerging non-model yeasts for the production of biofuels. Life 10 (11). https://doi.org/10.3390/life10110299.

Lachance, M.-A., Rosa, C.A., Carvajal, E.J., Freitas, L.F.D., Bowles, J.M., 2012. *Saccharomycopsis fodiens* sp. Nov., a rare predacious yeast from three distant localities. Int. J. Syst. Evol. Microbiol. 62 (Pt_11), 2793–2798. https://doi.org/10.1099/ijs.0.043109-0.

Li, J., Zhuang, N., Xie, J., Zhu, Y., Lai, H., Qin, W., Javed, M.S., Xie, W., Mai, W., 2019. Carboxymethyl cellulose binder greatly stabilizes porous hollow carbon submicrospheres in capacitive K-ion storage. ACS Appl. Mater. Interf. 11 (17), 15581–15590. https://doi.org/10.1021/acsami.9b02060.

Li, W., Wang, H., Xu, X.G., Yu, Y., 2020. Simultaneous nanoscale imaging of chemical and architectural heterogeneity on yeast cell wall particles. Langmuir 36 (22), 6169–6177. https://doi.org/10.1021/acs.langmuir.0c00627.

Maire, T., Allertz, T., Betjes, M.A., Youk, H., 2020. Dormancy-to-death transition in yeast spores occurs due to gradual loss of gene-expressing ability. Mol. Syst. Biol. 16 (11). https://doi.org/10.15252/msb.20199245.

Marín, F., Navarrete, H., Narvaez-Trujillo, A., 2018. Total petroleum hydrocarbon degradation by endophytic fungi from the Ecuadorian Amazon. Adv. Microbiol. 08 (12), 1029–1053. https://doi.org/10.4236/aim.2018.812070.

McCluskey, K., 2017. A review of living collections with special emphasis on sustainability and its impact on research across multiple disciplines. Biopreserv. Biobanking 15 (1), 20–30. https://doi.org/10.1089/bio.2016.0066.

Ojima, Y., Kosako, S., Kihara, M., Miyoshi, N., Igarashi, K., Azuma, M., 2019. Recovering metals from aqueous solutions by biosorption onto phosphorylated dry baker's yeast. Sci. Rep. 9 (1), 225. https://doi.org/10.1038/s41598-018-36306-2.

Orlean, P., 2012. Architecture and biosynthesis of the *Saccharomyces cerevisiae* cell wall. Genetics 192 (3), 775–818. https://doi.org/10.1534/genetics.112.144485.

Pellerin, C., Booker, S.M., 2000. Reflections on hexavalent chromium: health hazards of an industrial heavyweight. Environ. Health Perspect. 108 (9), A402–A407.

Petrovič, U., 2015. Next-generation biofuels: a new challenge for yeast: next-generation biofuels: a new challenge for yeast. Yeast 32 (9), 583–593. https://doi.org/10.1002/yea.3082.

Piló, F.B., Carvajal-Barriga, E.J., Guamán-Burneo, M.C., Portero-Barahona, P., Dias, A.M.M., Freitas, L. F. D. de, Gomes, F. de C.O., Rosa, C.A., 2018. *Saccharomyces cerevisiae* populations and other yeasts associated with indigenous beers (chicha) of Ecuador. Braz. J. Microbiol. 49 (4), 808–815. https://doi.org/10.1016/j.bjm.2018.01.002.

Pokrivcak, J., Chovanova Supekova, S., Lančarič, D., Savov, R., Tóth, M., Vašina, R., 2019. Development of beer industry and craft beer expansion. J. Food Nutr. Res. 58, 63–74.

Portero Barahona, P., Bastidas Mayorga, B., Martín-Gil, J., Martín-Ramos, P., Carvajal Barriga, E.J., 2020. Cellulosic ethanol: improving cost efficiency by coupling semi-continuous fermentation and simultaneous saccharification strategies. Processes 8 (11), 1459. https://doi.org/10.3390/pr8111459.

Portero-Barahona, P., Carvajal-Barriga, E.J., Martín-Gil, J., Martín-Ramos, P., 2019. Sugarcane bagasse hydrolysis enhancement by microwave-assisted sulfolane pretreatment. Energies 12 (9), 1703. https://doi.org/10.3390/en12091703.

Ram, B.K., Han, Y., Yang, G., Ling, Q., Dong, F., 2019. Effect of hexavalent chromium [Cr(VI)] on phytoremediation potential and biochemical response of hybrid napier grass with and without EDTA application. Plants 8 (11). https://doi.org/10.3390/plants8110515.

Riley, P., Moore, D., Ahmed, F., Sharif, M.O., Worthington, H.V., 2015. Xylitolcontaining products for preventing dental caries in children and adults. Cochrane Database Syst. Rev. https://doi.org/10.1002/14651858.CD010743.pub2.

Rodríguez, M.E., Infante, J.J., Molina, M., Rebordinos, L., Cantoral, J.M., 2011. Using RFLP-mtDNA for the rapid monitoring of the dominant inoculated yeast strain in industrial wine fermentations. Int. J. Food Microbiol. 145 (1), 331–335. https://doi.org/10.1016/j.ijfoodmicro.2010.11.035.

San Sebastián, M., Karin Hurtig, A., 2004. Oil exploitation in the Amazon basin of Ecuador: a public health emergency. Rev. Panam. Salud Públic 15 (3). https://doi.org/10.1590/S1020-49892004000300014.

Singh, P., Itankar, N., Patil, Y., 2021. Biomanagement of hexavalent chromium: current trends and promising perspectives. J. Environ. Manag. 279, 111547. https://doi.org/10.1016/j.jenvman.2020.111547.

Smith, D., 2003. Culture collections over the world. Int. Microbiol. 6 (2), 95–100. https://doi.org/10.1007/s10123-003-0114-3.

Smith, D., McCluskey, K., Stackebrandt, E., 2014. Investment into the future of microbial resources: culture collection funding models and BRC business plans for biological resource centres. SpringerPlus 3. https://doi.org/10.1186/2193-1801-3-81.

Spagnuolo, M., Yaguchi, A., Blenner, M., 2019. Oleaginous yeast for biofuel and oleochemical production. Curr. Opin. Biotechnol. 57, 73–81. https://doi.org/10.1016/j.copbio.2019.02.011.

Supraja, K.V., Behera, B., Paramasivan, B., 2020. Optimization of process variables on two-step microwave-assisted transesterification of waste cooking oil. Environ. Sci. Pollut. Contr. Ser. 27 (22), 27244–27255. https://doi.org/10.1007/s11356019-05384-8.

Teng, C.-H., Cai, M., Shin, S., Xie, Y., Kim, K.-J., Khan, N.A., Di Cello, F., Kim, K.S., 2005. *Escherichia coli* K1 RS218 interacts with human brain microvascular endothelial cells via type 1 fimbria bacteria in the fimbriated state. Infect. Immun. 73 (5), 2923–2931. https://doi.org/10.1128/IAI.73.5.2923-2931.2005.

The Yeasts, (n.d.). Retrieved January 28, 2021, from https://theyeasts.org/.

Torres, L.E.G., Koch, A., Portero, P., Carvajal, E.J., 2018. Dos nuevas especies de levaduras pertenecientes a la biodiversidad Ecuatoriana con potential actividad probiótica. Revista del Instituto de Salud Pública de Chile 2 (2). https://doi.org/10.34052/rispch.v2i2.37. Article 2.

Troilo, A., De Francesco, G., Marconi, O., Sileoni, V., Turchetti, B., Perretti, G., 2020. Low carbohydrate beers produced by a selected yeast strain from an alternative source. J. Am. Soc. Brew. Chem. 78 (1), 80–88. https://doi.org/10.1080/03610470.2019.1682887.

Türker, M., 2014. Yeast Biotechnology: Diversity and Applications.

Tyakht, A., Kopeliovich, A., Klimenko, N., Efimova, D., Dovidchenko, N., Odintsova, V., Kleimenov, M., Toshchakov, S., Popova, A., Khomyakova, M., Merkel, A., 2021. Characteristics of bacterial and yeast microbiomes in spontaneous and mixed fermentation beer and cider. Food Microbiol. 94, 103658. https://doi.org/10.1016/j.fm.2020.103658.

Ur-Rehman, S., Mushtaq, Z., Zahoor, T., Jamil, A., Murtaza, M.A., 2015. Xylitol: a review on bioproduction, application, health benefits, and related safety issues. Crit. Rev. Food Sci. Nutr. 55 (11), 1514–1528. https://doi.org/10.1080/10408398.2012.702288.

Wölnerhanssen, B.K., Meyer-Gerspach, A.C., Beglinger, C., Islam, M.S., 2020. Metabolic effects of the natural sweeteners xylitol and erythritol: a comprehensive review. Crit. Rev. Food Sci. Nutr. 60 (12), 1986–1998. https://doi.org/10.1080/10408398.2019.1623757.

World Federation for Culture Collections, (n.d.). Retrieved January 27, 2021, from http://www.wfcc.info/guidelines/.

Xu, Y., Chi, P., Bilal, M., Cheng, H., 2019. Biosynthetic strategies to produce xylitol: an economical venture. Appl. Microbiol. Biotechnol. 103 (13), 5143–5160. https://doi.org/10.1007/s00253-019-09881-1.

Further reading

Carvajal Barriga, E.J., Barahona, P.P., Tufiño, C., Bastidas, B., Guamán-Burneo, C., Freitas, L., Rosa, C., 2014. An overview of the yeast biodiversity in the Galápagos islands and other Ecuadorian regions. In: Grillo, O. (Ed.), Biodiversity—The Dynamic Balance of the Planet. InTech. https://doi.org/10.5772/58303.

Delgado-Ospina, J., Triboletti, S., Alessandria, V., Serio, A., Sergi, M., Paparella, A., Rantsiou, K., Chaves-López, C., 2020. Functional biodiversity of yeasts isolated from Colombian fermented and dry cocoa beans. Microorganisms 8 (7), 1086. https://doi.org/10.3390/microorganisms8071086.

Papalexandratou, Z., Falony, G., Romanens, E., Jimenez, J.C., Amores, F., Daniel, H.-M., De Vuyst, L., 2011. Species diversity, community dynamics, and metabolite kinetics of the microbiota associated with traditional Ecuadorian spontaneous cocoa bean fermentations. Appl. Environ. Microbiol. 77 (21), 7698–7714. https://doi.org/10.1128/AEM.05523-11.

Chapter 12

Yeast as a model organism for teaching biotechnology and human cell biology leading to sustainable futures

Ian G. Macreadie
Honorary Professor of Biotechnology, School of Science, RMIT University, Melbourne, VIC, Australia

Chapter outline

Yeast, as an inexpensive, convenient and excellent teaching resource

Teaching resources are increasingly being impacted by budget reductions and increasing costs associated with safety compliance. In addition, in 2020, COVID-19 added additional complexities and even further budget cuts with reduced enrolments that will cause impacts and uncertainties into the coming years. Whilst many course resources have been put totally on-line, there is nothing as good as real hands-on learning. Hands-on learning offers an

Importance of Microbiology Teaching and Microbial Resource Management for Sustainable Futures
https://doi.org/10.1016/B978-0-12-818272-7.00002-X

authentic experience, along with opportunities for real mistakes that can also be incorporated into teaching and learning. I emphasise here that it is important to perform appropriate risk assessments so that mistakes that would cause any injury are to be avoided. The mistakes that I refer to relate to experimental procedures that may lead to unwanted outcomes such as contamination of a culture.

The use of yeast as a model in teaching offers a very cost-effective means of providing a relevant laboratory training experience that also achieves some important learning objectives. For a considerable time, yeast has been employed as powerful research tools for research and biotechnology applications. The yeast that has pioneered the way is *Saccharomyces cerevisiae*, commonly referred to as brewers' or bakers' yeast. Not only has yeast proven itself as being safe, but yeast has also occupied niches in biotechnology for millennia. These niches include wine and beer production as well as bread production. However, in more recent times *Saccharomyces cerevisiae* has been used for the production of life-saving drugs such as human insulin, and vaccines for hepatitis B and human papilloma virus.

Low costs and the ease of maintaining and growing yeast make it highly attractive. Yeast cultures can be stored in frozen glycerol stocks at −80°C for a decade or more, whilst lyophilisation and storage on slants is reliable for storage at 4°C for 6 months or more. On solidified media in Petri dish plates yeast can also survive for weeks. A common practice amongst yeast researchers in distributing strains used to involve spreading the yeast on a sterile square cm of filter paper that was moistened with evaporated milk. This was then wrapped in sterile aluminium foil and placed into a regular envelope for mailing with a letter.

In comparison to many other common microorganisms used in research (e.g., *Escherichia coli*), yeast is safer, more robust and offers more advantages, but with similar culture costs. Again, in comparison to insect and mammalian cells, yeast culture is much less expensive, far more robust and much easier.

Yeast diversity

There are over 2000 species of yeast (www.theyeasts.org), but only a few species have been employed in biotechnology. Such yeasts include the budding yeast *Saccharomyces cerevisiae*, *Kluyveromyces lactis* and a yeast of major biotechnology importance, the methylotrophic yeast *Komagataella pastoris* (more commonly known by its previous name *Pichia pastoris*).

A few yeast species are significant human pathogens and intensely studied. These yeasts include *Candida albicans*, *Cryptococcus* spp. and the opportunistic pathogen, *Candida glabrata*, an interesting yeast which also serves in the food industry (Tam et al., 2015). The variety of yeast species provides much useful information about basic biology, genetics, pathogenicity and environmental responses, including drug susceptibility.

Whilst most yeast divide by budding, some yeast, most notably *Schizosaccharomyces pombe*, divides by fission. *Schizosaccharomyces pombe* also has a role in traditional beverages but is probably more notable for its contributions to the study of cell division. Its contributions to studies on human cell division provided insights into cancer, a Nobel Prize–winning achievement.

Saccharomyces cerevisiae continues to feature as a prominent species for research and indeed many basic discoveries, including many Nobel Prize–winning discoveries that have utilised the basic science discoveries made in *Saccharomyces cerevisiae*. For example, since the early 2000s, there have been five Nobel Prizes related to yeast studies:

- 2001: Leland H. Hartwell, R. Timothy (Tim) Hunt and Paul M. Nurse won the Nobel Prize in Physiology or Medicine for their discoveries of key regulators of the cell cycle (discussed in Macreadie, 2017). The research for this Nobel Prize was performed in *Schizosaccharomyces pombe.*
- 2006: Roger D. Kornberg won the Nobel Prize in Chemistry for his studies of the molecular basis of eukaryotic transcription (discussed in Macreadie, 2007a).
- 2009: Elizabeth H. Blackburn, Carol W. Greider and Jack W. Szostak won the Nobel Prize in Physiology or Medicine for the discovery of how chromosomes are protected by telomeres and the enzyme telomerase.
- 2013: James E. Rothman, Randy W. Schekman and Thomas C. Südhof won the Nobel Prize in Physiology or Medicine for their discoveries of machinery regulating vesicle traffic, a major transport system in our cells.
- 2016: Yoshinori Ohsumi won the Nobel Prize in Physiology or Medicine for discovery and elucidation of mechanisms underlying autophagy.

The importance of yeast studies is exemplified by the on-going Cold Spring Harbor courses on yeast genetics. A roll of honour displays instructors and attendees every year from 1970 to 2019, with the 2020 meetings and courses being cancelled due to COVID-19 [https://meetings.cshl.edu/alumni.aspx?course=C-YEAS&year=18]. The scientific output of instructors, assistants and 'students' is highly significant in its quality and volume, attesting to the power of yeast genetics and genomes.

Some major advantages of *Saccharomyces cerevisiae* are safety, its ability to be genetically transformed through homologous recombination, single copy and multicopy plasmids and the widespread collegiate behaviour of yeast researchers over many decades, enabling the sharing of information and resources. The impact of the latter should not be underestimated. Yeast research has greatly benefitted from the free sharing of knowledge, yeast strains, plasmids and other reagents.

The use of *Saccharomyces cerevisiae* in research has seen it used to study human genes, organelles and biological processes including those implicated with disease. The *Saccharomyces cerevisiae* genome was the first eukaryotic genome to be sequenced, with the effort being systematically co-ordinated through 74 labs around the globe (reviewed in Macreadie 2017). This is an example of the collegiate nature referred to above. It revealed that the 12 Mb genome was compact and contained 6000 genes: there was a gene for every 2 kb of chromosome! Unlike human genes, introns in yeast are rare. Whilst the functions of many yeast genes were known from yeast genetics, many were unknown through genetic studies. In order to reveal their functions, reverse genetics took place, also in a systematic approach of targeted gene knockouts being made in separate European and USA efforts. In order to make the collection complete, knockouts were made in diploid strains as well as haploid strains. If an essential gene was disrupted in a heterozygous diploid, this would be revealed after sporulation of the diploid strain. The sporulation would result in their being two viable spores and two inviable spores.

Observations of the yeast (*Saccharomyces cerevisiae*), as a microbe involved in fermentation were made by Louis Pasteur in the 1850s (reviewed in Barnett, 2000): in reality yeast has been used for biotechnology purposes for millennia. However, Pasteur's contributions including sterilisation and the observation that life did not spontaneously emerge led to the building foundations of microbiology as a discipline.

In the subsequent century, there were contributions to yeast genetics, reviewed by Roman (1986) and Dixon and Pretorius (2020). The earliest observations, from Winge (1935), were made in the Carlsberg Labs in Copenhagen, Denmark. Meanwhile, in the United States of America, Carl and Gertrude Lindgren worked on meiosis and mating type in yeast, sporulating diploids and characterising the four spores by micro dissection.

Yeast genetics

Some of the earliest studies leading to cytoplasmic genetics in yeast studies focused on the observations of variations in yeast colony size. Discovered in 1948 by Boris Ephrussi in France, regular-sized colonies were called 'grande', whilst a low frequency of colonies were small and called 'petite' (reviewed by Bernardi, 1979). The frequency of the petite colonies could be increased by treatment with mutagenic agents such as acriflavine. Petites were later found to have deletions in their mitochondrial genome, and in some instances, a loss of the entire mitochondrial genome. The petite phenomenon caused respiratory deficiency and malfunctions in energy production that led to the small size. It is worth noting that single nuclear gene mutations could affect one mitochondrial function (PET mutants), whilst single mitochondrial gene mutations (mit-) also led to respiratory deficiency but large colony size.

In the 1970s, the Monash University laboratory where I made my start in yeast genetics (and later yeast molecular biology) focussed on mitochondrial genetics, discovering the structure and function of the mitochondrial genome. The laboratory was headed by Antony Linnane who was complemented by

Philip Nagley, Bruce Lukins and Sangot Marzuki, ensuring a high-powered team to lead the Australian involvement in the world-wide competition to uncover the biogenesis of mitochondria. By the 1970s, there was a rudimentary map of the five genetic markers on the circular mitochondrial genome. These markers were most notably drug resistance determinants and included resistance determinants for erythromycin, chloramphenicol, paromomycin, mikamycin and oligomycin. The first three were in mitochondrial rRNA genes, somewhat like in bacteria, giving rise to popular speculation that mitochondria were evolved from bacteria. They presumably decreased binding of the drugs to the ribosomes allowing respiratory growth in the presence of the drugs. The oligomycin resistance mapped to a gene encoding subunit 9 of the mitochondria ATP synthetase, whilst mikamycin resistance mapped to the cytochrome *b* gene (Fig. 12.1).

In subsequent years, efforts focussed on other point mutations in the mitochondrial genome that lead to loss of respiratory function, *mit*-mutations. Other genes included subunits six and eight of the mtATPase, three subunits of cytochrome oxidase, a ribosomal protein, plus all of the tRNA genes required for mitochondrial translation. In elucidating the potential of the mitochondrial genome, there was a wealth of training for students, involving all aspects of genetics and biochemistry.

In the 1980s molecular biology, cloning and gene sequencing were emerging and I was fortunate in being able to train with those technologies to make more sense of the genetic knowledge. This was long before there was

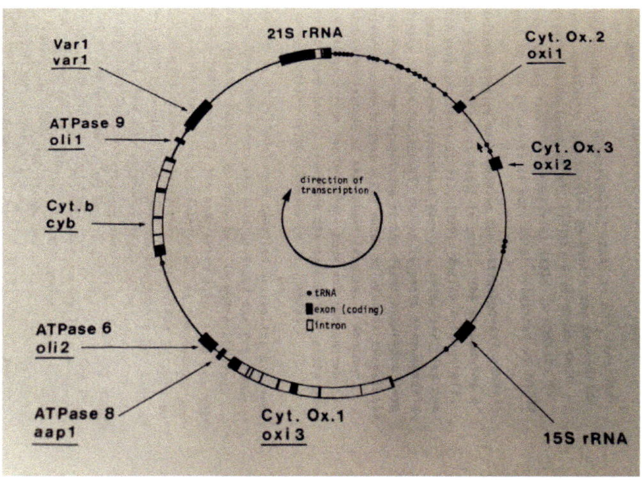

FIGURE 12.1 Map of the 75 kb yeast mitochondrial genome (1983). Note the direction of transcription, tRNA genes, rRNA genes and genes encoding the ribosomal protein Var1, cytochrome *b*, three subunits of cytochrome oxidase and three subunits of the ATP synthetase. Note also the presence of introns within many of genes, including the large rRNA gene. *Reproduced from Ian Macreadie (1983) PhD Thesis, Monash University, Australia.*

extensive information on gene sequences to truly complete gene structure and function elucidation. Nowadays, it is common to find a novel microorganism, determine its whole genome sequence and then to annotate the genome using bioinformatics tools. Such tools are based on similarities to known genes, and most often this information has come from model microbial organisms, including *Escherichia coli* and *Saccharomyces cerevisiae*. This is very noteworthy and stresses the importance of understanding the gene structure and function relationship unravelled in the model organisms.

Following the completion of my PhD studies, I enlisted in a post-doctoral position in Ron Butow's laboratory in Texas, where I continued mitochondrial studies. A major achievement in that lab was discovery of the function of a 'gene within a gene'. The 21S rRNA gene existed in two forms: forms with an intron (omega+) or forms without the intron (omega-), and within the intron, there was a gene of unknown function. When strains with the intron were crossed to strains without the intron almost all progeny received the intron. This was unlike other yeast mitochondrial genes where in a genetic cross of haploids of the opposite mating type, each haploid contributed genes equally and mitochondrial genes actively segregated, resulting in a 50:50 distribution of alleles. In an omega + to omega- cross, almost all of the resulting progenies were omega+. My work in Butow's laboratory operated on the premise that the gene within the intron encoded a factor for intron transmission, leading to its subsequent naming as the *fit1* gene. The idea was that there was a process of gene conversion, involving gap repair, whereby the intron DNA was actively being synthesised during a genetic cross, so that all progenies would receive it. I set out to specifically mutate the *FIT1* gene by adding the mitochondrial mutagen, manganese, to an active mating mixture of the two yeast types. It was hypothesised that the gene within the intron encoded an endonuclease that cut specifically at the site where the intron was to be placed, and the gap in the DNA would be repaired using the intron-containing mtDNA. The procedure turned out to be very successful and eight *fit1* mutants were isolated that were no longer active in high-frequency intron transmission (Macreadie et al., 1985).

This completed an era in mitochondrial gene discovery. Meanwhile, much was going on to elucidate the function of the yeast nuclear genes, and to some extent, this continues to the present day, with some gene functions still awaiting identification.

Another genomic element of *Saccharomyces cerevisiae* is the 2 μm plasmid, named after the plasmid's circumference. The plasmid is optional and is absent in some strains of *Saccharomyces cerevisiae*. Its purpose remains unknown and its advantage to yeast remains unclear. However, it has been a basis for obtaining yeast transformation with plasmids containing the origin of DNA replication from the 2 μm plasmid. This origin of replication allows for multiple copies of engineered plasmids (~50–200 copies per cell) to be replicated. This multi-copy expression can allow for high level expression of foreign genes.

Candida glabrata and medically important yeast

Whilst there is considerable research on other yeast, it is often to study infections caused by those yeast, or for their biotechnological application. In my own situation, I have also used *Candida glabrata*, which has previously been called *Torulopsis glabrata*. My first encounters with handling *Candida glabrata* were to study its mitochondrial *var1* gene in Ron Butow's lab (Ainley et al., 1985). This gene remains one of the most AT-rich known. In Dennis Winge's laboratory, we developed the first cloning system for *Candida glabrata* (Mehra et al., 1992); however, there have been improved and more versatile systems made in more recent times.

Candida glabrata is an obligate haploid that has uses in the food and beverage industry, but it is increasingly being found to be an opportunistic pathogen (Tam et al., 2015) and maybe this was a driver in re-naming it *Candida glabrata*. It has special new interest because of its intrinsically high resistance to drugs; it remains important to understand its mechanisms of drug resistance to enable improved and alternate drug treatment options (Almshawit and Macreadie, 2017a,b). Despite its new name, *Candida glabrata* bears closer similarity to *Saccharomyces cerevisiae* than any other *Candida* species (Tam et al., 2015).

For studies on the effects of drugs and other chemicals, *Candida glabrata* is very versatile for toxicity studies because of its extremely rapid growth: it can form colonies after overnight growth. This means that one can observe cell killing by the reduction in colony forming units (cfu) one day after treatments. Some examples of drugs and biogenic amines that have been studied by myself alone and with students are listed below.

There is considerable research on other yeast to study diseases caused by those yeast. Much of it focussed on *Candida* and there are on-going efforts to determine the role of all genes. A Candida Genome Database (www.candidagenome.org) has been established to catalog the functions of the *Candida* genes. At the moment, this database houses genomic information for *Candida glabrata*, *Candida albicans*, *Candida parapsilosis*, *Candida dubliniensis* and *Candida auris*. Information on the functions encoded by genes, phenotype of deletion constructs and links to scientific literature are reported.

The other yeast species of major medical significance are *Cryptococcus neoformans* and *Cryptococcus gattii*. These yeasts can cause meningitis, lung infection and disease of the central nervous system. They stand apart due to their severe disease, high mutation rates and being encapsulated.

Other yeast – important yeast for biotechnology

The most utilised yeast in biotechnology is *Pichia pastoris* (now known as *Komagataella pastoris*). Its use is popular because high levels of foreign proteins can be achieved when the open reading frame encoding the foreign protein replaces the ORF encoding alcohol oxidase which can comprise approximately half of the total protein in *Pichia pastoris* when growth is on methanol. The gene replacement procedure involves homologous recombination with a linear DNA molecule that can be cut from a plasmid or generated by PCR. The procedure was well established in *Saccharomyces cerevisiae* and many students and researchers have embarked on expression in *Pichia pastoris* due to the commercial development of kits enabling easy engineering of the yeast. Other methylotrophic yeasts including *Hansenula polymorpha* have also been developed for the biotechnology industry. The usual aims are high level production of valuable proteins for human use.

Kluyveromyces lactis is another yeast of special interest and worthy of mention. It has been used for some highly specific tasks such as production of human serum albumin. It benefits from abilities to secrete well and although it normally has no endogenous plasmid, it can maintain a multi-copy plasmid from *Kluyveromyces drosophilarum*, enabling foreign genes to be held at high copy with concomitant higher expression levels.

In next part of this chapter, I will outline yeast some simple uses of *Saccharomyces cerevisiae* for undergraduate teaching.

Basic microbiology training: testing of sterile techniques

The handling of yeast tests the skills of students in microbiology competence and sterile technique. For example, the formation of yeast colonies on solidified rich media takes up to 3 days allowing plenty of time for contaminants to grow and to be observed. Yeast media are particularly favourable for the germination of fungal spores that are frequently airborne and the most common contaminant. If students do not maintain sterile practices, this will be readily noticeable, usually with the growth of a spreading mould.

With to regard to genetically manipulating yeast, the yeast culture medium that is employed is also usually prepared without antibiotics, so sterile technique is again tested, perhaps even more, since plates may be incubated for 7–10 days.

Food and beverage production

Yeast plays a role in the production of a variety of foods and beverages (e.g., wine, beer, cheese, bread), and many students want to gain knowledge of the principles and practices used in the food industry. The production of beer using a 'home brew' kit is an excellent teaching example for students of food science and industrial microbiology. They can practice food hygiene, gowning up according to food safety practices and preparing reagents in a 'clean' environment.

Once set up, students then can monitor the beer fermentation, observing and recording the production of carbon dioxide, the drop in pH and the drop in specific gravity. These developments are all related and are explained by the chemistry equation demonstrating the reaction converting glucose to ethanol and carbon dioxide. Most of the carbon dioxide bubbles from the fermenter are lost, leading to a mass and density loss in the fermenter. Some carbon dioxide dissolves forming carbonic acid, which is observed as a drop in pH. This can also be related to atmospheric CO_2 increases which are currently leading to ocean acidification.

The drop in density offers an opportunity for student to revise some chemistry stoichiometry calculations.

$$\underset{\text{Glucose}}{C_6H_{12}O_6} \rightarrow \underset{\text{Ethanol}}{2(CH_3CH_2OH)} + \underset{\text{Carbon Dioxide}}{2(CO_2)} + \text{Energy(ATP)}$$

It is worthwhile calculating the mass lost to CO_2 bubbling from the vessel and using the equation to determine the mass of ethanol produced. Since ethanol levels are usually given as v/v%, it is also worth converting the w/w% to v/v% which is done by considering that water is 1 g/mL and ethanol has a density of 0.79 g/mL. Bottling the beer and adding a carbonation drop (glucose/sucrose mix) leads to further fermentation in the bottle, with bubbles being preserved.

This practical exercise has been employed in teaching at RMIT University for over a decade. Alternate processes that might be more inclusive of students who object to alcohol production have been considered. For example, the production of kombucha was considered; however, its production is much slower, making it difficult to fit into a semester. The beer production remains the easiest task to achieve and remains the popular exercise. It is complemented by the real-life fact that this is the world's largest industrial microbiology process. This does not mean this is where the jobs are: far from it. Big breweries employ very few microbiologists; however, the increasing numbers of microbreweries producing craft beers provide many new employment opportunities as highlighted by Rogers (2017). Some students enrolled in biotechnology courses have pursued careers in breweries, including managing their own breweries. For students enrolled in a food science program, this exercise is highly appropriate.

Transforming yeast and producing a protein that can be readily observed

Recombinant organisms, including yeast, are now a common part of our lives and students need to be fully aware of what goes on in the use of genetically engineered organisms, from the construction of the recombinant organism to the processing of the valuable end products.

Products such as human insulin, hepatitis B vaccine and human papillomavirus vaccine protect people from type I diabetes, and viral infections that can lead to liver and cervical cancers (Macreadie, 2007b; 2017). Demystifying genetic engineering (of yeast) can be performed readily in teaching laboratories and can teach the principles of transformation, selection and analysis of recombinant organisms. Using a freshly grown yeast colony from a plate, a yeast transformation kit and plates containing media that can be used to select for transformants, students can perform a yeast transformation within an hour. A good strain for yeast transformation can be W303-1a (*MATα leu2-3112 trp1-1 ura3-1 ade2-1 his3-11,15*), an auxotrophic strain that has requirements for leucine, uracil, tryptophan, adenine and histidine. Note that the red (cf. white) version of W303-1a offers the best teaching for this exercise!

Good plasmids for teaching include pYEX-S1 (Fig. 12.2) and pYEX-S1-βla. The reasons for this will be made clear from the following discussion. After transformation, the plasmid vectors pYEX-S1 and pYEX-S1-βla (which carry the *LEU2* and *URA3* genes) enable growth in the absence of leucine and uracil. The starting strain is grown in rich media prior to transformation.

A week later, the outcomes can be observed. Students should observe that the occurrence of colonies is dependent on a 'transforming plasmid': a 'no added plasmid' transformation serves to illustrate this and also serves to check on sterile technique. The pYEX plasmids are particularly useful for teaching since they have two selectable markers. pYEX-S1 was originally called pYEULS1 (Castelli et al., 1994). It was renamed by AMRAD for commercial use.

Students typically find 10–100 transformants when selecting in the absence of uracil, but in the absence of leucine they find none! This is due to phenotypic lag. The *LEU2* gene on the plasmid is defective and contains only 30 nucleotides upstream of its native start codon. To get enough of the *LEU2* gene product to enable growth in the absence of leucine, the plasmid needs to be amplified and this can eventually happen. Students should do a full phenotyping of transformants to see that they have produced what they expected, and to also experience genetic rearrangements. The phenotyping can be done by checking growth on a range of omission media. This should include complete minimal medium, as well as complete lacking uracil, complete lacking leucine, complete lacking tryptophan, complete lacking histidine and complete lacking adenine. Transformants will typically be LEU^+ URA^+ ADE^- HIS^- TRP^-, growing on all media except for media lacking tryptophan, histidine and adenine. This indicates that they have taken up and expressed *URA3* and *LEU2* genes from the pYEX plasmids. In some cases, there may be no growth on complete lacking leucine. This can be indicative of a plasmid rearrangement. It can be worth considering various options of how rearrangements could occur. Examples include plasmid rearrangement or possibly integration of the plasmid into the chromosome. This would mean just one copy of the defective *LEU2* gene and this probably would not allow growth in the absence of leucine.

With the phenotyping, it is also worth noting colony size and including a YEPE plate which is a rich medium but has ethanol as the carbon source. Many events, including transformation, can generate damage to the

pYEX™-S1 Yeast secretion expression vector

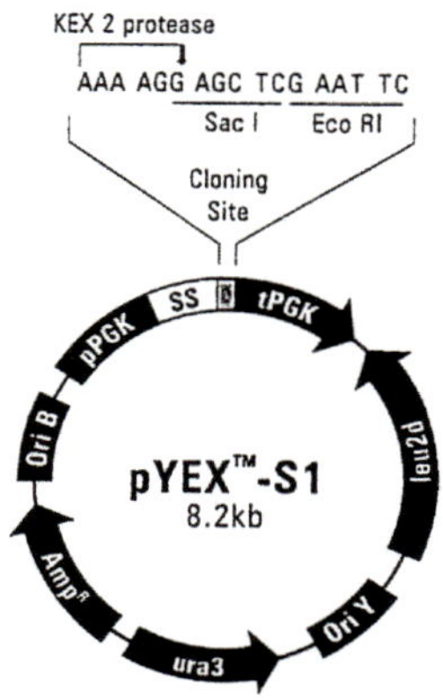

pPGK Phosphoglycerate kinase promoter
SS Secretion signal
tPGK Phosphoglycerate kinase terminator
Amp^R ß-lactamase gene for selection in E coli
leu2d Leucine selectable marker in yeast
ura3 Uracil selectable marker in yeast
Ori B E coli origin of replication
Ori Y Yeast origin of replication

Secretion Signal Sequence

pYEX™-S1 GAA TTC (Eco RI) TAT AAA ATG (met) AAT ATA TTT TAC ATA TTT TTG TTT TTG CTG TCA TTC GTT CAA GGT TTG GAG CAT ACT CAT CGA AGA GGC TCC TTA GTC AAA AGG (KEX 2 protease) AGC TCG (Sac I) AAT TC (Eco RI)

pYEX™-S1 Vector Control Regions and Restriction Enzyme Site Details:

PGK promotor: Start codon (ATG): 5247; Stop codon (TAA): 6847.

CS: *Sac*I (6433).

ß-lactamase gene region: Promoter: -10 2654-2659; -35 2631-2636; Start codon (ATG): 2701; Stop codon (TAA): 3559.

Plasmid *E. coli* replication region: Site of replication initiation: 4319; Region necessary for replication: 3626-4322.

Sequencing primers: pYEX™-S1 Forward sequencing primer binds nucleotides 6234-6254; pYEX™-S1 Reverse sequencing primer binds nucleotides 6483-6502.

ura3: Start codon (ATG): 2264.

leu2d: Start codon (ATG): 7928; stop codon (TAA): 6799.

No sites: *Acc*III, *Aoc*I, *Asp*718, *Avr*II, *Bal*I, *Bbe*I, *Bcl*I, *Bgl*II, *BspM*II, *BssH*II, *Bsu*361, *Ccr*I, *Cvn*I, *Eag*I, *Eco*521, *Eco*811, *Esp*I, *Kpn*I, *Mlu*I, *Mro*I, *Mst*II, *Nae*I, *Nar*I, *Nhe*I, *Not*I, *Nru*I, *PaeR*71, *Pfim*I, *PflM*I, *Pst*I, *Rsr*II, *Sac*II, *Sal*I, *Sau*I, *Sfi*I, *Sma*I, *Spe*I, *Sph*I, *Spl*I, *Sst*II, *Tth*111I, *Xba*I, *Xho*I, *Xma*I, *Xma*III.

One site: *Aat*1 (1827), *Aat*2 (2569), *Afl*2 (7667), *Apa*I (1889), *Aqu*I (295), *AspA*I (7947), *Ava*I (295), *BamH*1 (6714), *Bgl*1 (3370), *BstE*2 (7947), *Cfr*101 (3403), *Hind*3 (4735), *Hpa*1 (3), *Nco*1 (2054), *Nde*I (2348), *Ppum*1 (2107), *Pvu*2 (4556), *Sac*I (6330), *Sst*I (6330), *Stu*1 (1827).

Two sites: *Acc*I (1431, 1936), *Acy*1 (2566, 2948), *Aha*2 (2566, 2948), *AlwN*1 (1585, 3967), *ApalL*1 (2816, 4062), *Ava*3 (293, 1440), *Ban*1 (3535, 4632), *EcoN*I (7026, 7716), *EcoR*5 (2075, 7336), *EcoT*22I (293, 1440), *Nsi*1 (293, 1440), *PpuM*I (2107, 5813), *Sca*I (1951, 3007), *SnaB*7 (645, 6541).

Note: Restriction enzyme analysis prepared using MacVector™ 3.5. Vectors have not been tested with all of these enzymes; tables should serve as a guide only.

The absence of a particular restriction enzyme in the tables indicates more than two recognition sites in the vector.

MacVector is a trademark of International Biotechnologies, Inc.

FIGURE 12.2 pYEX-S1 plasmid. Product information sheet provided by AMRAD who distributed the plasmid. The plasmid was also available from CLONTECH. *Reproduced from Ian Macreadie personal collection.*

mitochondrial genome. This is not lethal for yeast but does lead to loss of respiratory function, petite colony size and lack of growth on YEPE. Students will get to see this first-hand.

Another mutation event leads to large white colonies. The red colour in W303-1a is due to the *ade2* mutation, but this can be suppressed through chromosomal mutation, which results in no red pigment. This upstream mutation in the adenine biosynthetic pathway leads to slightly faster growth and white colonies. The red colour accumulates with time and consumption of adenine in the media.

Finally, the next phase of the practical work is to look for foreign gene expression. The secretion of β-lactamase is easy and convenient in a teaching lab. Transformants are cultured for ideally 1–3 days in rich and/or selective media. A sample of the supernatant is removed and then run on an SDS-PAGE gel along with size markers. Pre-stained markers are ideal since students can visualise the running of the gel. Gels are then stained with Coomassie brilliant blue to visualise all proteins in the gel. With pYEX, it is observed that yeast does not secrete proteins that are readily observed; however, a couple of protein bands are usually observed in the supernatants from pYEX-S1 transformants. They correspond to β-lactamase produced in these transformants.

Beta-lactamase activity in supernatants can readily be detected by nitrocefin assay: the activity results in a purple to yellow colour change. This assay is extremely sensitive and works even when levels are so low that they produce weakly stained bands on gels.

On occasion, there may be no β-lactamase activity. This can be due to a lack of the β-lactamase in the transformant, due to a genetic rearrangement, or it could be due to degradation of the β-lactamase in the culture medium, or to loss of the plasmid in rich media. It really does not matter why in this instance as it is more important for students to consider the reasons for possibly not seeing β-lactamase in some samples.

Some questions that can test student learning are as follows:

1. How were yeast transformants obtained?
2. Could the plasmids be used to transform a wild-type strain?
3. How did the number of transformants compare on plates lacking leucine versus plates lacking uracil? Why are the numbers of transformants different?
4. Was there a difference in phenotype between the URA+ and LEU + transformants?
5. Why were there were some small colonies that lacked growth on YEPE?
6. There are two genes encoding β-lactamase in pYEX-S1-βla. Was the other β-lactamase gene expressed in yeast? Explain why.
7. Estimate the size of any product(s) secreted by the yeast. How do you know if β-lactamase was secreted?

The pYEX plasmids used in the exercise are examples of yeast episomal plasmids. They have an advantage of being capable of existing at high copy in *S. cerevisiae*. This means the foreign gene cloned into this plasmid will be at high level and this can improve foreign gene expression.

It is also worth noting that there are many other types of plasmids that give other options. They are listed in Table 12.1, and they all have useful purposes.

TABLE 12.1 Various options to produce genetic engineered yeast starting with a plasmid.

Plasmid or other DNA	Elements for selection and maintenance in *E. coli*	Elements for maintenance in *S. cerevisiae*	Elements for selection in *S. cerevisiae*	Promoter for foreign gene expression
Yeast episomal plasmids	Amp^r oriB	2 μm ori	*LEU2, URA3, TRP1, CUP1*	*CUP1, PGK, GPD*
Yeast centromeric plasmid	Amp^r oriB	Centromere and chromosomal ARS (autosomal replicating sequence)	*LEU2, URA3, TRP1, CUP1*	*CUP1, PGK, GPD*
Yeast integrating plasmids	Amp^r oriB	None	*LEU2, URA3, TRP1, CUP1*	*CUP1, PGK, GPD*
Yeast artificial chromosome	Amp^r oriB (but they are eliminated before going into yeast)	Centromere and chromosomal ARS and telomeres	*URA3, TRP1.* *HIS3* is placed between telomeric sequences and is eliminated before going into yeast	

Amp^r encodes beta lactamase gene which degrades ampicillin. Its presence causes resistance to ampicillin in *E. coli*. oriB is an origin of DNA replication for *E. coli*. ARS is an autonomous replication sequence. 2 μm ori is the origin of DNA replication from the *Saccharomyces cerevisiae* endogenous plasmid known as the 2 μm circle. The products encoded by the yeast genes are listed here: *CUP1* encoding metallothionein which confers copper resistance; *GPD* NAD-dependent glycerol-3-phosphate dehydrogenase and whose gene has a strong constitutive promoter; *HIS3*, encoding imidazole glycerol-phosphate dehydratase involved in histidine biosynthesis; *LEU2*, encoding beta-isopropylmalate dehydrogenase involved in leucine biosynthesis; *PGK* encodes phosphoglycerate kinase and whose gene has a strong constitutive promoter; *TRP1*, encoding phosphoribosylanthranilate isomerase involved in tryptophan biosynthesis; *URA3*, encoding orotidine-5′-phosphate (OMP) decarboxylase, involved in uracil biosynthesis.

For example, a centromeric plasmid has one or two copies per cell and is ideal for gene complementation studies. Integrating plasmids may also be useful. In the case of the biotechnologically important yeast, *Pichia pastoris,* integration is the only option: this yeast, like most others, does not allow any plasmid replication. Finally, Yeast Artificial Chromosomes were once essential for large-scale cloning of DNA for sequence analysis. This cloning/sequencing technology is now superseded. However, YACs can still have utility in assembling novel pathways in yeast.

Demonstrating mitochondrial function

Mitochondria are an essential important organelle for our existence, and as previously mentioned, mitochondrial function can be well studied in yeast, where its defects may only lead to conditional lethality. Whilst no eukaryote survives without mitochondria, yeast is an example of an organism that can live without mitochondrial function so long as a fermentable carbon source is available.

Yeast is an ideal organism to test for inhibitors of mitochondrial function. Using biogenic amines my students have had a high hit rate in looking for inhibitors of mitochondrial function. In small research projects, they have made their own media and titrated biogenic amines and various pharmaceuticals into growth media to examine effects on general growth as well as respiratory growth. Inhibition of mitochondrial function could be readily ascertained by growth inhibited on the latter, but not the former media, whilst inhibition of both indicated a general growth inhibition.

In the case of tryptamine, there was general growth inhibition and literature suggesting that tryptamine could bind tryptophan tRNA synthetase. There are two of these enzymes, one for the mitochondrion and one for the cytoplasm. Yeast studies (Kadkol and Macreadie, 2020) were consistent with tryptamine competitively inhibiting the tryptophan tRNA synthetases of the cytoplasm and mitochondria.

Another good example of an important biogenic amine affecting mitochondrial function is dopamine, an essential neurotransmitter. The effect on mitochondrial function was indicated by dopamine's inhibition of respiratory growth (Macreadie et al., 2010). The growth inhibition could be overcome by the addition of glutathione or ascorbate (Macreadie et al., 2010). In addition, dopamine can also cause cell death (Macreadie et al., 2010). Whilst seeming bizarre, these observations also match the effects on neuronal cells, and may provide insights into Parkinson's disease where dopaminergic neurons are depleted. Parkinson's disease treatment can involve supplementation with L-DOPA, which fortunately is not toxic to cells, including yeast (Iriate and Macreadie, 2020).

Chromosomal genetic engineering

It is useful to learn other approaches to genetic engineering, including chromosomal gene knockouts and replacements. Comprehensive libraries of gene knockouts have been made in *Saccharomyces cerevisiae* and every gene knockout has been analysed to elucidate functions. Knockouts in *Saccharomyces cerevisiae* can be readily performed by obtaining PCR products that comprise a selectable marker such as *URA3* or KAN^R and flanking sequences to allow double strand crossovers into a suitable strain. The KAN^R encodes kanamycin resistance in *E. coli* and G418 resistance in *Saccharomyces cerevisiae* and transformants can be readily selected on solidified media containing G418. Thus, KAN^R is a dominant selectable marker and can be used in many laboratory strains. A 'greener' (not involving the use of an antibiotic) approach is to use the *URA3* marker, but the host strain needs to be a *ura3* mutant.

Another, but more challenging, exercise is to perform a gene knockout in *Candida glabrata* where there is still much gene discovery to be made. *Candida glabrata* is an obligate haploid so a knockout of an essential gene will lead to death and no transformants. In *Saccharomyces cerevisiae,* essential genes are knocked out in diploids where one copy remains intact. Sporulation of the diploid results in half of the spores being inviable. The other big challenge with *Candida glabrata* is that homologous recombination is not as prevalent as in *Saccharomyces cerevisiae.* This means considerable illegitimate recombination occurs and integration may occur outside the targeted region.

Yeast as a model organism

As previously mentioned, there has been much success in using yeast as a model to uncover the basic biology of eukaryotic life. How far can we extend this? Can yeast be used to study complex human diseases, including neurodegenerative diseases like Alzheimer's disease? The answer is yes, and it also serves in excellent training, especially for higher degree studies.

The use of yeast as a model for a particular disease may require frequent refinement, re-working of the hypothesis under study and understanding of the limitations. I will again provide some examples from the work of my research students as well as some work of colleagues.

Anti-folate resistance

Folates are essential vitamins for growth and anti-folates are important drugs that target cell growth. The pathway for folate synthesis and utilisation is shown in Fig. 12.3. Understanding the importance of folic acid, its utilisation and synthesis, as well as the importance of anti-folates in health and disease is

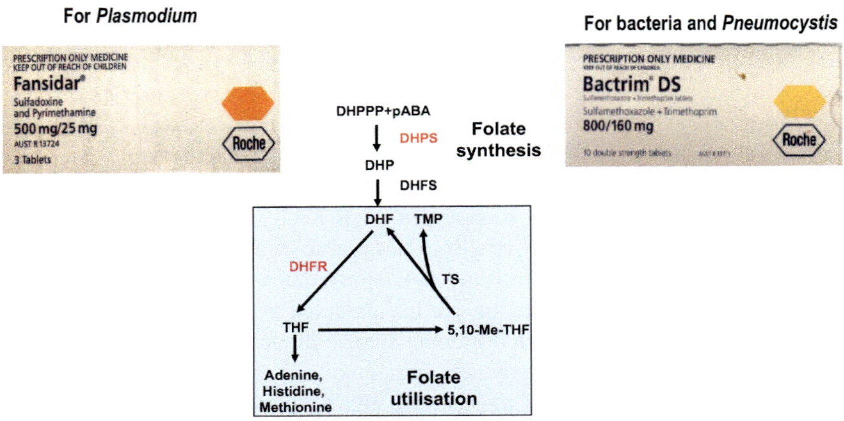

FIGURE 12.3 Representation of pathway for folate synthesis, utilisation and inhibition. Folate synthesis is unique to plants and microbes. It involves enzymic condensation of *p*-amino benzoic acid (pABA) with dihydropterin pyrophosphate (DHPPP) catalysed by dihydropteroate synthase (DHPS) to produce dihydropteroate (DHP). This is converted to dihydrofolate (DHF) by dihydrofolate synthase (DHFS). Dihydrofolate is utilised by all cells as indicated by the shaded area. It is reduced by dihydrofolate reductase (DHFR) to tetrahydrofolic acid (THF) which is used in production of three essential amino acids. THF is methylated and then converted back to DHF by the enzyme thymidylate synthase (TS), involving production of thymidine monophosphate (TMP) which is essential for DNA synthesis. Anti-folate therapy frequently simultaneously targets folate synthesis and utilisation. For example, Bactrim is used for bacterial infections and pneumocystis pneumonia, whilst Fansidar is used for malaria treatment. *Reproduced from Ian Macreadie personal collection.*

an essential part of learning for students of food science, biotechnology and medicine. Further, anti-folate resistance provides some excellent examples on the molecular mechanism of drug resistance due to competitive inhibition, target variation and gene amplification. As such, this pathway is highly recommended for student learning. Yeast can offer excellent insights into all aspects of folate synthesis and utilisation. It can also be engineered to study the impact of foreign genes involved in folate synthesis and utilisation, as well as anti-folate resistance.

One group of anti-folates target the enzyme dihydrofolate reductase (DHFR), an enzyme found in all cells. Methotrexate is an example of a DHFR inhibitor for humans that is used to arrest problematic cell proliferation such as occurs in cancer, psoriasis and rheumatoid arthritis. Methotrexate is an analogue of dihydrofolic acid and effectively binds to DHFR in the place of dihydrofolic acid, resulting in the blockage of folic acid utilisation, ultimately leading to cell death. Drugs like methotrexate should be prescribed carefully since they can target any cells, including healthy human cells, causing death of healthy epithelial cells, resulting in hair loss and other side effects: the disease treatment needs to be weighed up against the side effects.

DHFR is also targeted in microbial diseases including malaria, bacterial infection and some fungal infections. Fortunately, the differences in human DHFR and those of pathogens mean that some DHFR inhibitors can specifically target a pathogen DHFR. These include drugs like pyrimethamine for control of *Plasmodium falciparum* that causes malaria. However, resistance can occur if the DHFR gene becomes amplified (as occurs in some cancers) or artificially in a yeast expression system (Patel et al., 2004a) or if there are amino acid changes leading to drug resistance. The effects of DHFR inhibitors have been modelled in yeast studies by Carol Sibley and colleagues where the yeast DHFR was deleted and replaced with DHFR from a drug-resistant organism. As an example, Sibley and colleagues expressed mutant alleles from *Plasmodium falciparum* DHFR in yeast to observe their effects on resistance to a number of DHFR inhibitors (reviewed in Macreadie and Hopkins Sibley, 2001).

Another target of anti-folates is dihydropteroate synthase (DHPS) which is involved in folate synthesis. Animals do not have this enzyme and rely on obtaining exogenous folate which is produced by plants and microbes. Like DHFR inhibitors, inhibitors of DHPS have also been around for many decades. In fact, DHPS inhibitors were the first chemically synthesised antibiotics, the sulfa drugs, having their origins in the 1940s. The sulfa drugs are the textbook example of competitive inhibitors: they compete with the vitamin *p*-amino benzoic acid (pABA). Observing the sensitivity to numerous sulfa drugs is relatively easy if the correct precautions are taken; for example, it is important to deplete any small reserves of pABA (Castelli et al., 2001). One can then also observe the lack of inhibition by competition with pABA through exogenous addition or by endogenous production through overexpression of the pABA synthase (Castelli et al., 2001).

An interesting finding in yeast studies is that some sulfa drugs can replace pABA in the condensation with DHPP, to produce sulfa-DHP that can inhibit DHFR (Patel et al., 2003). Thus, sulfa drugs have more than one target (Patel et al., 2004b).

Learning mechanisms of drug resistance is of great importance for students and yeast can offer a means of studying some of the mechanisms of drug resistance without needing to set up complicated cell culture for organisms like *Plasmodium falciparum* that require cell culture. In some cases, organisms like *Pneumocystis jiroveci* which causes pneumocystis pneumonia in humans cannot be cultured so suspected mutant alleles involved in resistance to sulfa drugs were initially established by expressing the mutant alleles DHPS in yeast and by complementation of the *Pneumocystis jiroveci* DHPS gene in yeast (Iliades et al., 2004; Hauser and Macreadie, 2006).

Unravelling functions of HIV-1 auxiliary proteins

Yeast also offers the possibility of analysing the function of viral genes. For my research on HIV-1, yeast was particularly useful for the study of the four auxiliary proteins, Vpr, Nef, Vif and Vpu, which at the time of study were of unknown function.

The example of HIV-1 Vpr is of particular interest. In yeast, Vpr displayed toxicity so it was important to use a copper-inducible system for Vpr production. Without copper addition, very little Vpr was produced, and cells grew normally. However, with the addition of 0.5 mM copper, Vpr production was increased, and cell growth was arrested (Macreadie et al., 1995). This function of Vpr in causing a growth arrest was also observed in HIV-1 infected cells (Re et al., 1995). This provides a good example of how well protein functions are preserved in yeast and human cells. It is noteworthy that the isolation of HIV-1 from infected individuals into laboratory culture, where growth was selected, led to many isolates with mutated versions of Vpr.

Another HIV-1 protein of interest, Nef, was expressed at high levels in yeast where it was myristylated and was able to be harvested from cell culture by differential centrifugation (Azad et al., 1994). Post-translational modifications in yeast are generally as good as those in mammalian cells. However, one exception is N-linked glycosylation which is complex in humans but of the high mannose type in yeast. In order to expand the utility of yeast to produce correctly glycosylated human proteins, the biotechnology company GlycoFi utilised the yeast *Pichia pastoris* to replace yeast glycosylation machinery with that of human cells (Wildt and Gerngross, 2005).

Amyloid beta and Alzheimer's disease

A possibly unexpected big jump into a new area of yeast research was in Alzheimer's disease (AD). I was the first person to start expressing amyloid beta in yeast. Amyloid beta is a small peptide of most commonly 40 or 42 amino acids that is derived from a cellular protein called Amyloid Precursor Protein. Amyloid beta is abundant in senile plaques that are highly associated with AD. At the time there was a general feeling that AD occurred as a result of extracellular plaque in the brain, and this made it harder to apply to yeast technology. However, I reasoned that there would be merit in examining the properties of amyloid both inside the cell and on the outside. My first publication on this was in 2007 where the 42 amino acid long version of amyloid beta was expressed in the yeast cytoplasm where it was fused to green fluorescent protein (GFP) (Caine et al., 2007). Our earlier attempts at producing amyloid beta on its own had initially been unsuccessful. Populations of yeast transformed to produce GFP or amyloid beta fused to GFP appeared quite similar with respect to growth. The populations expressing amyloid beta fused to GFP grew at 95% normal growth rate. A genome wide expression analysis comparing genes up- and down-regulated was highly revealing and still gets re-examined a decade later. The gene array analysis is publicly available at RMIT University's research repository [https://researchrepository.rmit.edu.au/discovery/fulldisplay/alma9921861365401341/61RMIT_INST:ResearchRepository?tags=scholar]. A stand-out effect was the heat shock stress response. This response occurs in yeast when cells are exposed to heat shock, reactive oxygen species or misfolded proteins. In response to this, a heat shock factor (HSF1) trimerises and binds to heat shock elements (HSEs) that are just upstream of around 50 genes encoding proteins involved in the heat shock response. To further demonstrate this phenomenon, a heat shock reporter plasmid was inserted into the yeast. This plasmid contained an HSE upstream of *lacZ*, which encodes β-galactosidase. Yeast producing amyloid beta fused to GFP had significantly increased levels of β-galactosidase activity, demonstrating that amyloid beta caused a stress response. More recently, an improved reporter has been made. It comprises an mCherry reporter which can exhibit red fluorescence in live cells (Luu and Macreadie, 2018), allowing an examination of the cells without cell disruption and biochemical assay. It also allows individual cell analyses.

Also, in the original study (Caine et al., 2007) it was noted that not all cells exhibited green fluorescence. The numbers were much lower than expected, around 20%. It subsequently became apparent that this was a very interesting phenomenon that was informing us about ageing. Only the old cells fluoresced: young cells got rid of their amyloid beta that was fused to GFP and no longer fluoresced. This has led to my great interest in ageing and proteostasis, which declines in people also, leading to accumulation of toxic proteins like amyloid beta and α-synuclein which is associated with Parkinson's disease.

Also noted by Caine et al. (2007) was the appearance of amyloid beta fused to GFP inside cells. In the fluorescent cells it appeared in novel punctate patches. It is likely that this is associated with increased membrane synthesis and disruption in these cells, although further work is required to prove that it is so.

Another interesting outcome of the genome-wide expression was that the most increased gene expression was that of *AHP1*, which encodes a lipid peroxidase. *AHP1* has counterparts in humans that are suggested to be playing the same role in mitochondria and in the cytoplasm (Macreadie and Dhakal, 2019). Amyloid beta has a very hydrophobic C-terminus that embeds into membranes, and it is likely that the punctate patches of GFP fused to amyloid beta include membrane that has been disrupted. It is consistent that there is increased expression of this lipid peroxidase to deal with oxidative damage to membranes. In order to study in more detail other effects of amyloid beta in cells, a number of studies looking for agents providing chemo protection have been performed in an *ahp1* deleted strain.

Yeast, like neuronal cells, is also sensitive to the 42 amino acid amyloid beta peptide and can be killed by it in a dose-dependent manner when it is freshly prepared and added to yeast cells suspended in water (Bharadwaj et al., 2008).

If left overnight to form fibrils, the amyloid beta is no longer toxic (Bharadwaj et al., 2008), an observation also made on neuronal cells. The yeast assays have been performed in *Candida glabrata* yeast where colony counts can be made a day after plating out cells.

Studies on statins including their protection against Alzheimer's disease

Yeast is also an excellent model for studying the effects of the cholesterol lowering drugs known as statins, since yeast have essentially the same pathway, with the exception of some different enzymes and reactions in the final steps of the biosynthetic pathway, so that in yeast the end product is ergosterol, rather than cholesterol. However, ergosterol and cholesterol are functionally equivalent. In yeast, the synthesis of ergosterol can be blocked and replaced by the addition of cholesterol. Statins bind to inhibit the enzyme HMGCR (3-hydroxy-3-methylglutaryl-CoA reductase) lowering the production of mevalonate. Complete blocking of this step is lethal as cholesterol and ergosterol are essential for cell membrane fluidity. In yeast, inhibition by statins can lead to inhibition of growth and the loss of the mitochondrial genome (Westermeyer and Macreadie, 2007). In humans, the loss of mtDNA is lethal but there certainly are indications of effects mitochondrial function. It also turns out that one statin, simvastatin, provides high protection against AD and Parkinson's disease (Wolozin et al., 2007).

Yeast studies offered an opportunity to investigate the mechanism of simvastatin's effect. Incubating cells producing GFP-Aβ or native Aβ with three statins showed that whilst all caused some reduction in GFP-Aβ or native Aβ, the treatment with simvastatin led to a >99% reduction in levels of GFP-Aβ and more than 80% of Aβ (Dhakal et al., 2019). Thus, yeast studies imply the protective effect of simvastatin operates by reducing levels of Aβ.

Further use of recombinant yeast in testing effects on Aβ has shown that there can be toxic synergies with other factors such as tyramine (Dhakal and Macreadie, 2020) and aluminium (Mcdonald et al., 2021). This is an important area for the future and may lead to unravelling of the complexities and sporadic nature of AD.

Conclusions

Yeast offers a plethora of opportunities for teaching undergraduate students, higher degree students and for the further teaching of senior researchers. I have briefly discussed the use of yeast in our daily lives providing important food and beverages, as well as important medicines to save lives and prevent cancer. In addition, I have provided examples of where yeast has been used as model organism, providing insights into aspects of human biology and diseases like AIDS, malaria and AD. It is important that students of bioscience have exposure to this knowledge and processes including good training in genetic engineering to understand its benefits and possible abuses and misuses. The safety, convenience and low cost of yeast research enable yeast to be a highly convenient teaching resource.

References

Ainley, M., Macreadie, I.G., Butow, R.A., 1985. *var1* gene on the mitochondrial genome of *Torulopsis glabrata*. J. Mol. Biol. 184, 565–576.

Almshawit, H., Macreadie, I., 2017a. Fungicidal effect of thymoquinone involves generation of oxidative stress in *Candida glabrata*. Microbiol. Res. 195, 81–88.

Almshawit, H., Macreadie, I., 2017b. Doxycycline induces SLT2 phosphorylation in *Candida glabrata*. Int. J. Clin. Med. Microbiol. 2, 122.

Azad, A.A., Failla, P., Lucantoni, A., Bentley, J., Mardon, C., Wolfe, A., Fuller, K., Hewish, D., Sengupta, S., Sankovich, S., Grgacic, E., McPhee, D., Macreadie, I.G., 1994. Large scale production and characterisation of recombinant human immunodeficiency virus type 1 Nef. J. Gen. Virol. 75, 651–655.

Barnett, J.A., 2000. A history of research on yeasts 2: Louis Pasteur and his contemporaries, 1850-1880. Yeast 16, 755–771.

Bernardi, G., September 1979. The petite mutant of yeast. Trends Biochem. Sci. 197–200.

Bharadwaj, P., Waddington, L., Varghese, J., Macreadie, I., 2008. A new method to measure cellular toxicity of non-fibrillar and fibrillar Alzheimer's Aβ using yeast. J. Alzheim. Dis. 13, 147–150.

Caine, J., Sankovich, S., Antony, H., Waddington, L., Macreadie, P., Varghese, J., Macreadie, I., 2007. Alzheimer's Aβ fused to green fluorescent protein induces growth stress and a heat shock response. FEMS Yeast Res. 7, 1230–1236.

Castelli, L.A., Mardon, C.J., Strike, P.M., Azad, A.A., Macreadie, I.G., 1994. High-level secretion of correctly-folded -lactamase from *Saccharomyces cerevisiae*. Gene 142, 113–117.

Castelli, L.A., Nguyen, N.P., Macreadie, I., 2001. Sulfa drug screening in yeast: fifteen sulfa drugs compete with *p*-aminobenzoate in *Saccharomyces cerevisiae*. FEMS Microbiol. Lett. 199, 181–184.

Dhakal, S., Macreadie, I., 2020. Tyramine and amyloid beta 42: a toxic synergy. Biomedicines 8, 145. https://doi.org/10.3390/biomedicines8060145.

Dhakal, S., Subhan, M., Fraser, J.M., Gardiner, K., Macreadie, I., 2019. Simvastatin efficiently reduces levels of Alzheimer's amyloid beta in yeast. Int. J. Mol. Sci. 20, E3531. https://doi.org/10.3390/ijms20143531.

Dixon, T.A., Pretorius, I.S., 2020. Drawing on the past to shape the future of synthetic yeast research. Int. J. Mol. Sci. 21, 7156. https://doi.org/10.3390/ijms21197156.

Hauser, P.M., Macreadie, I.G., 2006. Isolation of the Pneumocystis carinii dihydrofolate synthase gene and functional complementation in *Saccharomyces cerevisiae*. FEMS Microbiol. Lett. 256, 244–250.

Iliades, P., Meshnick, S.R., Macreadie, I.G., 2004. DHPS mutations found in Pneumocystis jiroveci can affect sulfamethoxazole resistance in a *Saccharomyces cerevisiae* model. Antimicrob. Agents Chemother. 48, 2617–2623.

Iriate, C.E., Macreadie, I., 2020. Comparison of cytocidal activities of L-DOPA and dopamine in *S. cerevisiae* and *C. glabrata*. Curr. Bioact. Compd. 16, 90–93.

Kadkol, C., Macreadie, I., 2020. Inhibition of yeast growth by tryptamine and recovery with tryptophan. Curr. Bioact. Compd. 16, 48–52. https://doi.org/10.2174/1573407214666180713094152.

Macreadie, I., 2007a. Nobel Prize in Chemistry: molecular basis of eukaryotic transcription. Chem. Aust. 74, 11.

Macreadie, I., 2007b. Therapeutic products from yeast. Microbiol. Australia 28, 82–84.

Luu, Y.N., Macreadie, I., 2018. Development of convenient system for detecting yeast cell stress, including that of amyloid beta. Int. J. Mol. Sci. 19, 2136.

Macreadie, I.G., 2017. Yeasts, molecular and therapeutic applications. In: Kirk-Othmer Encyclopedia of Chemical Technology (2017). John Wiley & Sons, Inc., pp. 1–24

Macreadie, I., Hopkins Sibley, C., 2001. Yeast enlists in the fight against malaria. Today Life Sci. 13, 46–50.

Macreadie, I., Dhakal, S., 2019. Insights from yeast on oxidative stress in Alzheimer's Disease, focusing on Ahp1p/Prx5. OBM Geriat. 3 (3), 10. https://doi.org/10.21926/obm.geriatr.1903064.

Macreadie, I.G., Scott, R.M., Zinn, A.R., Butow, R.A., 1985. Transposition of an intron in yeast mitochondria requires a protein encoded by that intron. Cell 41, 395–402.

Macreadie, I.G., Castelli, L.A., Hewish, D., Kirkpatrick, A., Ward, A.C., Azad, A.A., 1995. A domain of human immunodeficiency virus type-1 Vpr protein containing H(S/F)RIG amino acid motifs causes growth arrest and structural defects. Proc. Natl. Acad. Sci. U.S.A. 92, 2770–2774.

Macreadie, I.G., Bartone, N., Sparrow, L., 2010. Inhibition of respiratory growth and survival in yeast by dopamine and counteraction with ascorbate or glutathione. J. Biomol. Screen 15, 297–301.

Mcdonald, J.B., Dhakal, S., Macreadie, I.G., 2021. A toxic synergy between aluminium and amyloid beta in yeast. Int. J. Mol. Sci. 22, 1835.

Mehra, R.K., Thorvaldsen, J.L., Macreadie, I.G., Winge, D.R., 1992. Cloning system for Candida glabrata using elements from the metallothionein-IIa-encoding gene that confer autonomous replication. Gene 113, 119–124.

Patel, O., Satchell, J., Baell, J., Fernley, R., Coloe, P., Macreadie, I., 2003. Inhibition studies of sulfonamide-containing folate analogs in yeast. Microb. Drug Resist. 9, 139–146.

Patel, O., Karnik, K., Macreadie, I.G., 2004a. Over-production of dihydrofolate reductase leads to sulfa-dihydropteroate resistance in yeast. FEMS Microbiol. Lett. 301–305.

Patel, O., Mberu, E., Nzila, A., Macreadie, I., 2004b. Sulfa drugs strike more than once. Trends Parasitol. 20, 1–3.

Re, F., Braaten, D., Franke, E.K., Luban, J., 1995. Human immunodeficiency virus type 1 Vpr arrests the cell cycle in G2 by inhibiting the activation of p34cdc2-cyclin B. J. Virol. 69, 6859–6864.

Rogers, P., 2017. Beer: old's now new again. Microbiol. Aust. 38 (2), 80–81. https://doi.org/10.1071/MA17034.

Roman, H., 1986. The early days of yeast genetics. A personal narrative. Annu. Rev. Genet. 20, 1–12.

Tam, P., Kee, G., Piechocinski, M., Macreadie, I., 2015. *Candida glabrata*: friend and foe. J Fungi 1, 277–292.

Westermeyer, C., Macreadie, I.G., 2007. Simvastatin reduces ergosterol levels, inhibits growth and causes loss of mitochondrial DNA in *Candida glabrata*. FEMS Yeast Res. 7, 436–441.

Wildt, S., Gerngross, T.U., 2005. The humanisation of N-linked glycosylation pathways in yeast. Nat. Rev. Microbiol. 3, 119–128.

Winge, Ö., 1935. On haplophase and diplophase in some Saccharomycetes. C. R. Trav. Lab. Carlsberg Ser. Physiol. 21, 77–109.

Wolozin, B., et al., 2007. Simvastatin is associated with a reduced incidence of dementia and Parkinson's disease. BMC Med. 5, 20.

Index

'*Note*: Page numbers followed by "f" indicate figures those followed by "t" indicate tables and "b" indicate boxes. Numbers indicate the page ranges for the mentioned words.'

C

D

E

N

O

P